WILLIAM F. MAAG LIBRARY
YOUNGSTOWN STATE UNIVERSITY

TOPICS IN
INORGANIC AND ORGANOMETALLIC STEREOCHEMISTRY

VOLUME 12

ADVISORY BOARD

STEPHEN J. ANGYAL, *University of New South Wales, Sydney, Australia*
ALAN R. BATTERSBY, *Cambridge University, Cambridge, England*
GIANCARLO BERTI, *University of Pisa, Pisa, Italy*
F. ALBERT COTTON, *Texas A&M University, College Station, Texas*
JOHANNES DALE, *University of Oslo, Oslo, Norway*
DAVID GINSBURG, *Technion, Israel Institute of Technology, Haifa, Israel*
JAN MICHALSKI, *Centre of Molecular and Macromolecular Studies, Polish Academy of Sciences, Lodz, Poland*
KURT MISLOW, *Princeton University, Princeton, New Jersey*
SAN-ICHIRO MIZUSHIMA, *Japan Academy, Tokyo, Japan*
GUY OURISSON, *University of Strasbourg, Strasbourg, France*
VLADIMIR PRELOG, *Eidgenössische Technische Hochschule, Zurich, Switzerland*
GÜNTHER SNATZKE, *Ruhruniversität, Bochum, Federal Republic of Germany*
HANS WYNBERG, *University of Groningen, Groningen, The Netherlands*

TOPICS IN

INORGANIC AND ORGANOMETALLIC STEREOCHEMISTRY

EDITED BY

GREGORY L. GEOFFROY

Pennsylvania State University
University Park, Pennsylvania

VOLUME 12

of

TOPICS IN
STEREOCHEMISTRY

Series Editors

NORMAN L. ALLINGER
ERNEST L. ELIEL

AN INTERSCIENCE® PUBLICATION

JOHN WILEY & SONS

New York · Chichester · Brisbane · Toronto

An Interscience® Publication

Copyright © 1981 by John Wiley & Sons, Inc.

All rights reserved. Published simultaneously in Canada.

Reproduction or translation of any part of this work beyond that permitted by Sections 107 or 108 of the 1976 United States Copyright Act without the permission of the copyright owner is unlawful. Requests for permission or further information should be addressed to the Permissions Department, John Wiley & Sons, Inc.

Library of Congress Catalog Card Number: 67–13943

ISBN 0–471–05292–2

Printed in the United States of America

10 9 8 7 6 5 4 3 2 1

FOREWORD

This volume, devoted entirely to inorganic stereochemistry, is not a standard volume and so it does not bear our standard introduction. When Professor F. A. Cotton joined our editorial advisory board, he expressed dissatisfaction with the paucity of chapters dealing with *in*organic stereochemistry in previous volumes of this series. He suggested that the only appropriate way to make up for past deficiencies in this area was to devote an entire volume of Topics in Stereochemistry to the inorganic aspects of the subject. This suggestion intrigued the editors; however, from past experience we doubted whether we would be able to put together an entire volume made up of contributions by inorganic chemists. Dr. Cotton had a suggestion in this regard: he recommended that we ask Professor Gregory Geoffroy of Pennsylvania State University to edit the proposed venture. Dr. Geoffroy was kind enough to accept our invitation, and the present volume is the result. We hope it will, in a tangible way, signal to the inorganic community the editors' awareness of stereochemistry as more than a prerogative of organic chemists and of Topics in Stereochemistry as a good vehicle for the dissemination of information in the area of inorganic stereochemistry. We hope that there will be many inorganic contributions in future volumes, and we encourage offerings of potential chapters in this area.

At the same time, after reading the chapters Dr. Geoffroy has acquired, we are convinced that there is much information here of considerable interest to our regular readers with organic backgrounds. Indeed, in view of the close relationship between organic and inorganic chemistry that has recently developed affecting both synthesis and mechanism, it is only appropriate that stereochemistry be included in the confluence of these two areas. We hope that our readers, including those who consider themselves organic chemists, will agree with that sentiment.

We are pleased to welcome Professor Alan R. Battersby of Cambridge, Professor Günther Snatzke of the University of Bochum, and Professor Jan Michalski of the Polish Academy of Sciences to our circle of editorial advisors, and we take this opportunity to announce that Professor Samuel H. Wilen of the City University of New York will join us as a co-editor beginning with the next volume (Volume 13).

N. L. ALLINGER
E. L. ELIEL

February 1981

PREFACE

Volume 12 of Topics in Stereochemistry is devoted entirely to inorganic and organometallic stereochemical subjects.

The volume begins with a chapter by T. E. Sloan of Chemical Abstracts Service on the notation and nomenclature used to denote stereochemistry in inorganic and organometallic compounds. The need for a rational and systematic stereochemical nomenclature has become increasingly important as research interest in organometallic and asymmetric compounds has grown during the past two decades. The systematic notations that have been developed to satisfy these needs are reviewed and discussed in this chapter, with specific attention given to the use of Cahn-Ingold-Prelog ligand priority numbers for denoting complex stereochemistry.

The ubiquitous involvement of metal-carbon bonds in changes of bonding at carbon centers induced by metals, either catalytically or stoichiometrically, is well established. In Chapter 2 T. C. Flood reviews the stereochemical aspects of these reactions and importantly puts the stereochemistry into mechanistic perspective where possible. Reactions that form and disrupt metal-carbon σ bonds are discussed by reaction type: oxidative addition, reductive elimination, insertion, electrophilic cleavage, transmetalation, and addition.

An important area of current stereochemical research is the synthesis of asymmetric compounds mediated by transition metals. The incorporation of chiral ligands or the use of chiral complexes can lead to asymmetric synthesis where the chiral discrimination rivals that of enzymes. In Chapter 3 B. Bosnich and M. D. Fryzuk review the major achievements in this field. The authors attempt to give a mechanistic basis for the reactions they discuss, in spite of the notable lack of rationality and systematization in these studies.

Metal nitrosyl complexes represent an important class of organometallic compounds. The nitrosyl ligand can adopt linear and bent geometries conforming to the valence-bond structures $|N{\equiv}O|^+$ and $\langle N{=}O\rangle^-$, respectively, depending on the electronic nature of the complex. This area of organometallic chemistry has grown rapidly in recent years and previous reviews are severely outdated. Thus the review of the structural aspects of nitrosyl complexes by R. D. Feltham and J. H. Enemark in Chapter 4 is timely. This review is restricted to a discussion of metal nitrosyl complexes whose structures have been determined by X-ray crystallography, neutron diffraction, electron diffraction or microwave spectroscopy, that is, of compounds whose stereochemistry is known with certainty.

The stereochemistry of compounds of germanium and tin is the subject of Chapter 5 by M. Gielen. This article nicely complements Sommer's earlier review of the stereochemical aspects of silicon chemistry (L. Sommer, *Stereochemistry, Mechanism, and Silicon*, McGraw-Hill, New York, 1965); it thus completes coverage of all the heavier group IV elements, since few stereochemical studies have been conducted with lead compounds. The optical stability of stereoisomers of germanium and tin compounds is discussed, and there is a summary of the stereochemistry of substitution reactions and intermolecular rearrangements.

In Chapter 6 B. F. G. Johnson and R. E. Benfield discuss the structures of transition metal carbonyl clusters, summarizing recent progress in this rapidly expanding area and providing a semi-quantitative rationalization of the structures for these clusters. Focus is on the bonding considerations of the central metal framework as well as the influence of the ligand envelope on the structures of such clusters.

G. L. GEOFFROY

University Park, Pennsylvania
February 1981

CONTENTS

STEREOCHEMICAL NOMENCLATURE AND NOTATION IN INORGANIC CHEMISTRY
by Thomas E. Sloan, Chemical Abstracts Service, Columbus, Ohio 1

STEREOCHEMISTRY OF REACTIONS OF TRANSITION METAL–CARBON SIGMA BONDS
by Thomas C. Flood, Department of Chemistry, University of Southern California, Los Angeles, California 37

ASYMMETRIC SYNTHESIS MEDIATED BY TRANSITION METAL COMPLEXES
by Bryce Bosnich, Lash Miller Chemical Laboratories, University of Toronto, Toronto, Canada, and Michael D. Fryzuk, University of British Columbia, Vancouver, Canada 119

STRUCTURES OF METAL NITROSYLS
by Robert D. Feltham and John H. Enemark, Department of Chemistry, University of Arizona, Tucson, Arizona 155

THE STEREOCHEMISTRY OF GERMANIUM AND TIN COMPOUNDS
by Marcel Gielen, Vrije Universiteit Brussel and Université Libre de Bruxelles, Brussels, Belgium 217

STEREOCHEMISTRY OF TRANSITION METAL CARBONYL CLUSTERS
by Brian F. G. Johnson and Robert E. Benfield, University Chemical Laboratory, Cambridge University, Cambridge, England 253

SUBJECT INDEX **337**

CUMULATIVE INDEX, VOLUMES 1–12 . . . . **349**

TOPICS IN
INORGANIC AND ORGANOMETALLIC STEREOCHEMISTRY

VOLUME 12

Stereochemical Nomenclature and Notation in Inorganic Chemistry

THOMAS E. SLOAN

Chemical Abstracts Service, Columbus, Ohio

I.	Introduction	2
II.	Historical Development of Stereochemical Nomenclature	2
III.	Ligand Indexing in Stereochemical Notation	8
	A. The Cahn-Ingold-Prelog (CIP) Notation	9
	B. Computer Assignment of Ligand Priorities and Notations	9
IV.	Application of CIP Ligand Priorities to Coordinated Ligands	10
V.	Systematic Stereochemical Notation for Coordination Compounds Using CIP Priority Numbers	17
	A. The Stereochemical Descriptor	17
	B. Ligand Degeneracy and Multidentate Ligands	19
	1. The trans Maximum Difference Subrule for Coordination Numbers 4 to 6	19
	2. Priming Subrule a	19
	3. Priming Subrule b	19
	C. Assignment of Configuration Numbers	20
	D. Assignment of Chirality Symbols	25
	E. Examples	27
	1. Tris(1,2-propanediamine-N,N')cobalt(3+)	27
	2. [[N,N'-Ethylenebis(glycinato)] (2—)] [oxalato(2—)] cobalate(1—)	28
	3. Bromochloro(2-ethyl-13-methyl-3,6,9,12,18-pentaazabicyclo-[12.3.1]octadeca-1(18),2,12,14,16-pentaene-N^3, N^6, N^9, N^{18})-iron(1+)	29
	4. Bis[N-(2-aminoethyl)-1,2-ethanediamine-N,N',N'']-bromochloropraseodymium(1+)	30
	5. Tetraaqua(2,4,5,6($1H,3H$)-pyrimidinetetrone-5-oximato-N^5,O^4)(2,4,5,6,($1H,2H$)-pyrimidinetetrone-5-oximato-O^4,O^5)-strontium	30
VI.	Hapto Complexes	31
VII.	Polynuclear and Cluster Complexes	34
VIII.	Conclusion	35
	References	35

I. INTRODUCTION

The synthesis of urea in 1828 by Wöhler and the founding of structural chemistry in the mid-nineteenth century by Kekulé, Pasteur, van't Hoff, and Le Bell mark the origins of stereochemistry (1). The development of structural carbon chemistry would have been difficult without the experimental clues and disclosures derived from the development of stereochemistry. It was the number of isomers of disubstituted benzene that provided the proof in distinguishing between the several possible structural formulations. The then new structural concepts and insights gained led to the rapid development of carbon (organic) chemistry and for the next 60 years overwhelmed the more traditional subjects in chemical investigations of "mineral" and "pneumatic" chemistry. The residue of this preponderance of carbon chemistry still exist in the language and notation found in today's chemical literature.

II. HISTORICAL DEVELOPMENT OF STEREOCHEMICAL NOMENCLATURE

Three of the great precepts of classical carbon chemistry: (*a*) the constant valency of four for the carbon atom, (*b*) the ability of carbon to form homogeneous chains, and (*c*) the tetrahedral configuration of four coordinate carbon, were so successful and pervasive that the same concepts were thought to be the basis of structural chemistry for all chemical systems. Following these precepts and preserving an ideal valency, scientists set forth the chain-type structures to explain the structures and properties of many inorganic systems. Experimental evidence that challenged these theories continued to accumulate, and finally Werner formulated and popularized the more complex and less certain bonding theory of coordination compounds.

The first nomenclature applied to isomers of inorganic compounds was derived from the colors of compounds with the same stoichiometric proportions (2). Cobalt trichloride tetraammoniate had two forms known as the cobalt *praseo* or green complex and the cobalt *violeo* or violet complex. Often within the same family of compounds the color nomenclature was able to group together molecules of similar characteristics. Since there was no agreement among inorganic chemists of the period on the constitutional structures of inorganic species, it would have been difficult to develop any type of rational nomenclature.

As the Werner-Jorgensen controversy was actively pursued (11), experimental evidence of structures accumulated, and the classification by color broke down. The color nomenclature became less meaningful. Finally, as Werner's structural postulates became widely accepted, it became possible to develop a stereochemical nomenclature and notation based on the accepted and experimentally accurate structure.

Werner recognized geometric similarities of disubstituted platinum(II) planar compounds and disubstituted cobalt(III) octahedral compounds to the geometric isomerism of ethylene compounds and used the terms "cis" and "trans" to describe the adjacent and opposite or corner and axial isomeric forms of these compounds (3). Werner also developed the first stereochemical numbering for the ligand positions on the octahedral chromium and cobalt complexes (Fig. 1). The cis compounds were designated 1,2 and the trans compounds 1,6.

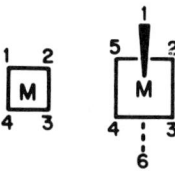

Fig. 1. Werner numbering for square planar and octhedral structures.

Stereochemical nomenclature did not change appreciably in the next 50 years. The Report of the Committee for the Reform of Inorganic Chemical Nomenclature (1940) does not mention stereochemical nomenclature (4). A comprehensive review of coordination compound nomenclature by Fernelius reaffirms the use of the terms "cis" and "trans", recognizes a standard numbering for the square planar structure, gives the Werner numbering of the octahedral structure, and proposes standard numbering practices for binuclear structures of the fused square planar and fused tetrahedral types (5). This review also notes the well-established use of the (+), (−), or, alternatively, d or l to indicate the sign of rotation for optical isomers.

The numbers and complexities of isomeric inorganic compounds increased rapidly following the renewed interest in inorganic and coordination chemistry fostered by the development of the atomic bomb during World War II. Stereochemical research had focused predominantly on coordination numbers 4 and 6. The cis and trans terminology was found to be inadequate for the relatively well-studied six-coordinate octahedral compounds (Fig. 2). In structure I of Figure 2, "cis" describes the adjacent relationship of the tricarbonyl and the trichloro ligands, and the trans isomer is the only other structure possible. However in structure II of Figure 2 the tricarbonyl ligands are mutually opposite or trans, but the chlorine ligand is adjacent or cis to the phosphine and pyridine ligands. The terms "fac" and "mer" have been used to describe the facial and meridional distributions of ligands in trisubstituted octahedral complexes. Although commonly used, these terms are not adequate because there are four possible geometric arrangements, one mer and three fac isomer for octahedral [Ma_3bcd]

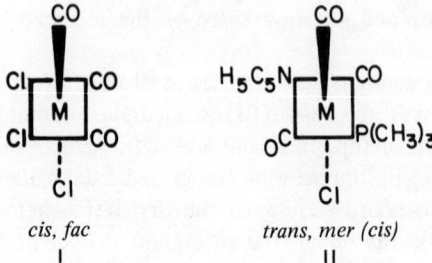

Fig. 2. Stereochemical terms for tris monodentate octhedral complexes.

compounds. Similarly, there are 15 geometric isomers of the octahedral [Mabcdef] compounds, and numerous isomers for multidentate chelating ligands such as the octahedral structure with the quadridentate ligands in Figure 3, which exists in three geometric forms. Structures **I** and **II** are commonly designated α-*cis* and β-*cis*, respectively, and structure **III** as trans.

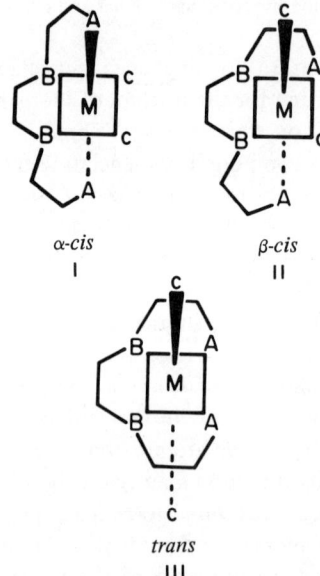

Fig. 3. Geometric isomers of octhedral complexes with linear tetradentate ligands.

These deficiencies in stereochemical terminology were compounded by the synthesis of compounds having coordination numbers 2 through 9 with many different geometries. The inadequacy of stereochemical nomenclature was recognized by several investigators. In a review of compounds forming higher coordination number polyhedra and molecular clusters, Muetterties and Wright proposed rules for numbering polyhedra and produced a table of numbered polyhedra with 5 to 20 vertices, representing 29 different geometries (6). The numbering rules were developed from the symmetry proporties of the polyhedra. They define a principal or preferred axis, number a position on this axis, and number positions in belts or planes perpendicular to the preferred axis, in a clockwise direction. This generates the Werner numbering of the octahedron and is very similar to the numbering of closo polyhedral boranes adopted by IUPAC (7).

In developing a linear notation for coordination compounds, McDonnell and Pasternack proposed a ligand locant designator notation that follows the numbering pattern proposed by Muetterties (8). The McDonnell locant designators substitute lowercase Roman characters to designate the ligand positions. This was done to eliminate any confusion that would occur between the substitution numbering within a ligand and the numbering of the ligand position. Another innovation in the McDonnell notation is the class symbol (Arabic number, capital Roman letter) to indicate the coordination geometry about the central atom (Fig. 4).

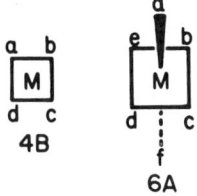

Fig. 4. McDonnell locant designators and class symbols for square planar and octahedral structures.

The traditional stereochemical descriptors cis, trans, fac, and mer, along with the McDonnell locant designators, were codified in the *IUPAC Nomenclature of Inorganic Chemistry, 1970* (9). The IUPAC nomenclature elaborates extensively on the McDonnell notation to allow for chelating ligands, develops ligating atom priorities to eliminate ambiguities, and extends the notation to polynuclear compounds.

A method for unambiguous, structurally exact stereochemical nomenclature is provided by specific ligand numbering or locant designators. These have been employed very sparingly in the chemical literature, however, perhaps because of the rather complicated rules for the unambiguous application of locants to the complicated structures for which they are most useful. Also, in the case of directionally specific locants, it is difficult to distinguish enantiomeric pairs of compounds (Table 1). A further notation or symbol is required when the geo-

Table 1
Chiral $(NbCl_2F_5)^{2-}$ Compounds.

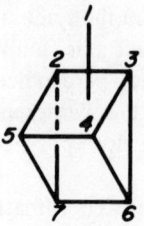

Muetterties-Wright Numbering

1,2-dichloro-3,4,5,6,7-pentafluoroniobate(2-) I
1,3-dichloro-2,4,5,6,7-pentafluoroniobate(2-) I
2,4-dichloro-1,3,5,6,7-pentafluoroniobate(2-) VII
2,6-dichloro-1,3,4,5,7-pentafluoroniobate(2-) VIII
2,7-dichloro-1,3,4,5,6-pentafluoroniobate(2-) IV
3,5-dichloro-1,2,4,6,7-pentafluoroniobate(2-) VII
3,6-dichloro-1,2,4,5,7-pentafluoroniobate(2-) IV
3,7-dichloro-1,2,4,5,6-pentafluoroniobate(2-) VIII

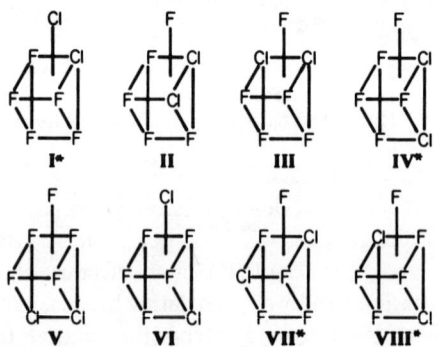

Fig. 5. Isomers of $[NbCl_2F_5]^{2-}$ (chiral structures indicated by asterisk).

metric configuration of an asymmetric compound is known, but not the absolute configuration, since it is not possible to draw and number such a structure ambiguously (structure **IV**, Fig. 5).

The stereochemical nomenclature for optically active compounds has also abounded with attempts to provide rational and exact notation. The sign of optical rotation, + or −, was first used to indicate an optically pure or nearly optically pure isomer, and the terms "dextrorotatory" and "levorotatory" were used interchangeably with + and —. As the tetrahedral carbon atom hypothesis of van't Hoff and Le Bel gained acceptance and experimental data on optically active compounds accumulated, it became obvious that the sign of optical rotation bore no fundamental relationship to the inherent configuration of a molecule or to the relative configurations of related molecules.

Fischer assigned one of the two possible structures to dextrorotatory (+)-tartaric acid, after investigating the structural similarities of a set of similar compounds (10). The configuration of (+)-glyceraldehyde was related to (+)-tartaric acid and then by various chemical means to many other molecules. The molecules related to (+)-glyceraldehyde were said to be the D series and those related to (—)-glyceraldehyde formed the L series. Many naturally occurring amino acids were assigned to the L series. Simple sugars having several asymmetric carbon atoms were related to the D series through (+)-glucose. The convention evolved by assigning the configuration of a carbohydrate based on the highest asymmetric carbon atom in the chain. Multicentered amino acids were assigned a configuration based on the lowest numbered carbon or carbon α to the carboxy group. This set of conventions resulted in the assignment of some molecules with the same ligand configuration to different configuration series; for example, the naturally occurring amino acid threonine is assigned to the L series and the natural sugar threose is assigned to the D series (Fig. 6). The absolute configuration of (+)-glyceraldehyde was not known with certainty until Bijvoet et al. determined the absolute configuration of tartaric acid in 1951 (11). (The Fischer assignment in 1896 amounted to a very astute guess with a 50–50 chance of being correct).

With the establishement of the absolute configuration of (+)-glyceraldehyde, Cahn, Ingold, and Prelog circumvented the conflicting carbohydrate–amino acid notation dilemma by developing the CIP sequence rule and the R,S notation (12). Blackwood et al. extended the application of the CIP sequence rule to olefin geometric isomers and proposed the E,Z notation (13).

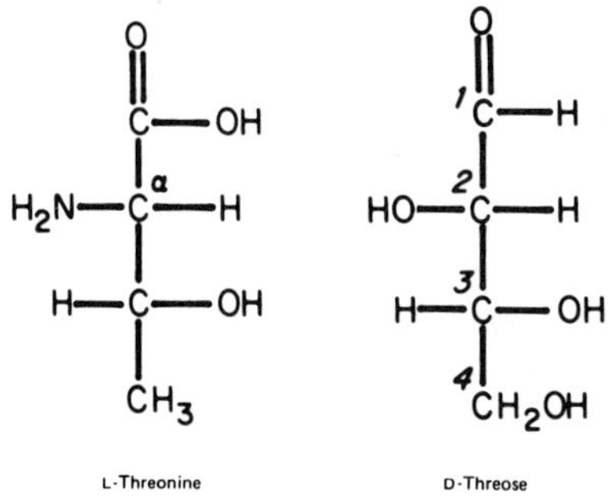

Fig. 6. Amino acid and carbohydrate conventions for configuration assignment.

Coordination and inorganic chemists used the sign of rotation of the optically pure isomer, for example, (+), dextro, d or D, and (−), levo, l or L, to describe optically active compounds, until Saito et al. determined absolute configuration of (+)-tris(ethylenediamine)cobalt(3+) in 1955 (14). Not to be outdone by organic chemists, Piper first proposed the use of a Δ to indicate the right-handed helical arrangement, as viewed on the C_3 axis, of (−)-tris(ethylenediamine)cobalt(3+) and Λ for the (+) isomers (15). Other workers noted that another helical configuration in the bis(ethylenediamine) octahedral complexes can be related to tris-(ethylenediamine)cobalt^{3+} by reference to the C_2 axis. The symbols M and P were proposed for denoting this configuration based on the C_2 axis (16). Numerous examples followed of the use of Δ and Λ based on either the C_3 or the C_2 axis and less commonly of the designations M and P. Cahn, Ingold, and Prelog proposed the use of R and S based on comparison of an octahedral molecule to a nondirectional Werner numbering and determining the right- or left-handed path of the sequence 1,2,3 as viewed from the side of the model remote from 4,5,6 (12). This resulted in the assignment of R to (+)-tris(ethylenediamine)cobalt(3+). During this period, depending on the preference of the individual, chelated tris bidentate and bis bidentate octahedral compounds were given the notation $\Delta\,(C_3)$, $\Lambda\,(C_2)$, $P\,(C_3)$, $M\,(C_2)$, S, L, l or (−), all for the same configuration.

The 1970 second edition of the *IUPAC Nomenclature of Inorganic Chemistry* redefined Δ and Λ on the basis of the helical system generated by two nonorthogonal skew lines (17). The symbol Δ is assigned to the right-handed helix and Λ to the left-handed helix. This definition is independent of symmetry axes in the octahedral structure but is coincident with Piper's original proposal. This notation has been widely adopted, and the designation of tris bidentate or bis bidentate octahedral compounds by Δ or Λ can be accepted with a high degree of confidence.

III. LIGAND INDEXING IN STEREOCHEMICAL NOTATION

The need for a rational and systematic stereochemical nomenclature has become increasingly acute since the establishment of the absolute configurations of (+)-tartaric acid by Bijvoet and (+)-tris(ethylenediamine)cobalt(3+) by Saito. The importance of stereochemical configuration in living systems, the influence of stereoconfiguration on the course of reactions, and the worldwide distribution of stereochemical research reinforces the need for a systematic nomenclature to effectively present and communicate the results of this research. All the systematic notations that have been developed to satisfy these needs are based on a ligand index number derived from procedures for structural constitutional numbering.

A. The Cahn-Ingold-Prelog (CIP) Notation

The Cahn, Ingold, and Prelog notation is based on a priority ranking of ligands in a three-dimensional, valence-bond molecular representation (12). The priority numbers are assigned on the basis of atom atomic number, and procedures are given to explore the valence-bond structures by path, atomic number, and other criteria to determine ligand priority. The procedures presented are stylized and schematic because of the deficiencies of conventional valence-bond representations for certain chemical structures.

To reduce the problems associated with valence-bond representations and to provide simplifications useful to computer analysis of graph properties and stereochemical characteristics, several investigators have studied ligand indexing in the quest of computer-assisted solutions to problems in chemistry and chemical documentation. The computer studies have been based on graph representations of chemical formulas similar to the Morgan algorithm. The Morgan algorithm is the basis of the Chemical Abstracts Service Chemical Registry System (18).

B. Computer Assignment of Ligand Priorities and Notations

Wipke developed the stereochemical extension of the Morgan algorithm (SEMA), which utilizes the Morgan numbers in a stereochemically exact representation of the molecular graph (19). From the combination of the three-dimensional representations and the Morgan numbers, a parity is established that is sufficient to represent the graph node under consideration. This is similar to the CIP R, S nomenclature, but there is no correspondence whatsoever of the parity to R or S, since there is no relationship between or translation of the Morgan numbers to the CIP priority numbers (20). Wipke et al. have modified SEMA to treat five-coordinate, trigonal bipyramidal phosphorus compounds (20).

The canonical topological structure description (TSD) developed by Davis has been influenced by the CIP procedures but has utilized the more computer-compatible atomic sequencing index (ASI) and parity vectors of Professor Ugi in a more efficient system (21). The ASI numbers are related to the atomic numbers of atoms at the molecular graph nodes. In the canonical TSD the atomic sequence numbers are placed on a three-dimensional structure of the molecule to determine the parity vector of the node. The canonical TSD-parity vector analyzes structures node by node in such a way that the orientation of one node does not influence the representations of the other nodes. In both the CIP methodology and the permutational isomer notation, the molecule must be analyzed globally.

The CANON algorithm and permutational isomer notation employ in turn a relative ranking of ligands based on the atomic number of the atom under consideration, the atomic numbers of its α neighbors, and the permutational graph properties of their molecular structure (22,23). The CANON ligand indexing is specifically designed to give a relative ranking that is the same as the CIP relative priority for most compounds. A structure with the CANON index numbers is compared to a structure of the same geometry with a previously determined standard numbering. The permutations of the ligand positions on the candidate structure to those of the standard representation are determined, and a notation if derived for geometric and enantiomeric forms of the structure.

IV. APPLICATION OF CIP LIGAND PRIORITIES TO COORDINATED LIGANDS

The systematic stereochemical notation that has gained the greatest acceptance and usage is the Cahn-Ingold-Prelog (CIP) R, S notation, and it is described below in some detail with application to coordinating ligands.

The rules and procedures for determining the ligand index number and the chirality symbol comprise the CIP sequence rule. The CIP sequence rule states in part (12):

Ligand Complementation. All atoms other than hydrogen are complemented to quadriligancy (or higher for some inorganic molecules) by providing respectively one or two duplicate representations of any ligands (replica atoms) to which (they) are doubly or triply bonded, and any necessary number of phantom atoms of atomic number zero.

Sequence Rule. The ligands associated with an element of chirality are ordered by comparing them at each step in bond-by-bond explorations of them, from the element, along the successive bonds of each ligand, and where the ligands branch, first along branched paths providing highest precedence to their respective ligands, the explorations being continued to total ordering by use of the following Standard Subrules, each to exhaustion in turn, namely:

0. Nearer end of axis or side of plane precedes further.
1. Higher atomic number precedes lower.
2. Higher atomic mass-number precedes lower.
3. Seqcis precedes seqtrans. (Z precedes E).
4. Like pair R,R or S,S precedes unlike R,S or S,R,. . . .
5. (R) precedes (S), . . .

Chloromethyl(trimethylsilyl)stannanamine, (Fig. 7) provides an example of ranking by atomic number with the Cl, Si, N, and C atoms directly bonded to tetrahedral tin. The ranking by atomic number is $17 > 14 > 7 > 6$, which gives

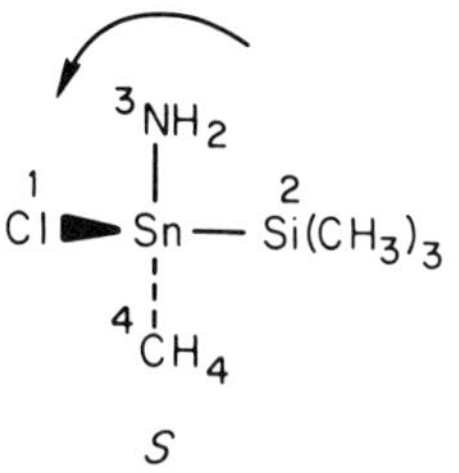

$$\text{Cl} \blacktriangleright \overset{\overset{3}{\text{NH}_2}}{\underset{\underset{4}{\text{CH}_4}}{\overset{1}{\text{Sn}}}} - \overset{2}{\text{Si}(\text{CH}_3)_3}$$

S

Fig. 7. Cahn-Ingold-Prelog ligand relative priority determined by atomic number.

the relative seniority $Cl > Si > N > C$, and the relative priority numbers $Cl = 1$, $Si = 2$, $N = 3$, and $C = 4$ as shown. Note the higher the seniority, the lower the CIP priority number.

Compounds with polyatomic ligands frequently require differentiation of constitutionally different donor atoms of the same atomic number. The expansion of the six amine ligands in the [(dimethylamine)(methylamine)(methylamino)(N-methylethylamine)(pyridine)(pyrimidine)metal] complex (Fig. 8) is shown in Table 2. The ligands are explored from the metal center outward, starting with the donor atom, until a priority is established based on the atomic number or the next appropriate subrule in sequence. The ligand atoms are representative by their respective atomic numbers in the exploration table and are compared at each step. All atoms with coordination numbers less than four (generally with double or triple bonds in organic ligands) are complemented with a replica atom. Electron pairs are shown as phantom atoms of atomic number zero (see the expansion of the methylamino and pyrimidine ligands in Table 2). The complementation of double bonds in pyridine and pyrimidine are shown by the replica atoms of atomic number 6 shown in parentheses in Table 2. The replica atoms on the ortho carbon positions of the pyridine ligand and position 6 on the pyrimidine ligand are assigned the average atomic number 6.5 $[(6 + 7)/2 = 6.5]$, to allow for the equivalency due to the delocalized π orbitals or resonance with the adjacent nitrogen and carbon atoms.

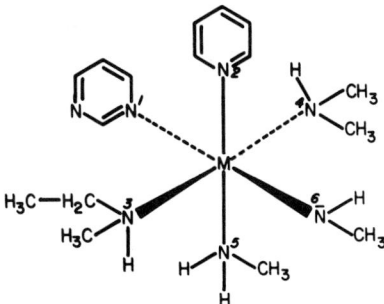

Fig. 8. Illustration of assignment of CIP relative priority numbers for ligands.

Table 2
Exploration Table (by Atomic Number).

The determination of the ligand CIP priorities for [nitratonitro(pyridine 1-oxide)(1,1,1-trifluoro-2,4-pentanedionato)metal] complex (Fig. 9) is shown in Table 3. The nitro, nitrate, and pyridine 1-oxide are complemented to quinqueligancy at the nitrogen atom and quadriligancy at the oxygen atoms. The resonance of the nitrate ligand results in a replica atom of average atomic number 3.5 [(7 + 0)/2 = 3.5], complemented to the nitrate nitrogen. The chelating 1,1,1,-trifluoro-2,4-pentanedione ligand is explored from each donor oxygen atom independently to give the oxygen of the carbonyl group adjacent to the trifluoromethyl group a higher relative CIP priority than the remaining oxygen donor.

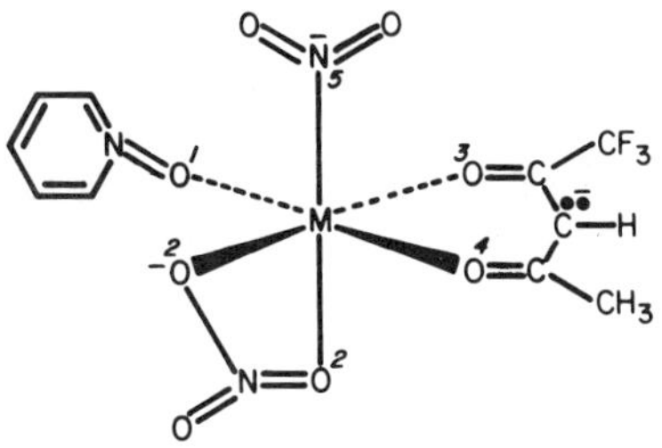

Fig. 9. Illustration of assignment of CIP relative priority numbers for ligands.

The exploration and assignment of CIP priority numbers for ammine(3-chloro-4-methoxyphenolato)(3-chloro-5-methoxyphenolato)-(N,N-dimethylcyclobutylamine)(pyridine)cobalt (Fig. 10) is examined in Table 4. The higher priority of the 3-chloro-5-methoxyphenol versus the 3,4 isomer is determined at the fifth level by comparing oxygen atomic number 8 to the carbon atomic number 6 in the second branch of the exploration tree.

The cyclopropyl and cyclobutylamine ligands illustrate that the nature of the exploration path, not the group size, is a critical feature in determining CIP

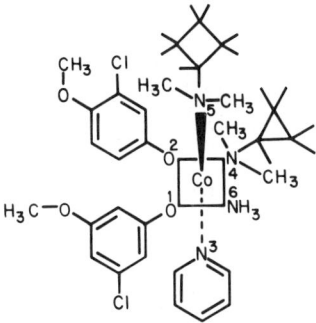

Fig. 10. Assignment of CIP priority numbers for ligands with donor atoms of the same atomic number.

Table 3
Exploration Table (by Atomic Number).

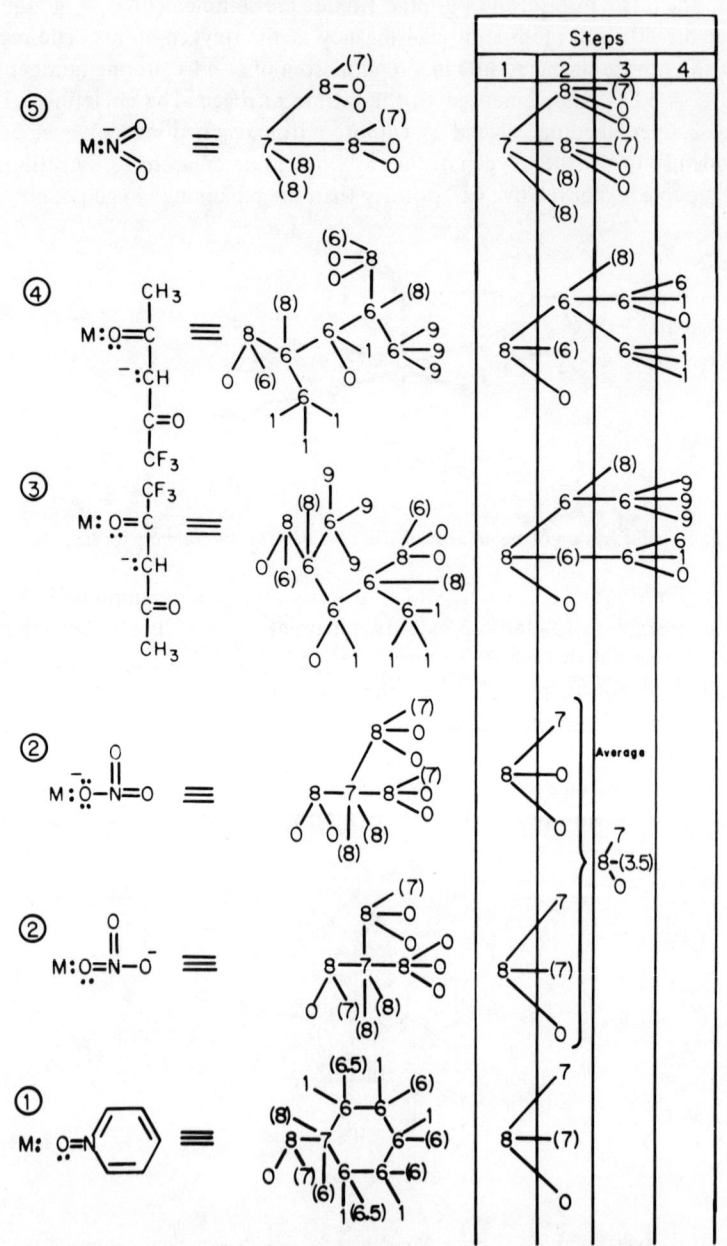

Table 4
Ligand CIP Exploration Table (by Atomic Number).

Table 4 (*continued*)

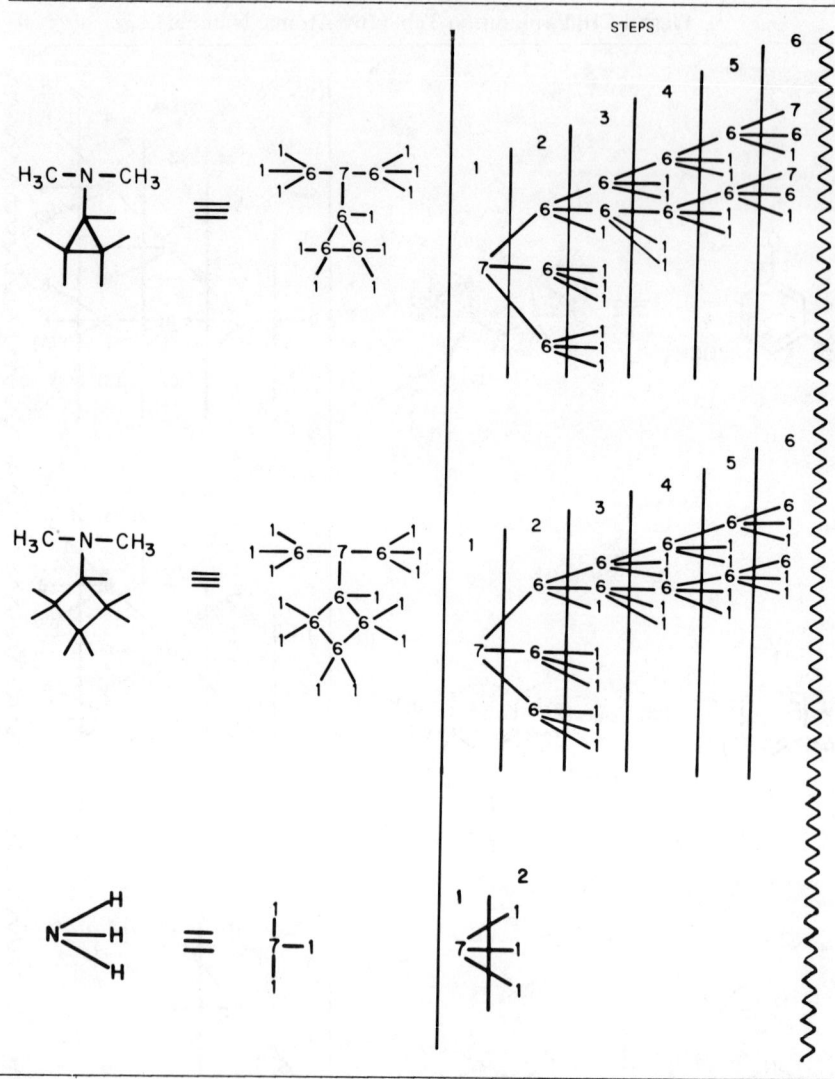

priorities. The cyclopropyl group results in the highest priority by returning to the amine nitrogen before the cyclobutyl group at the sixth level of the exploration.

The relative ranking of ligands is the most accepted and utilized principle for the description of the stereochemistry of organic molecules. This is not the accepted practice in the description of the stereochemistry of coordination and inorganic compounds. These compounds are described either by the traditional terminology or more often by means of stereospecific structural representations.

V. SYSTEMATIC STEREOCHEMICAL NOTATION FOR COORDINATION COMPOUNDS USING CIP PRIORITY NUMBERS

Brown, Cook, and Sloan reported a stereonotation system that is based on CIP priorities for ligands and is designed specifically to define the facets of stereochemistry that are unique to coordination and inorganic structures (24,25). This notation was introduced into the Chemical Abstracts Service Chemical Registry System and Chemical Abstracts Indexes in 1972 and was extended to higher coordination numbers in 1977.

A. The Stereochemical Descriptor

The CIP priority number-based notation uniquely defines the geometry of the central atom, the relative geometric distribution of ligands within the basic coordination structure, the chirality of the central atom, and the chirality associated with the ligands. This notation is achieved with a stereochemical descriptor that is composed of four parts: (a) a symmetry site term, (b) a configuration number, (c) a central atom chirality symbol, and (d) a ligand stereochemical lable. The symmetry site terms for the most common coordination polyhedra of coordination numbers 4 through 9 are given in Table 5.

The configuration number is a one- to nine-digit number that identifies ligand atoms on symmetry elements of a structure and by this means distinguishes the geometric isomers. The configuration numbers are composed of the CIP priority numbers for donor atoms.

Three different chirality lables are employed to designate the chirality at a coordination center. Although these symbols all convey essentially the same information (R, C, and Δ indicating right-handed or clockwise, and S, A, and Λ indicating left-handed or anticlockwise), the principles defining right- and left-handed are different for each set of symbols. Because of the differing assignment principles, there is no exact translation of R to C or C to Δ. For this reason the three different sets of chirality symbols are retained to emphasize the different set of defining principles.

Table 5
Symmetry Site Terms.

Four-Coordinate Polyhedra

T-4	tetrahedron
SP-4	square plane

Five-Coordinate Polyhedra

TB-5	trigonal bipyramid
SP-5	square pyramid

Six-Coordinate Polyhedra

OC-6	octahedron
TP-6	trigonal prism

Seven-Coordinate Polyhedra

PB-7	pentagonal bipyramid
OCF-7	octahedron face monocapped
TPS-7	trigonal prism square face monocapped

Eight-Coordinate Polyhedra

CU-8	cube
SA-8	square antiprism
DD-8	dodecahedron
HB-8	hexagonal bipyramid
OCT-8	octahedron trans-bicapped
TPT-8	trigonal prism triangular face bicapped
TPS-8	trigonal prism square face bicapped

Nine-Coordinate Polyhedra

TPS-9	trigonal prism square face tricapped
HB-9	heptagonal bipyramid

Stereochemical symbols and terms associated with ligand atoms may be given in the ligand portion of the name, for example, tris[*L*-aspartato(2−)] cobaltate-(3−) or more preferrably after the central atom chirality label, for example, [*OC*-6-21-Δ-(*R*,*R*,*R*)]-tris(1,2-propanediamine)cobalt(3+). In the latter cases the CIP priority number of the ligating atom may be used as a locant for defining the position of the closest ligand stereocenter in an ambiguous context. The lowest CIP priority number should be used for multidentate ligands.

B. Ligand Degeneracy and Multidentate Ligands

The CIP sequence rule provides numbers for ligating atom ranking based on constitutional and stereochemical differences. Therefore many ligating atoms receive the same or degenerate CIP priority numbers. To use the CIP priority numbers in an efficient and exact notation for highly symmetrical structures such as the octahedron, the following subrules were devised.

1. The trans Maximum Difference Subrule for Coordination Numbers 4 to 6

The same relative priority numbers are assigned to all constitutionally equivalent atoms. For example, in the system $Ma_2b_2c_2$ (where M is a central atom and a, b, and c represent monodentate ligands), the relative priority numbers are 1, 1, 2, 2, 3, 3. Whenever a choice exists in distinguishing between constitutionally equivalent donor atoms, the preference is given to the donor atom trans to the donor of highest CIP priority number. In this way configuration numbers can be chosen based on the symmetry characteristics of a particular geometry to distinguish between the position isomers of any geometry through coordination six by giving a number no more than three digits long.

2. Priming Subrule a

When there are two or more equivalent bidentate or tridentate ligands [e.g., $M(AA)_3$, $M(AB)_3$, and $M(BAB)_2$, where M is a central atom and AA, AB, and BAB represent multidentate chelating ligands] and the same priority numbers thus occur in equivalent ligands, the ties are broken by identically priming, double priming, and so on, all the CIP priority numbers of ligating atoms within a ligand to determine both the configuration number and the chirality symbol.

3. Priming Subrule b

In the cases of symmetrical quadridentate, quinquedentate, and sexidentate, ties between equivalent ligating atoms are broken, where necessary, by priming ligating atom priority numbers in half of the ligand. When two or more nonequivalent tie-breaking choices exist, the tie is broken by application of the trans maximum difference subrule. Primed ligating atom priority numbers are less preferred (higher value) than those that are unprimed but are of the same absolute value; doubly primed priority numbers are less preferred than primed, and so on. The primes are not included in the configuration number except when absolutely necessary, for example, for octahedral complexes containing two identical tridentate ligands and trigonal prismatic complexes containing two or more identical multidentate ligands.

C. Assignment of Configuration Numbers

Tetrahedral complexes (*T*-4) have no geometric isomers, therefore no configuration number is given. Square planar complexes (*SP*-4) exhibit planar geometric isomerism. The geometric isomerism is indicated by giving the CIP priority number for the ligating atom trans to the most preferred donor atom of CIP priority 1 (Table 6).

Octahedral complexes (*OC*-6) may have up to 15 geometric isomers. The first digit of the two-digit configuration number is the CIP priority number of the ligating atom trans to the most preferred atom of CIP priority 1. These two atoms define the principal axis of the octahedron.* The second digit of the configuration number is the CIP priority number for the atom trans to the most preferred ligating atom in the plane perpendicular to the principal axis (Table 7).

Table 6
Derivation of Configuration Numbers for Square-Planar Structures.

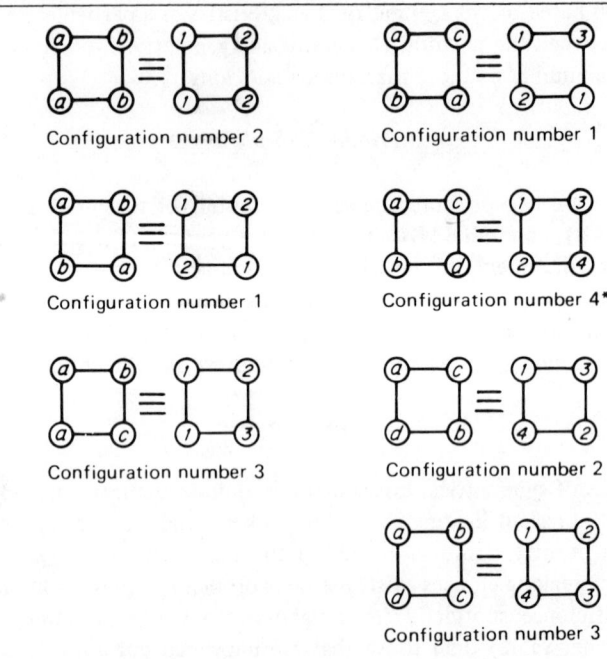

*(Note that cis-trans terminology is not adequate to distinguish the three isomers of an [Mabcd] square planar complex)

---

*If there are two ligating atoms of priority 1, the axis is picked on the basis of the principle of maximum trans difference (see p. 16); thus 1,3 has precedence over 1,2, etc.

Table 7
(Ma₂bcde) Octahedral Isomers.

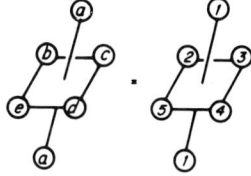

Configuration Number = 14

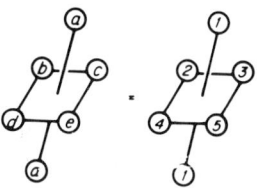

Configuration Number = 15

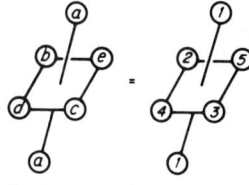

Configuration Number = 13

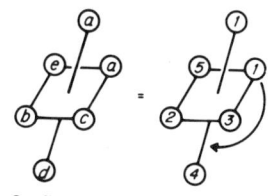

Configuration Number = 42
Chirality Symbol = C *

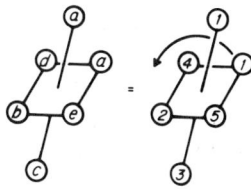

Configuration Number = 32
Chirality Symbol = A

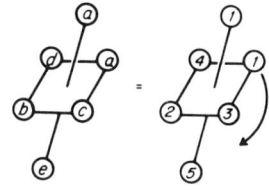

Configuration Number = 52
Chirality Symbol = C

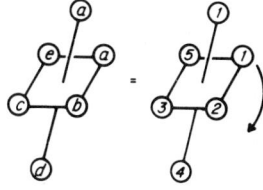

Configuration Number = 43
Chirality Symbol = C

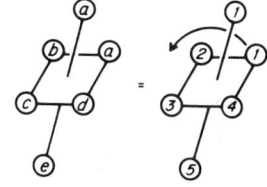

Configuration Number = 53
Chirality Symbol = A

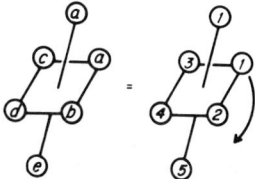

Configuration Number = 54
Chirality Symbol = C

*(Note preference is given to the axis with the highest CIP priority number opposite 1)

The configuration numbers for bipyramidal structures (*TB*-5, *PB*-7, *HB*-8, *HB*-9) and the square pyrimidal structure (*SP*-5) consist of the CIP priority numbers of the atoms on the highest order axis given in lowest sequential order[*] followed by the CIP priority numbers of the atoms in the plane perpendicular to this axis. These numbers are also cited in the lowest sequential order starting with the priority number of the highest CIP priority donor atom in the plane, then proceeding either right or left to give the lowest numerical sequence at the first point of difference. The priority numbers for the axial ligands in the bi-pyrimidal structures (*PB*-7, *HB*-8, *HB*-9) are set off from the planar priority numbers by a hyphen (e.g., *HB*-8-xx-xxxxxx; Table 8).

The configuration numbers for the five-coordinate structures *TB*-5 and *SP*-5 consist of two digits. The trigonal bipyramidal structure (*TB*-5) is denoted in the configuration number by giving the priority number for the atoms on the C_3 axis in lowest sequential order. The atoms in the trigonal plane define the chirality, and their sequential order determines the chirality symbol. The configuration number for the square pyramidal structure (*SP*-5) consists of the priority number of the atom on the C_4 axis and the priority number of the atom trans to the atom of highest CIP priority in the plane perpendicular to the C_4 axis.

Eclipsed structures (*TP*-6, *CU*-8, *DD*-8, *TPT*-8, *TPS*-9) are viewed from the atom of highest priority on the highest order axis or from a point on the highest order axis above the preferred facial plane. The configuration number consists of the priority numbers for the ligating atoms on the highest order axis, when present, cited in ascending sequential order, followed by the CIP priority number of the highest priority atom in the first plane and its eclipsed partner in a lower plane. The configuration number is completed by continuing either right or left to the next-most-preferred atom in the top plane and giving its priority number with its eclipsed partner, continuing in the same direction and manner until the priority numbers for all ligating atoms have been noted.

The configuration number for the trigonal prism, square face, tricapped structure is constructed in the same manner; however the priority numbers for the atoms in the intervening staggered trigonal plane are given as they are encountered, proceeding eight right or left to the next sequential priority number by viewing the projection of the structure from the C_3 axis.

The configuration number for the trigonal prism (*TP*-6) consists of three numbers. It is determined by viewing the structure from above the preferred trigonal face on the C_3 axis and giving the priority numbers for the eclipsed atoms in the lower trigonal face only. The numbers are cited in the order from lowest to highest as the priority numbers occur in the preferred or upper triangular face. Priority numbers for atoms on the highest order axes in the *TPT*-8 structure are set off by hyphens (e.g., *TPT*-8-xx-xxxxxx; Table 9).

Staggered structures (*OCF*-7, *TPS*-7, *SA*-8, *OCT*-8, *TPS*-8) are viewed from

[*](e.g., 1,2,4 > 1,2,5 > 1,3,4),

Table 8
Derivation of Configuration Numbers for Pyramidal and Bipyramidal Structures.

1. Trigonal bipyramid

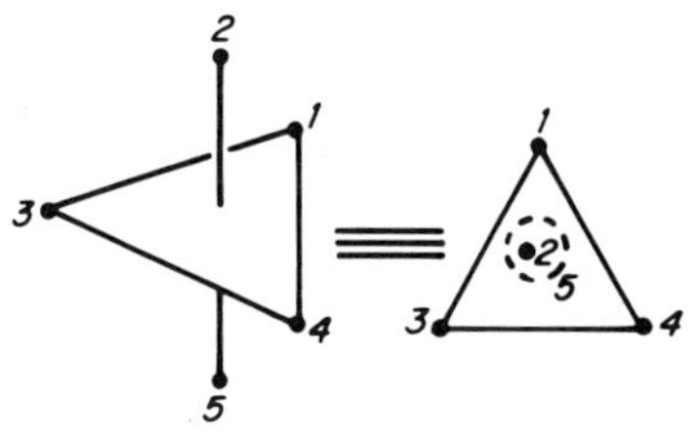

TB-5-25

2. Square pyramid

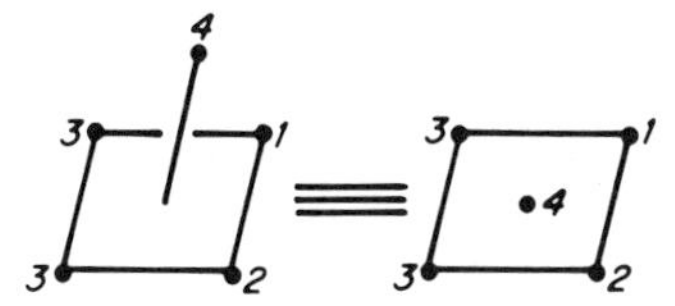

SP-5-43

3. Pentagonal bipyramid

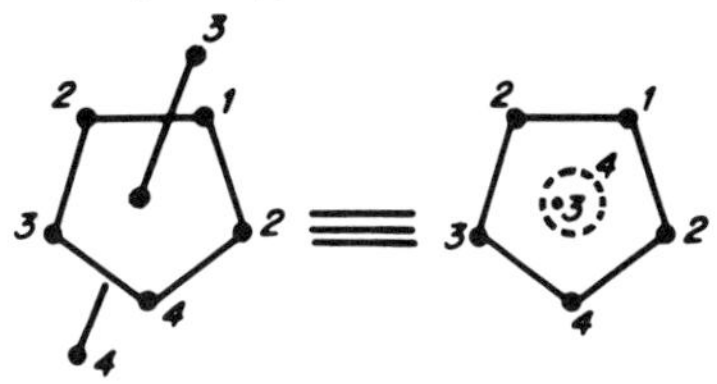

PB-7-34-12342

4. Hexagonal bipyramid

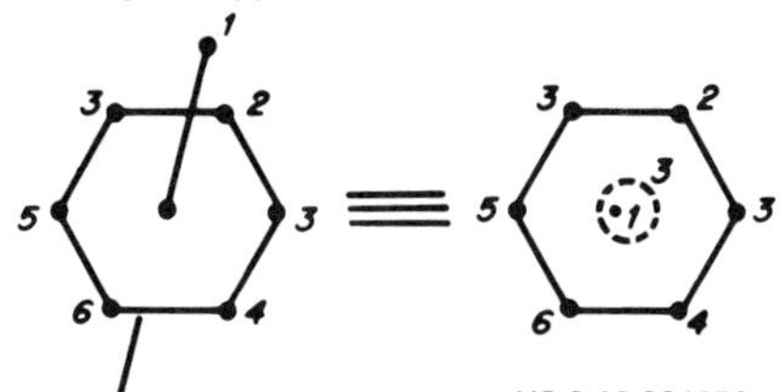

HB-8-13-234653

5. Heptagonal bipyramid

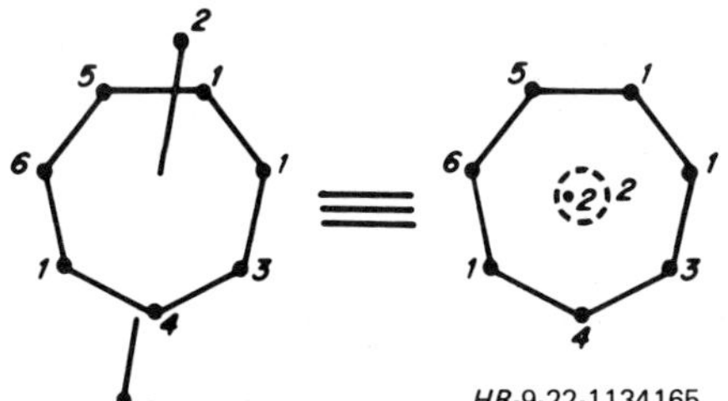

HB-9-22-1134165

---

the atom of highest priority on the highest order axis or from a point on the highest order axis above the most-preferred plane. The configuration number begins with the priority number(s) of the atom(s) on this axis given in ascending sequential order. This priority number is separated from the remaining portion of the configuration number by a hyphen (e.g., *TPS*-7-x-xxxxxx). The remaining portion of the configuration number for staggered structures is derived by viewing the structure from the highest priority atom on the preferred axis or from a position on the highest order axis above the preferred facial plane and by citing the CIP priority numbers of the ligating atom with the highest priority in the top

Table 9
Derivation of Configuration Numbers for Eclipsed Ligand Structures.

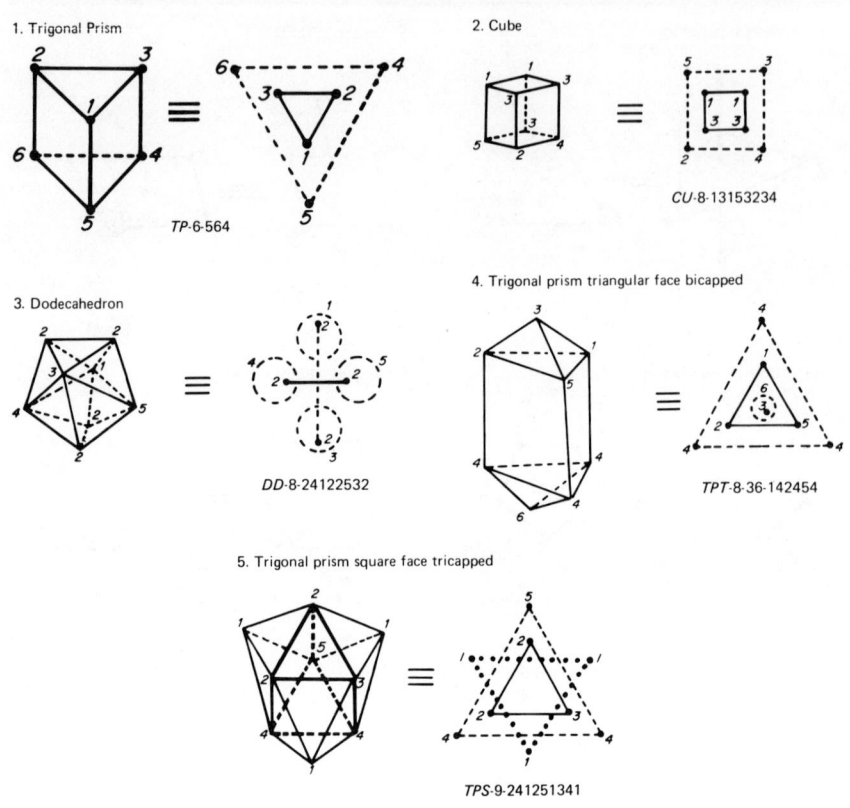

plane. The configuration number is then completed by continuing to give the priority numbers of donor atoms in sequence as they are encountered, either clockwise or anticlockwise around the projection of the model structure, alternating between planes to give the lowest numerical sequence determined at the first point of difference (Table 10).

The assignment of the configuration numbers for complexes with chelating ligands follows the same procedures as outlined above, with the addition that primed priority numbers are employed to remove the degeneracy of the CIP priority numbers assigned to equivalent donor atoms in identical bidentate, tridentate, and in equivalent groupings of symmetrical multidentate chelating ligands (Table 11). The priming procedure is not necessary for octahedral cis-bis bidentate and tris bidentate structures because of the helical assignment of the Δ and Λ chirality symbols. The priming subrule a and the C and A chirality symbols

Table 10
Derivation of Configuration Numbers for Staggered Ligand Structures.

2. Octahedron face monocapped   **3. Trigonal prism square face monocapped**

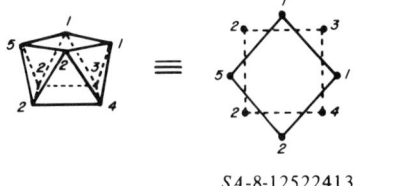

OCF-7-4-213251   TPS-7-4-214345

5. Square antiprism   **8. Octahedron trans bicapped**

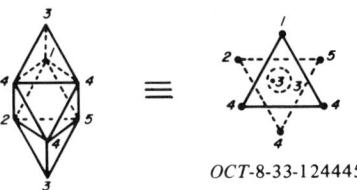

SA-8-12522413   OCT-8-33-124445

10. Trigonal prism square face bicapped

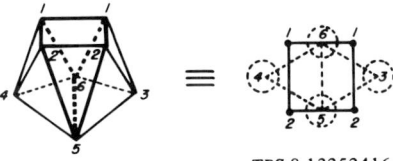

TPS-8-13252416

---

could be employed to provide the chirality assignment, but this approach is not felt be as useful given to the simplicity of the helical convention and the widespread use of the helical notation for octahedral structures.

D. Assignment of Chirality Symbols

The symbols R for a right-handed path and S for a left-handed path are the chirality symbols used for tetrahedral compounds. The tetrahedral center is viewed from the side opposite the highest CIP-numbered ligand priority number 4, (lowest priority), and as the path defined by the sequence 1,2,3 is right- or left-handed, the tetrahedral center is assigned the label R or S, respectively.

The chirality of cis-bis bidentate and tris bidentate octahedral compounds is indicated by Δ for a right-handed and Λ for a left-handed helix. The octahedral

Table 11
Derivation of Configuration Numbers for Chelated Structures.

PB-7-22'-1233'2'

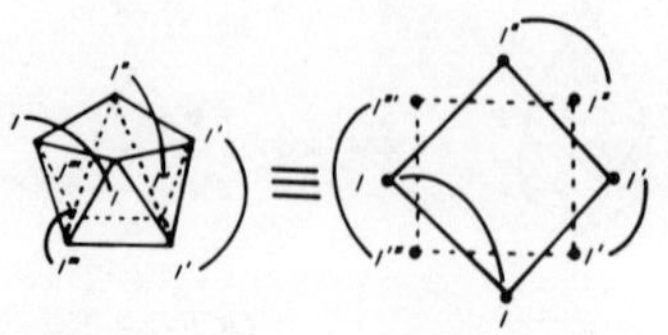

SA-8-11'1'1"1"'11'"

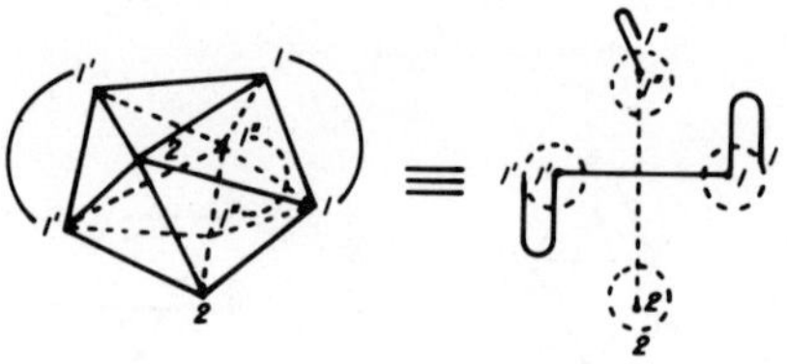

DD-8-111"1"1'1'22

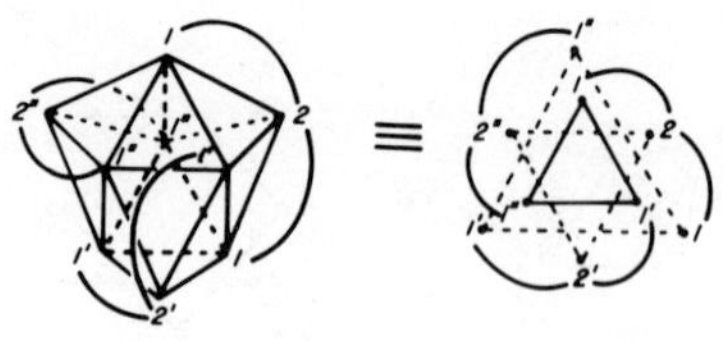

TPS-9-11"21'12'1"'1'2"

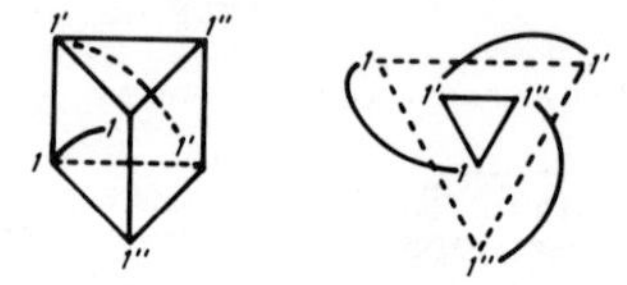

TPT-6-1"11'

structure is viewed such that two bidentate ligands define a pair of intersecting skew nonorthogonal lines. The intersecting nonorthogonal line segments in turn define either a right- or a left-handed helix and are symbolized Δ and Λ (Fig. 11).

The chirality symbols C for clockwise and A for anticlockwise are used for central atom geometries other than the tetrahedral and cis-bis bidentate and tris bidentate octahedral compounds. The structure is viewed either from the atom of lowest CIP priority number on a major symmetry axis or from a point on a major axis above the preferred facial plane of the structure. The preferred facial plane will contain the largest number of lowest priority numbers or will be adjacent to a plane with the larger number of lowest priority numbers. As the projection of the structure is viewed from this point, the CIP priority numbers will define a sequence that increases either clockwise or anticlockwise and is symbolized C or A, respectively.

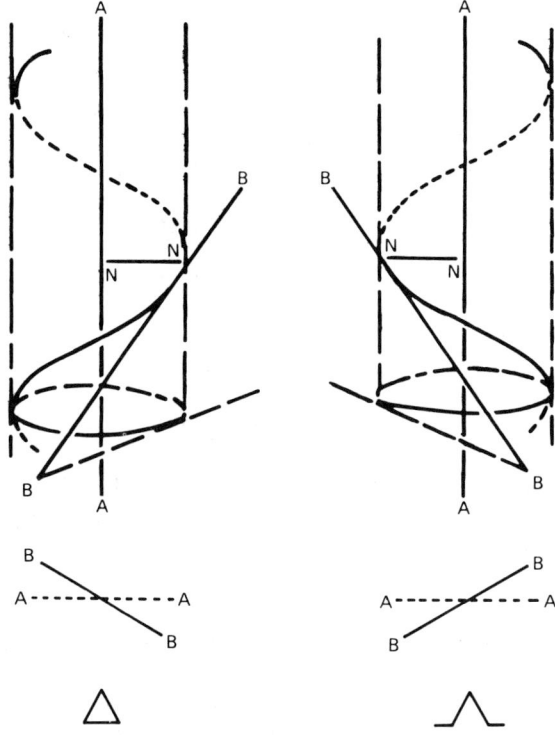

Fig. 11. A left handed helix with axis *AA* tangent *BB*.

E. Examples

The structures in Figures 12 through 16 illustrate the ultility of a ligand index-based stereochemical notation for coordination compounds.

1. Tris(1,2-propanediamine-N,N')cobalt(3+)

The tris(1,2-propanediamine-N,N')cobalt(3+) structure of Figure 12 has two geometric configurations each for the R,R,R and S,S,S ligand configuration, both of which have a Δ and a Λ form to give eight isomers. The R,R,S, and R,S,S ligand configurations have two facial and six meridional configurations, each in a Δ and a Λ form, to give 16 isomers. This results in a total of 24 possible isomers. In the structure shown, the asymmetric carbon atoms are in a meridional configuration. The principal axis is determined by the maximum difference in the trans CIP priority numbers, the 1−2 axis. The configuration number is determined by noting the priority numbers of the atoms trans to the highest CIP

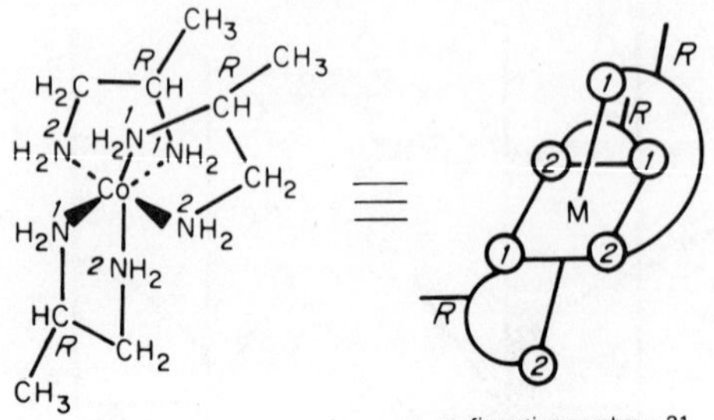

configuration number = 21
chirality symbol = Δ

[OC-6-21-Δ-(R),(R),(R)]-tris-(1,2-propanediamine-N,N')-cobalt(3+)

Fig. 12. [Co(1,2-pn)$_3$]$^{3+}$.

priority atoms on the principal axis and in the plane, 12. The helical configuration of the chelate rings is right-handed and is symbolized by Δ. The complete stereodescriptor is [OC-6-21-Δ-(R),(R),(R)]. The stereodescriptors for all 24 isomers are tabulated in Table 12.

2. [[N,N'-Ethylenebis(glycinato)] (2−)] [oxalato(2−)] cobalate(1−)

An octahedral structure with a quadridentate ligand is shown in Figure 13. The oxalate oxygens are CIP priority number 1, the glycine oxygens are priority number 2, and the nitrogens are priority number 3. The quadridentate ligand is

Table 12
Stereonotation for 24 Isomers of [Co(1,2-pn)$_3$]$^{3+}$.

[OC-6-21-Δ-(R),(R),(R)]	[OC-6-21-Λ-(S),(S),(S)]
[OC-6-22-Δ-(R),(R),(R)]	[OC-6-22-Λ-(S),(S),(S)]
[OC-6-21-Δ-(S),(S),(S)]	[OC-6-21-Λ-(R),(R),(R)]
[OC-6-22-Δ-(S),(S),(S)]	[OC-6-22-Λ-(R),(R),(R)]
[OC-6-43-Δ-(R),(R),(S)]	[OC-6-44-Λ-(R),(S),(S)]
[OC-6-44-Δ-(R),(S),(S)]	[OC-6-43-Λ-(R),(R),(S)]
[OC-6-32-Δ-(R),(R),(S)]	[OC-6-24-Λ-(R),(S),(S)]
[OC-6-13-Δ-(R),(R),(S)]	[OC-6-42-Λ-(R),(S),(S)]
[OC-6-42-Δ-(R),(R),(S)]	[OC-6-23-Λ-(R),(S),(S)]
[OC-6-24-Δ-(R),(S),(S)]	[OC-6-32-Λ-(R),(R),(S)]
[OC 6-42-Δ-(R),(S),(S)]	[OC-6-13-Λ-(R),(R),(S)]
[OC-6-23-Δ-(R),(S),(S)]	[OC-6-42-Λ-(R),(R),(S)]

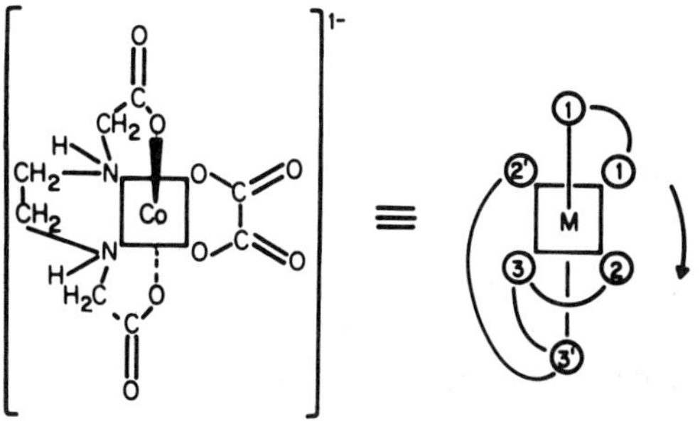

α—cis

Configuration number = 33
Chirality symbol = C

$(OC\text{-}6\text{-}33\text{-}C) - [[N,N'\text{-ethylenebis(glycinato)}](2-)]^{-1}$ [oxalato(2-)] cobalate(1-)

Fig. 13. $[Co[(CH_2)_2[NH(C_2H_2O_2)]_2(C_2O_4)]$.

primed to give the maximum difference on the principal axis, 1–3′. The configuration number is 33, with the glycine nitrogens trans to the oxalate oxygens. Primes are not generally necessary in the configuration numbers for octahedral complexes, but they are used to determine the chirality descriptor. The chirality is determined by the increasing sequence 1232′ or clockwise, C. The complete descriptor is (OC-6-33-C) for the α-cis isomer and (OC-6-32) for the β-cis isomer.

3. Bromochloro(2-ethyl-13-methyl-3,6,9,12,18-pentaazabicyclo[12.3.1]-octadeca-1(18),2,12,14,16-pentaene-N^3,N^6,N^9,N^{18})iron(1+)

A seven-coordinate pentagonal bipyramidal structure with a planar macrocyclic ligand is shown in Figure 14. The bromine atom is priority number 1, the chlorine is 2, the imine nitrogens are 3 and 4, the pyridine nitrogen is 5, and the amine nitrogens have priority numbers of 6 and 7 as shown. The configuration number starts with the priority numbers of the atoms on the C_5 axis, 12, and

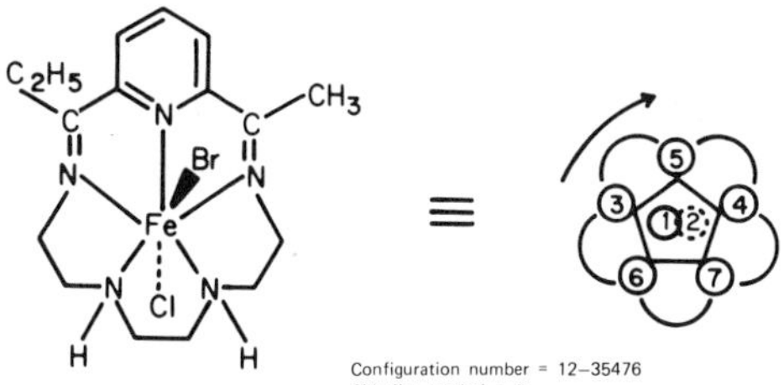

Configuration number = 12-35476
Chirality symbol = C

(PB-7-12-35476-C)-bromochloro (2-ethyl-13-methyl- = 3,6,9,12,18-pentaazabicyclo [12.31] octadeca-1(18),2, = 12,14,16-pentaene-N^3,N^6,N^9,N^{18},) iron (1+)

Fig. 14. $[FeBrCl(C_{16}H_{25}N_5)]$.

continues with the priority numbers for the atoms in the pentagonal plane in the increasing sequential order 35476. When viewed from atom priority number 1 on the C_5 axis, this sequence is clockwise or C. The complete descriptor is (*PB*-7-12-35476-*C*).

4. Bis[N-aminoethyl]-1,2-ethanediamine-N,N',N''] bromochloropraseodymium(1+)

An eight-coordinate, triangular face, bicapped trigonal prismatic structure is shown in Figure 15. The structure is viewed from the atom of highest priority on the C_3 axis, priority number 4. The configuration number begins with the priority numbers for the donor atoms on the C_3 axis given in lowest sequential order, 4,4', and continues with the priority number of the highest priority atom in the adjacent plane, 1, along with the priority number for the atom it eclipses, 3'. The configuration number sequence continues either right or left to the donor atom of next highest priority in the first plane. This is the atom of priority number 3. It and the atom it eclipses, 2, are cited next. The sequence continues in a like manner to give 44'-13'324'4. The sequence increases in clockwise direction when viewed from atom priority number 4 on the C_3 axis. The complete descriptor is (*TPT*-8-44'-13'324'4-*C*).

5. Tetraaqua(2,4,5,6(1H,3H)-pyrimidinetetrone 5-oximato-N^5,O^4)-(2,4,5,6,(1H,2H)-pyrimidinetetrone 5-oximato-O^4,O^5)strontium

The structure in Figure 16 is an eight-coordinate, square face, bicapped trigonal prism. The structure is viewed from a point on the C_3 axis above the

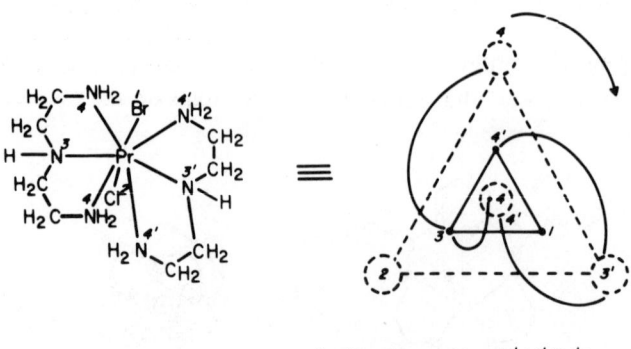

Configuration number = 44'-13'324'4
Chirality symbol   = C

(*TPT*-8-44'-13'324'4-*C*)-bis[*N*-(2-aminoethyl)-1,2-ethanediamine-=
N,N',N''] bromochloropraseodymium(1+)

Fig. 15.   [Pr($C_4H_{13}N_3$)$_2$BrCl]$^+$.

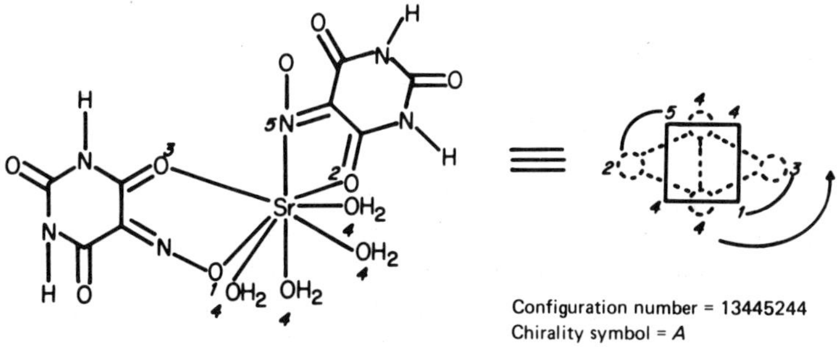

Configuration number = 13445244
Chirality symbol = A

(*TPS*-8-13445244-*A*)-tetraaqua(2,4,5,6(1*H*,3*H*)-pyrimidinetetrone 5-= oximato-N^5,O^4)(2,4,5,6(1*H*,3*H*)-pyrimidinetetrone 5-oximato-O^4,O^5) = strontium

Fig. 16. [Sr(H$_2$O)$_4$(C$_4$H$_2$N$_3$O$_4$)$_2$].

uncapped square or rectangular face. The first digit of the configuration number is the priority number of the most-preferred atom in the square face, 1. The second number is that of the atom encountered next on the structure projection as one proceeds either right or left to the atom in the plane adjacent to the square face, priority number 3. The priority numbers for the remaining atoms are given as they are encountered, continuing in the same direction on the projection of the structure to give 13445244. As viewed from above the square face, the sequential order increases in the anticlockwise direction, symbolized A. The complete descriptor is (*TPS*-8-13445244-*A*).

The Chemical Abstracts notation provides a vehicle for inorganic and coordination chemists to describe the gross symmetry of the central atom, the geometric distribution of ligands on the coordination polyhedron, and the stereochemical configurations at the central atom and at centers in the ligands. This notation is especially useful for describing the more complicated types of stereochemical configurations (Table 12), for the compounds of higher coordination numbers, and for lower symmetry coordination polyhedra that have potentially thousands of possible isomers.

VI. HAPTO COMPLEXES

The stereochemistry of bis(η^5-cyclopentadienyl)iron(ferrocene), tricarbonyl-(η^5-cyclopentadienyl)manganese, and (η^6-benzene)tricarbonylchromium has been extensively reviewed by Schlögl (26). His review is primarily concerned with ring substitution, and the stereonotation is the extension of the CIP sequence rule, R,S, to the cyclopentadiene ring. This is achieved by considering the hapto attachment to the metal equivalent to a single bond. This device results in a four-coordinate center (not tetrahedral), which can be ranked by the normal

CIP procedures. The ranking and notation for 1-acetyl-2-methylferrocene is shown in Figure 17. The atomic number of iron, 26, is the highest ranking atom, priority number 1. The methyl-substituted ring atom has a priority number of 2; the α-unsubstituted ring atom has a priority number of 3; and the carbonyl atom has a priority number of 4. The chirality symbol is given to the highest CIP priority cyclopentadienyl ring atom. When this atom is viewed from opposite the lowest priority ligand, number 4, the sequence 1,2,3 is right-handed and is given the chirality symbol R. This application of the CIP sequence rule has been generally accepted and is employed in Chemical Abstracts indexes to assign chirality symbols to polysubstituted bis(η^5-cyclopentadienyl) metal complexes. Before the concept of centrochirality was adopted in 1967 (26), cyclopentadienyl metal compounds were assigned R and S descriptors according to the concept of planar chirality. Following the planar chirality procedure, the molecule is viewed from a point above the substituted ring on the C_5 axis. The ring substituents are ranked by the CIP procedures and the R or S chirality symbol is assigned based on the clockwise or anticlockwise sense of the resulting sequence. According to the rules of planar chiralty, the ferrocene complex in Figure 17 would be assigned the symbol S, but this approach has been abandoned.

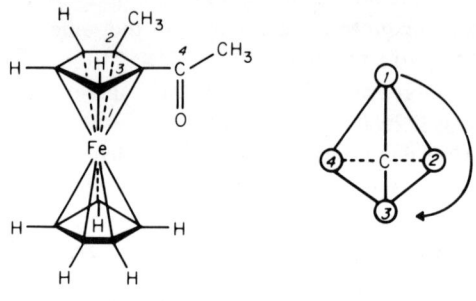

(R)-1-acetyl-2-methyl-ferrocene

Fig. 17. [Fe(η^5-C$_5$H$_5$)[η^5-(CH$_3$CO)(CH$_3$)C$_5$H$_3$]].

Optically active η^5-cyclopentadienyl metal complexes, in which the metal is a chiral center, are becoming increasingly available. Stanley and Baird have suggested an extension of the CIP sequence rule to be applied to the metal center (27). This proposal states that the polyhapto ligand "be considered pseudo-atoms of atomic weight equal to the sum of the atomic weight of all the atoms bonded to the metal atom." This proposal has been accepted and employed by several researchers in this field. It should be noted that to keep within the generally accepted priorities of the CIP sequence rule as quoted above, the first priority is "higher atomic number precedes lower." Thus the pseudo-atom proposal might be altered such that polyhapto ligands would be considered pseudo-atoms of atomic number equal to the sum of the atomic numbers of all atoms bonded to the metal atom, to be consistent with the CIP sequence rule. Thus η^8-C_8H_8, η^7-C_7H_7, η^6-C_6H_6, η^5-C_5H_5, η^4-C_4H_6, and η^3-C_3H_5 would be treated as pseudo-atoms of atomic number $48 > 42 > 36 > 30 > 24 > 15$. This places η^5-cyclopentadienyl, the most common polyhapto ligand, between the second and third periods of the periodic table (e.g., between Cl, atomic number 17, and As, atomic number 33), considering only common donor atoms. This is approximately the same relative priority obtained by comparison of atomic weights (e.g., η^5-C_5H_5, 60; Cl, 35.5; and As, 75), for the same donor atoms.

Applying the pseudo-atom convention to the iron complex in Figure 18, the iodine atom is priority 1, the η^5-$C_5H_3R_2$ ligand is priority number 2, the phosphorus atom is priority number 3, and carbon is priority number 4 (28). When the iron is viewed from the side opposite the priority number 4, the sequence is 1,2,3 in the anticlockwise or S direction. The highest priority carbon in the cyclopentadienyl ligand, indicated by an asterisk, is designated with the R chirality symbol by application of the extended CIP sequence rule.

Bernal, Brunner, et al. have further extended the pseudo-atom priorities to a series of five-coordinate molybdenum complexes, one of which is shown in Figure 19 (29). They assign the chirality symbol according to the octahedral sequence rule by treating the square pyramid as an octahedral with a vacant sixth position. Accordingly, the ligand positions on the octahedron are numbered 1 through 6 such that 1 and 6 are always trans and 2 through 5 will advance either right or left around the belt of the octahedron according to the priorities of the belt donor atoms. The structure is then viewed from opposite the triangular face 4,5,6. The sequence 1,2,3 is clockwise, and the chirality symbol R is assigned.

The assignment of descriptors and configuration is sometimes arbitrary, at

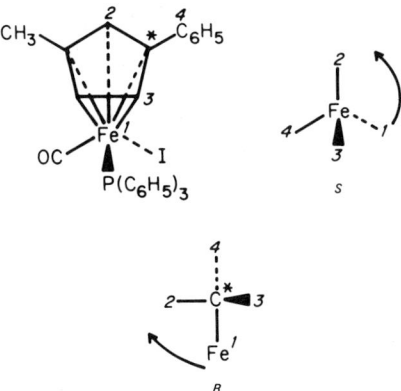

Fig. 18. [Fe[3-CH$_3$)-1-(C$_6$H$_5$)-η^5-C$_5$H$_3$] [P(C$_6$H$_5$)$_3$](CO)]].

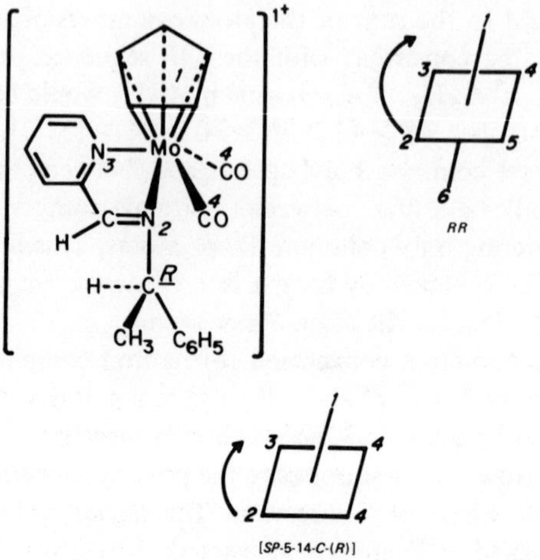

Fig. 19. [Mo(η^5-C$_5$H$_5$)(CO)$_2$ [2-[[[[2-(C$_6$H$_5$)]C$_2$H$_5$]N]CH]C$_5$H$_4$N]].

best, when based on model structures and pseudo-atom coordination numbers. A more explicit stereochemical notation is achieved by using the Chemical Abstracts notation, which states within the stereodescriptor the model structure on which the notation is based. In the Chemical Abstracts notation the pseudo–square pyramidal structure is [*SP*-5-14-*C*-(*R*)]. This structure can be expected to result in geometric isomers when one of the carbonyls is substituted by a ligand such as a phosphine. This is easily noted in the configuration number.

The definition of a pseudo-atom coordination number for polyhapto ligands will be a difficult one on which to achieve a consensus.* For example, η^5-cyclopentadienyl can be thought to be either five-coordinate (having five nearest neighbor carbon atoms to the coordination center), three-coordinate (due to the donation of six valence electrons and the usual replacement of three monohapto ligands), or monocoordinate (in the pseudo-atom concept above). Whatever coordination number or coordination geometry is chosen, the Chemical Abstracts notation is unambiguous because the necessary information is stated exactly as an integral part of the stereodescriptor.

VII. POLYNUCLEAR AND CLUSTER COMPLEXES

Bridged binuclear, trinuclear, and tetronuclear chelated octahedral structures are examined by Schaffer (30), who use the skew line helical definition for the

---

*The author solicits all comments concerning atomic number versus atomic weight and pseudo-atom (pseudomonocoordinate) definition.

chirality symbols Δ and Λ. The configurational isomers for a tetrakisbidentate, edge-fused bis-octahedral structure are ΔΔ, ΔΛ, ΛΔ, and ΛΛ.

Many polynuclear and cluster compounds with challenging stereochemical properties are frequently prepared and reported. For example, asymmetric metalocarboranes have been prepared by Miller et al. (31), and an asymmetric tetrahedral cluster with four different atoms at the tetrahedral apexes was recently reported by Richter and Vahrenkamp (32). Polynuclear and cluster compounds, especially those with polyhapto ligands and ligands that contain involved stereochemistry in and of themselves, have potentially tens of thousands, if not hundreds of thousands, of isomeric configurations. Intermediate size chemical systems with relatively simple and straightforward stereochemical relationships can be described by the various notations already outlined. These notations will require some extensions and further elaboration, such as an appropriate set of locants to place a stereodescriptor at an exact position in an extended structure and, undoubtedly, further refinements that cannot be foreseen until a particular configuration is demonstrated. However the larger compounds with the very complicated stereochemical relationships will, in all probability, require the more powerful and rapid analysis available only through the application of computer systems.

VIII. CONCLUSION

The contributions of stereochemistry to the overall development of structural chemistry can hardly be overstated. The study of the stereochemistry of coordination and inorganic systems has been a lively and rewarding topic for the last three decades. I have outlined the development of stereochemical nomenclature in the twentieth century and have included a description of the development of ligand ranking and ligand indexing procedures.

The application of CIP ligand priority numbers in the Chemical Abstracts stereochemical notation has been provided, with several illustrative examples. I have considered several recent proposals regarding the use of CIP priority numbers in a stereochemical notation for polyhapto compounds.

The implications of the large numbers of isomers and the many intricate stereochemical relationships involved in polynuclear and cluster compounds indicates that although a start has been made toward a rational stereochemical nomenclature for coordination and inorganic compounds, much work still remains.

REFERENCES

1. S. F. Mason, "The Foundations of Classical Stereochemistry," in *Topics in Stereochemistry*, Vol. 9, N. L. Allinger and E. L. Eliel, Eds., Wiley, New York, 1976, pp. 1–34.

2. E. Fremy, *Ann. Chim. Phys.* [3], **35**, 257 (1852); *J. Prakt. Chem.*, **57**, 95 (1852).
3. A. Werner, *New Ideas on Inorganic Chemistry*, 2nd ed., E. P. Hedley, Trans., Longmans, Green, New York, 1911, pp. 229–268.
4. W. P. Jorissen, H. Bassett, A. Damiens, F. Fichter, and H. Remy, *J. Chem. Soc.*, **1940**, 1404.
5. W. C. Fernelius, "Nomenclature of Coordination Compounds and Its Relation to General Inorganic Nomenclature," in *Chemical Nomenclature*, Advances in Chemistry Series, Vol. 8, American Chemical Society, Washington, D.C., 1953, pp. 27–30.
6. E. L. Muetterties and C. M. Wright, *Q. Rev. Chem. Soc.*, **21**, 109 (1967).
7. *IUPAC Nomenclature of Inorganic Chemistry, 1970*, 2nd ed., Butterworths, London, 1971, pp. 92–97. Rule 11.
8. R. F. Pasternack and P. M. McDonnell, *Inorg. Chem.*, **4**, 600 (1965); *J. Chem. Doc.*, **5**, 56 (1965).
9. *IUPAC Nomenclature of Inorganic Chemistry, 1970*, 2nd ed., Butterworths, London, 1971, pp. 55–71. Rules 7.5 and 7.6.
10. E. Fischer, *Ber. Dtsch. Chem. Ges.*, **29**, 1377 (1896).
11. J. M. Bijvoet, A. F. Peerdeman, and A. J. van Bommel, *Nature (London)*, **168**, 271 (1951).
12. R. S. Cahn, C. Ingold, and V. Prelog, *Angew. Chem., Int. Ed. Engl.*, **5**, 385 (1966).
13. J. E. Blackwood, C. L. Gladys, K. L. Loening, A. E. Petrarca, and J. E. Rush, *J. Am. Chem. Soc.*, **90**, 509 (1968); J. E. Blackwood, C. L. Gladys, A. E. Petrarca, W. H. Powell, and J. E. Rush, *J. Chem. Doc.*, **8**, 30 (1968).
14. Y. Saito, K. Nakatsu, M. Shiro, and H. Kuroya, *Acta Crystallogr.*, **8**, 729 (1955).
15. T. S. Piper, *J. Am. Chem. Soc.*, **83**, 3908 (1961).
16. A. J. McCaffery, S. F. Mason, and R. E. Ballard, *J. Chem. Soc.*, **1965**, 2883.
17. *IUPAC Nomenclature of Inorganic Chemistry, 1970*, 2nd ed., Butterworths, London, 1971, pp. 75–83. Rule 7.8.
18. H. L. Morgan, *J. Chem. Doc.*, **5**, 107 (1965).
19. W. T. Wipke and T. M. Dyott, *J. Am. Chem. Soc.*, **96**, 4834 (1974).
20. F. Choplin, R. Marc, G. Kaufmann, and W. T. Wipke, *J. Chem. Inf. Comput. Sci.*, **18**, 110 (1978).
21. H. W. Davis, *Computer Representation of the Stereochemistry of Organic Molecules*, Birkhäuser Verlag, Basel, Switzerland, 1976.
22. W. Schubert and I. Ugi, *J. Am. Chem. Soc.*, **100**, 37 (1978), and private communication.
23. I. Ugi, D. Marquarding, H. Klusacek, G. Gokel, and P. Gillespie, *Angew. Chem., Int. Ed. Engl.*, **9**, 703 (1970).
24. M. F. Brown, B. R. Cook, and T. E. Sloan, *Inorg. Chem.*, **14**, 1273 (1975).
25. M. F. Brown, B. R. Cook, and T. E. Sloan, *Inorg. Chem.*, **17**, 1563 (1978).
26. K. Schlögl, "Stereochemistry of Metallocenes," in *Topics in Stereochemistry*, Vol. 1, N. L. Allinger and E. L. Eliel, Eds., Wiley, New York, 1967, pp. 39–91.
27. K. Stanley and M. C. Baird, *J. Am. Chem. Soc.*, **97**, 6598 (1975).
28. T. G. Attig, R. G. Teller, S. M. Wu, R. Bau, and A. Wojcicki, *J. Am. Chem. Soc.*, **101**, 619 (1979).
29. T. Bernal, S. J. LaPlaca, J. Korp, H. Brunner, and W. Herrmann, *Inorg. Chem.*, **17**, 382 (1978).
30. U. Thewalt, K. A. Jensen, and C. E. Schäffer, *Inorg. Chem.*, **11**, 2129 (1972).
31. V. R. Miller, L. G. Sneddon, D. C. Beer, and R. N. Grimes, *J. Am. Chem. Soc.*, **96**, 3090 (1974).
32. F. Richter and H. Vahrenkamp, *Angew. Chem., Int. Ed. Engl.*, **17**, 864 (1978).

Stereochemistry of Reactions of Transition Metal–Carbon Sigma Bonds

THOMAS C. FLOOD

Department of Chemistry, University of Southern California, Los Angeles, California

I.	Introduction		38
II.	Reactions That Form Metal-Carbon Sigma Bonds		39
	A.	Oxidative Addition of Alkyl and Alkenyl Species	39
		1. Alkyl Species to Metal Anions	39
		2. Alkyl Species to Neutral Complexes	45
		3. Oxidative Addition of Alkenyl Species	51
		4. Oxidative Addition of Carbon-Carbon Bonds	56
	B.	Transmetalation	58
	C.	Additions to Complexed Alkenes and Alkynes	63
		1. Additions of Oxygen and Nitrogen Nucleophiles	63
		2. Additions of Carbon Species	69
		3. Addition of Hydride, β-Additions, and β-Eliminations	76
III.	Reactions That Disrupt Metal-Carbon Sigma Bonds		82
	A.	Insertion Reactions	82
		1. Carbon Monoxide Insertion	83
		2. Alkene, Alkyne, Sulfur Dioxide, and Other Insertions	90
	B.	Reductive Elimination	94
		1. Forming Carbon-Carbon Bonds	94
		2. Forming Carbon-Hydrogen Bonds	96
		3. Forming Carbon-Heteroatom Bonds	98
	C.	Electrophilic Cleavage	99
		1. Protonolysis	99
		2. Cleavage by Halogens	101
		3. Cleavage by Mercuric Salts	104
		4. Oxidative Cleavage and Autoxidation	107
IV.	Conclusions		109
	Acknowledgements		110
	References		111

I. INTRODUCTION

The ubiquitous involvement of metal-carbon σ bonds in changes of bonding at carbon centers induced by metals, either catalytically or stoichiometrically, is now well established. Consequently considerable effort has been expended in examining mechanism and stereochemistry of reactions of such σ-bonded species. This review emphasizes stereochemical aspects of these reactions, but incorporates enough data of other types to put the stereochemistry into mechanistic perspective where possible.

Reactions that form and those that break σ bonds of metals to sp^3 or sp^2 carbon are discussed, with the center of attention on species of reasonably well-defined or inferable structure. Thus although an extremely large body of data has arisen in the past 10 years on the chemistry of organocuprates, this is discussed relatively briefly: the structure of these species and the mechanisms by which they react are still particularly vaguely understood. Reactions of η^3-allyl and higher π-complexed systems are not discussed, since these generally result in interconversions of π systems rather than generation of σ-bonded alkyls.

It may be stated at the outset that the picture of the mechanisms of reactions of M—R compounds that has emerged is frequently one of significant complexity. For many reactions there appears to be a family of mechanisms that lie close to one another in energy. Which mechanism(s) will operate in a given case is therefore the result of a delicate balance among all the factors one can enumerate, and sooner or later any plausible path that can be imagined will evidence itself in some system or other. The stereochemistry of a particular type of reaction may therefore also show striking changes as a consequence of variation of subtle factors as well.

There are some exceptions to the last generalization. One of these is the insertion of carbon monoxide into M—C bonds. In every well-defined case examined to date, this reaction has appeared to proceed with complete retention of configuration at the migrating carbon. This result is assumed in the discussion in Sect. II and is documented in detail in Sect. III-A.

In practice, a stereochemical study usually involves a sequence of reactions, each one exhibiting its own stereochemical characteristics. Thus various parts of some studies are discussed in different sections of this review. Also, since any classification of reactions is somewhat artificial and arbitrary, closely related transformations may appear in different sections. Therefore liberal cross-referencing has been included. Naturally the division of reactions into those that form and those that disrupt metal-carbon bonds depends on the direction in which a given reaction proceeds under a certain set of conditions. Nevertheless, it forms a convenient and realistic basis for the organization of a large body of experimental information.

Reviews have appeared that deal with various aspects (usually mechanisms or synthetic applications) of individual topics discussed herein. Many of these are referenced in the appropriate sections. The general area of organo–transition metal stereochemistry was reviewed more briefly in 1974 (1).

The reader who is less familar with organo–transition metal chemistry may wish to consult some brief introductory reviews before continuing (2–4).

II. REACTIONS THAT FORM METAL–CARBON BONDS

A. Oxidative Addition of Alkyl and Alkenyl Species

Oxidative addition is formally defined as in eq. [1], where $M^m L_n$ is a coordinatively unsaturated metal complex (i.e., has fewer than 18 electrons in its valence shell) that adds the two fragments X and Y from rupture of the X–Y bond. The new ligands are formally considered to be negatively charged, so the oxidation state of the metal has increased by 2. In general usage, the situation of eq. [2a] has also come to be known as oxidative addition, where in this case ML_n is coordinatively saturated; the consecutive reactions

$$M^m L_n + X\text{-}Y \rightleftharpoons X\text{-}\underset{Y}{M^{m+2}}L_n \qquad (1)$$

$$ML_n + X\text{-}Y \rightleftharpoons [X\text{-}ML_n]^+ + Y^- \qquad (2A)$$

$$[X\text{-}ML_n]^+ + Y^- \rightleftharpoons X\text{-}\underset{Y}{M}L_{n-1} + L \qquad (2B)$$

$$[ML_n]^- + X\text{-}Y \rightleftharpoons X\text{-}ML_n + Y^- \qquad (3)$$

of eqs. [2a] and [2b] together have been called "oxidative elimination" by some authors. Clearly, the reactions of eqs. [2a] and [3] are formally the same; thus that all these processes (eqs. [1–3]) are grouped together here. A recent review discusses general aspects of the mechanisms of oxidative additions of organic halides to group 8 transition metal complexes (5).

1. Alkyl Species to Metal Anions

The first example demonstrating the stereochemistry of oxidative addition of an alkyl leaving group species to a metal anion was that of neohexyl brosylate to an iron anion, eq. [4] (6). Because of the strong preference for an anti arrangement of the t-butyl group and either the brosylate or the ligated iron group, the vicinal hydrogens of the neohexyl-d_2 moiety should be distinctly anti in the erythro diastereomer and gauche in the threo isomer. These are readily differen-

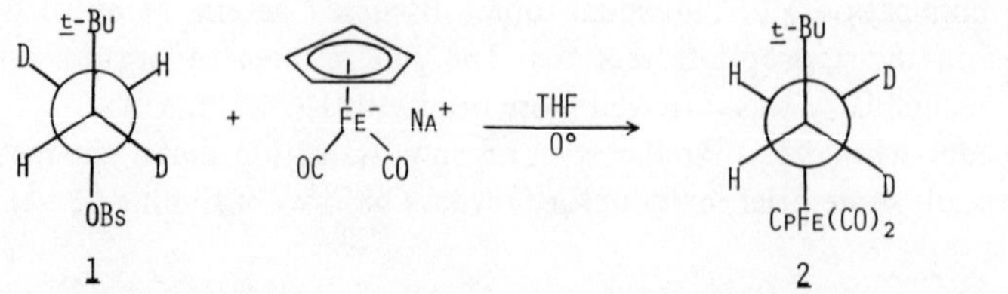

$$\text{1} \quad + \quad \text{CpFe(CO)}_2^- \text{Na}^+ \quad \xrightarrow[0°]{\text{THF}} \quad \text{2} \quad (4)$$

tiated by their coupling constants ($J_{\text{erythro}} > J_{\text{gauche}}$). It was found that *erythro*-1 and [CpFe(CO)$_2$]$^-$ yielded *threo*-2; therefore the alkylation of iron by neohexyl brosylate proceeds with greater than 95% inversion of configuration at carbon. The same result was found with neohexyl bromide in place of brosylate, and it was independent of whether the counterion of the iron anion was Li$^+$, Na$^+$, or [MgBr]$^+$ (6). Thus variation in stereochemistry as a function of the counterion is not in evidence here as it is in nucleophilic reactions of Me$_3$Sn$^-$M$^+$ with alkyl halides (7).

A second primary substrate, β-phenethyl-1,2-d_2 tosylate has more recently been examined in alkylation of [CpFe(CO)$_2$]$^-$ using the same nuclear magnetic resonance (NMR) technique as that developed by Whitesides et al. (6) and has exhibited the expected inversion of configuration (8). These results appear to be general for secondary substrates as well. Optically active *sec*-butyl bromide alkylates this iron anion with inversion, the configuration of product 3 being demonstated by PPh$_3$-induced CO insertion followed by oxidative cleavage and hydrolysis to (−)-2-methylbutanoic acid as shown in Scheme 1. This correlation

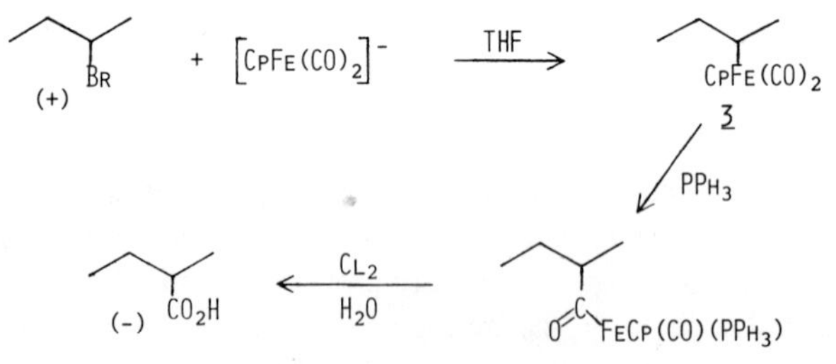

Scheme 1

relies on the now well-established generalization that CO insertion proceeds with retention of configuration at the migrating alkyl carbon (see Sect. III-A-1) (9). Similarly, *cis*- and *trans*-4-methylcyclohexyl benzenesulfonate alkylate [CpFe(CO)$_2$]$^-$ affords the secondary alkyliron species via an inversion path (eg. [5]) (10). Configurational assignment of 4 was made by comparison of the relative chemical shifts of the methyl group in *cis*- and *trans*-4 to those in 1-methyl-4-*t*-butylcyclohexane or in the low-temperature spectrum of 1,4-dimethyl-

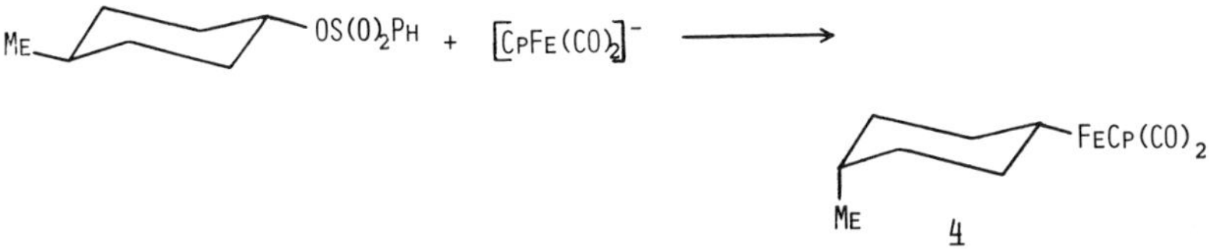

$$\text{Me} \overset{\displaystyle\frown}{\underset{\displaystyle\smile}{}} OS(O)_2Ph \;+\; [CpFe(CO)_2]^- \;\longrightarrow$$

$$\overset{\displaystyle\frown}{\underset{\displaystyle\smile}{\underset{Me}{}}} FeCp(CO)_2 \quad (5)$$

4

4-*t*-butycyclohexane or in the low-temperature spectrum of 1,4-dimethylcyclohexane. Since in this case there is some independent evidence for the carbon configuration in **4**, its reported subsequent oxidatively induced carbonylation (10) also constitutes independent evidence for the stereochemistry of that reaction (Sects. III-A-1).

A different iron anion, $Na_2Fe(CO)_4$, also exhibits classical nucleophilic behaviour toward carbon electrophiles. Reaction with (*S*)-2-octyl tosylate yielded the expected alkyliron anion that was subsequently treated as shown in Scheme 2 (11). Clean second-order kinetics were observed (12). Assignment of configuration of iron alkyl **5** was accomplished by invoking the well-known retention of configuration at carbon for both the CO insertion and Baeyer-Villiger oxidation. (11).

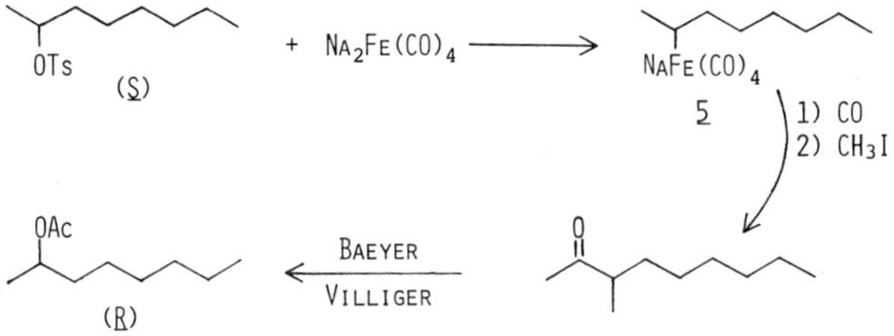

SCHEME 2

Other anionic metal complexes also undergo alkylation with inversion at carbon. Pentacarbonylmanganate(-I) was alkylated by (*S*)-(−)-α-phenethylbromide, carbonyl insertion was induced by triphenylphosphine, and the resulting acyl complex was cleaved by bromide (Scheme 3). The rotation of the α-phenyl-

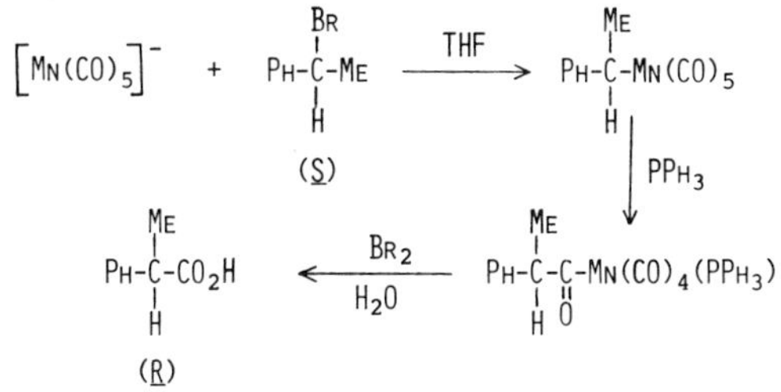

SCHEME 3

propionic acid was reported to correspond to greater than 75% stereospecificity overall, with inversion in the first step (9). Using the same NMR analysis as described before in the case of $CpFe(CO)_2CHDCHDPh$, (8) *erythro*-PhCHDCHDOTs was shown to suffer inversion of configuration upon attack by both $[Mn(CO)_4(PEt_3)]^-$ and $[CpW(CO)_2(PEt_3)]^-$ (13).

Perhaps the earliest indication of stereochemistry in nucleophilic reactions of metal anions was the report by Heck (14) that cyclohexene oxide was opened by $HCo(CO)_4$ to yield trans product **6** (Scheme 4). It was not determined whether

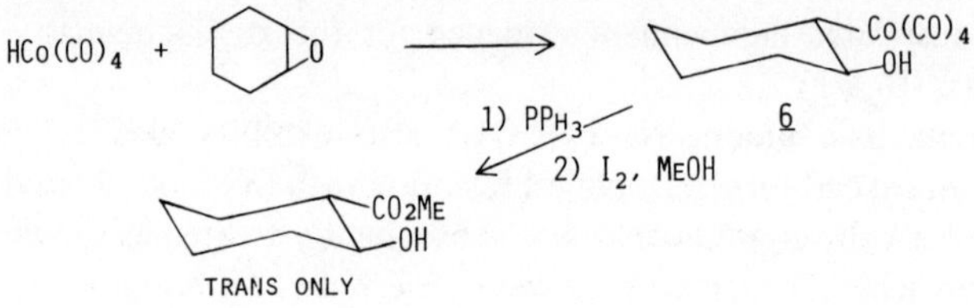

SCHEME 4

this involved general acid catalysis by the acidic cobalt hydride, or whether $[Co(CO)_4]^-$ was involved. Later, the opening of cyclohexene oxide and attack on a variety of other cyclohexyl substrates by bis(dimethylglyoximato)(pyridine)-cobaltate(I) [(pyridine)cobaloxime anion, **7**] were reported to proceed with inversion of configuration at carbon (e.g., eq. [6]) (15). Stereochemical assignments were made on the basis of NMR analysis utilizing the fact that the very bulky $Co(dmgH)_2(py)$ group will exhibit a strong preference for an equatorial position in the same way that a *t*-butyl group does. Confirmation of this

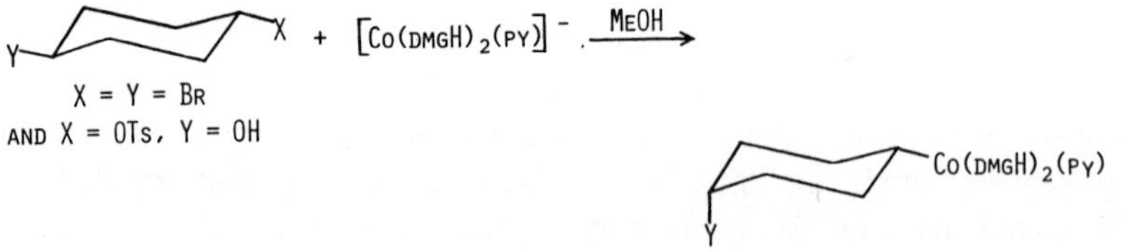

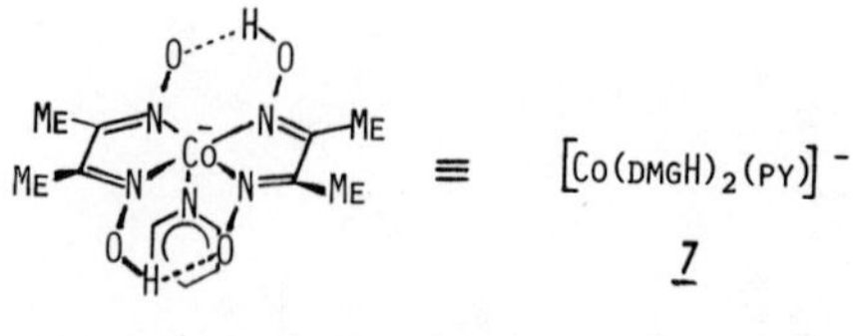

result was made by the report that the primary species *threo*-*t*-BuCHDCHDOTf alkylates cobaloxime anion with inversion of configuration at carbon (16). This triflate also alkylates $[CpMo(CO)_3]^-$ with the same stereochemistry.

Organocuprates, $[R_2Cu]Li$, have proved to be exceptionally useful reagents for organic synthesis. Part of this utility stems from their highly stereospecific coupling, particularly with vinylic halides (Sect. II-A-3). Cuprate conjugate addition to α,β-unsaturated carbonyl compounds has been reviewed (17a), and more recently an extensive review of cuprate coupling reactions with organic halides has appeared (17b). This chapter does not attempt to update the previous treatments, especially since no definitive mechanistic work or new stereochemical generalizations have emerged. Structures of $[R_2Cu]Li$ and all the related reagents are poorly defined, making it difficult to draw clear mechanistic conclusions.

Coupling of cuprates with saturated organic substrates proceeds with net inversion of configuration at the electrophilic carbon. For example, the coupling of lithium diphenylcuprate with (R)-(−)-2-bromobutane occurred with predominant inversion (84−92% e.e.) of configuration (18). Also, cis-1-bromo-4-t-butylcyclohexane in reaction with lithium dimethylcuprate afforded a 55% yield of 1-methyl-4-t-butycyclohexane, with a trans-cis ratio of only 4.5 : 1 (19). In contrast to these results, more recently highly stereospecific inversion of configuration was found upon alkylation of a variety of cyclic and acyclic secondary tosylates by both lithium diphenyl- and dimethylcuprate (20). Although this coupling is not well understood, if the initial step were alkylation of an anionic copper center, and if reductive elimination of the two particular organic groups were to proceed with retention (as it almost certainly would − see Sect. III-B-1), then these results would be seen as yet another typical S_N2 displacement by a metal with inversion at carbon. Partial racemization of secondary bromide substrates would be easily understood as the encroachment of a radical component on the reaction.

Having cited a number of examples from the chemistry of molybdenum, tungsten, manganese, iron, and cobalt metalate anions behaving as normal nucleophiles toward alkyl halides or sulfonates, we should reiterate one statement made in the introduction. Although the bulk of examples of eq. [3] proceed with inversion of configuration at carbon, there are clearly other mechanisms that lie very close in energy. For example, $[Co(dmgH)_2(py)]^-$, 7, and bromide 8 fail to react, whereas 7 and bromide 9 form 10 in 7% yield. Configurational assignments made by NMR appear to be reliable (21). Apparent retention in 10 and accompanying reduction of 9 under the alkylation conditions are consistent with some type of radical process wherein the "retention" upon formation of 10 would actually reflect the strongly preferred approach of the very bulky cobalt intermediate to the much less hindered side of the corresponding cyclopropyl radical. A similar result has been observed upon reaction of cobaltate anion 7 with 1-adamantyl and 1-norbonyl bromides (22). Again the yields were low (11 and 3%, respectively), with most of each alkyl halide being recovered unchanged. Optically active 1-methyl-2,2-diphenylcyclopropyl bromide in reaction with **7** yielded 60% of 1-methyl-2,2-diphenylcyclopropylcobaloxime. Cleavage of this by HBr afforded racemic 1-methyl-2,2-diphenylcyclopropane (23). Since it is very likely that protic cleavage would be stereospecific (see Sect. III-C-2), it is probable that the alkyl-cobalt bond formed stereorandomly. In each of the last two cases, some consideration was given by the respective authors to possible frontside associative mechanisms at carbon, either frontside S_N2, or a pentacoordinated carbon intermediate (22,23). In the absence of any bona fide example of a nucleophilic frontside associative substitution at a saturated carbon, perhaps a tentative mechanism such as that in eq. [7] (also suggested by these authors) seems more likely. The recent report that either 3α- or 3β-iodocholestane

$$R\text{-}X + [Co(I)]^- \longrightarrow R\cdot + Co(II) + X^- \longrightarrow R\text{-}Co(III) \quad (7)$$

yields a 3β cobalt-carbon bond upon reaction with an electrochemically generated Co(I) chelate complex (24) provides a further example of nonstereospecific behavior.

There is evidence that even when the alkylating agent is of a structure that does not preclude backside nucleophilic attack, the mechanism of eq. [7] or some similar radical mechanism may operate. Reaction of $[CpFe(CO)_2]^-$ with a variety of alkyl iodides (e.g., n-butyl, etc.) in THF at room temperature was found to yield strong electron spin resonance (ESR) signals of the corresponding alkyl radicals (25). Simple alkyl bromides and chlorides did not exhibit detectable signals, but alkyl and benzyl bromides in reaction with $[CpFe(CO)_2]^-$ did lead to observation of the corresponding radicals. Possibly more indicative of the quantitative role of radical intermediates was the reaction of the iron anion with cyclopropylcarbinyl iodide as compared to the bromide (eq. [8]) (25). Substantial amounts of rearranged product in the case of the iodide were attributed to nongeminate radical coupling processes rather than rearrangement of a carbanionic intermediate. Radical paths have similarly been implicated in reactions of $[CpFe(CO)_2]^-$ with 1,2-dihalobenzocyclobutenes (26).

Formation of racemic products, or a significant racemic product component, from optically active alkyl halides is not necessarily evidence for a radical mechanism in oxidative addition reactions. In the case of metalate anion nucleophiles, as well as in some neutral complexes, there can arise a metal-metal exchange reaction of the type shown in eq. [9] that is analogous to the Finkelstein reaction of organic halides with halide ions (see Sect. II-B). For example, the racemization of $MeCH(CO_2Et)Mn(CO)_5$ in the presence of excess $[Mn(CO)_5]^-$ has been attributed to such a reaction (9).

$$L_nM\text{-}R + [L_nM']^- \rightleftharpoons [L_nM]^- + L_nM'\text{-}R \qquad (9)$$

2. Alkyl Species to Neutral Complexes

Stereochemical investigations have been reported for complexes of Rh(I), Ir(I), and Ni-Pt(0), all of which can lead to well-defined products resulting from formal two-electron changes at the metal. This reaction has exhibited much more variation in its behavior than have the alkylations of anionic complexes discussed in Sect. II-A-1.

The earliest and, until recently, the most thoroughly examined example of oxidative addition of alkyl halides was the Ir(I) system shown in eq. [10]. This class of reactions is difficult to study: the reaction of eq. [10] was reported to proceed with inversion at carbon (27), with retention (28) and, in the case of **14**, not to proceed at all (29). The first two reports are now known to be incorrect (30,31), and this system is indeed reticent when the reagents are too scrupulously purified, but it is not totally unreactive (29,30).

Osborn et al. (30,31) observed that whichever of the two diastereomers of compounds **14** to **16** was used, the same mixture of diastereomeric alkyliridium-(III) products was formed. NMR spectroscopy was used to establish relative configurations within the diastereomers as described in Sect. II-A-1. Resolved α-haloester **13** yielded **12**, for which no significant optical rotation could be detected. The possibility of epimerization of the α-haloesters or their adducts **12** was ruled out, since no deuterium was incorporated into **12** when MeOD was employed as cosolvent for the oxidative addition reaction, and since one isolated diastereomeric adduct of **11** and **16** was not epimerized over long periods in the presence of excess **11**. All these additions could be initiated by oxygen or by standard radical initiators such as AIBN, and all were pronouncedly inhibited by species such as galvanoxyl. A free radical mechanism such as that in Scheme 5 is consistent with these data (30,31).

$$\text{INITIATION} \longrightarrow \text{Ir}^{II} \text{ OR } R\cdot$$

$$\text{Ir}^{I} + R\cdot \longrightarrow \text{Ir}^{II}\text{-R}$$

$$\text{Ir}^{II}\text{-R} + R\text{-X} \longrightarrow R\text{-Ir}^{III}\text{-X} + R\cdot$$

SCHEME 5

In contrast to the radical path for addition to iridium(I) generally exhibited by alkyl halides, a group of highly reactive halides such as methyl iodide, benzyl chloride, allyl chloride, and chloromethyl methyl ether is not inhibited by the usual radical scavengers in its oxidative additions (30,32). Furthermore, methyl iodide addition is clearly first order in complex **11** and first order in CH_3I, and has activation parameters very similar to those for the Menschutkin reaction (when L = PPh_3 in benzene, $\Delta H^{\ddagger} = 5.6$ kcal/mol, and $\Delta S^{\ddagger} = 51$ e.u.) (33,34). The kinetic product of the addition of CH_3I to **11** (L = PPh_2Me) is trans, as demonstrated by the sequence of reactions shown in Scheme 6, where structural assignments were based on analysis of far-infrared Ir—Cl stretching frequencies

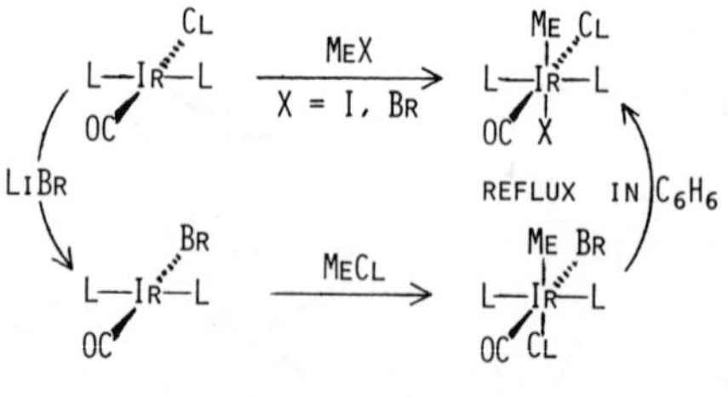

SCHEME 6

(35). The same stereochemistry has been assumed in virtually all other reported studies of alkyl halide oxidative addition to **11**, if it has been considered at all.

It is problematic that in spite of the trans stereochemistry and S_N2-like kinetic behavior of the CH_3I addition, when the reaction was carried out in the presence of $^{131}I^-$ (34), Cl^-, or SCN^- (28) *no* incorporation of these ligands into **12** in place of I^- occurred. This apparent discrepancy may be tentatively rationalized by proposing that a pentacoordinate low-spin d^6 intermediate such as $[IrCl(CH_3)(CO)L_2]^+$ would reasonably be expected to be square pyramidal based on empirical and theoretical considerations (36), and that tight ion pairs in nonpolar solvents such as those used in these oxidative additions might not undergo anion exchange as rapidly as bond formation. In any event, the critical test for a possible S_N2 mechanism, examination of the stereochemistry at carbon, has not yet been accomplished.

Square planar rhodium(I) complexes exhibit less reactivity toward oxidative addition than do their iridium analogs, and the chemistry observed is strongly dependent on the ligands employed. For example, *trans*-RhCl(CO)L$_2$ is highly unreactive when L = PPh$_3$. However when L = PEt$_3$, methyl iodide will oxidatively add, though α-bromoesters will not (31). Complex **17** has been thoroughly examined in its reactivity with alkyl halides, and all the data support an S_N2 mechanism. Nevertheless, the stereochemistry at carbon has not been determined because there is a rapid metal-metal interchange. Both primary and secondary alkyl adducts of **17** (R = Me) rapidly exchanged alkyl groups with **17** (R = Et) itself (37). Some stereochemical information is available for rhodium(I), but in the form of the microscopic reverse, reductive elimination; therefore it is discussed in Sect. III-B-3.

$X = BF_2$, H
$R = Me$, Et

17

More recently, significant effort has gone into elucidating pathways of oxidative addition to palladium(0) and platinum(0) complexes. Ethyl (S)-(−)-α-bromopropionate, **13**, undergoes oxidative addition to "Pd(*t*-Bu-NC)$_2$" as shown in eq. [11]. Although highly labile toward β-elimination, the product was stable enough

(11)

(S)-**13**

to allow attempts to measure its optical rotation, though none could be detected. Similarly, ethyl (S)-(+)-α-bromophenylacetate yielded an inactive palladium complex (38). It is true that rotations of highly colored complexes often cannot be measured, but the products in this case were yellow, and a specific rotation as low as −0.96° has been measured for the similar compound *trans*-PdCl(CHDPh)-(PEt$_3$)$_2$, the optical activity resulting only from the presence of the deuterium isotope (39). Nevertheless, the optical stability of the aforementioned products could not be determined, and an α-metalloester might be especially prone to racemization via paths such as those involving intermediates **18** or **19**.

$$\underset{\text{EtO}}{\text{HO}}C=C\underset{\text{PdXL}_2}{\text{Me}} \qquad \underset{\text{H}}{\text{Me}}C=C\underset{\text{O-PdXL}_2}{\text{OEt}}$$
$$\underline{18} \qquad \qquad \underline{19}$$

Very similarly, optically active **13** as well as its chloride analog in reaction with Pt(PEt$_3$)$_3$ yielded an addition product for which no rotation could be detected (31,40). In this study, however, the O-bonded platinum analog of **19** was observed by NMR to be a kinetic component of the product mixture at low temperature (41). Interpretation of the stereochemical result would therefore be particularly difficult, except that these authors have gathered a substantial and convincing body of data to establish the predominance of a free radical path. For example, oxidative additions to Pt(PEt$_3$)$_3$ were strongly inhibited by radical scavengers, and 6-bromo-1-hexene in reaction with this complex yielded both linear and cyclic alkylplatinum products (31,32,40). Many of these observations were duplicated using Pd(PEt$_3$)$_3$ (32). A radical chain, probably similar to that shown in Scheme 5 for iridium, thus seems indicated. Chemically induced dynamic nuclear polarization (CIDNP) effects were observed for certain products from especially reactive alkyl halide substrates, establishing the presence of one-electron paths in side product formation, but not necessarily in the formation of the alkylplatinum products themselves (41).

As was the case with Ir(I) and Rh(I), a second class of substrates undergoes oxidative addition to Pd(PEt$_3$)$_3$ and to Pt(PEt$_3$)$_3$ without rate retardation in the presence of radical scavengers. This includes allylic and benzylic halides, methyl halides, and in the case of palladium and platinum, aryl and vinyl halides (32,40). A radical pair mechanism might accommodate the lack of scavenging, but, at least in the case of benzylic substrates, stereochemical experiments have been definitive and have indicated a predominant $S_N 2$ pathway (39,42).

Initial experiments employed optically active PhCH(CF$_3$)Cl in reaction with Pd(PPh$_3$)$_4$, where the CF$_3$ group was used to circumvent the problem of rapid β-hydride elimination that generally plagues metal alkyls having β-hydrogens. Nevertheless, the product was racemic (43). It is possible that this oxidative addition proceeded by the same free radical path discussed above, but the cause of racemization was not determined.

In further work, β-elimination was avoided by the use of two techniques. The first was removal of the need for β-blocking altogether by employing optically active **20** or **21** — species that have been in use by organic chemists for mechanistic studies for some time. The second was the rapid trapping of any alkylmetal species that might have β-hydrogens, such as metallic derivatives of **22**, via CO

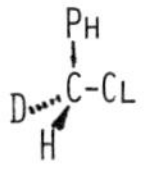

(R)-20 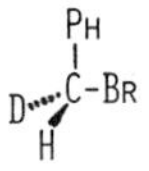   (R)-21 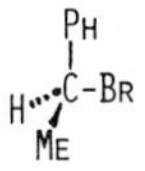   (S)-22

insertion (39,42,43). The second device also solves the problem of stereochemical assay of the alkylmetal species, since it is well documented that the CO insertion proceeds with retention of configuration at carbon (Sect. III-A-1).

Primary benzylic substrates such as **20** and **21** oxidatively add with net inversion of configuration as shown by a series of reactions outlined in Scheme 7 (39,42). In this case the alkyl group cannot β-eliminate, and intermediate **23** can be isolated. Subsequent addition of CO, followed by oxidative cleavage, ultimately led to ester **25**, whose optical purity indicated that the oxidative addition had proceeded with 74% net inversion. When addition of **20** was effected directly under CO atmosphere using either $Pd(CO)(PPh_3)_3$ or $Pd(PPh_3)_4$, complete inversion of configuration was observed. Incurrence of a racemic component in the

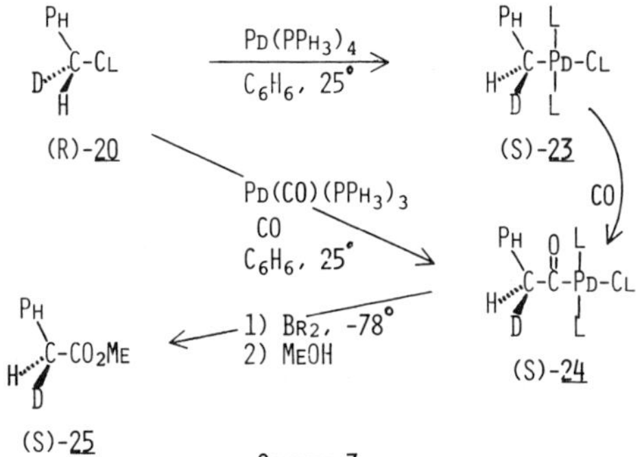

SCHEME 7

absence of CO was not associated with formation of an η^3-benzylic intermediate, since the analogous complex **26** was shown not to undergo racemization of the chiral methylene even though **27** exhibited substantial fluxional behavior (46).

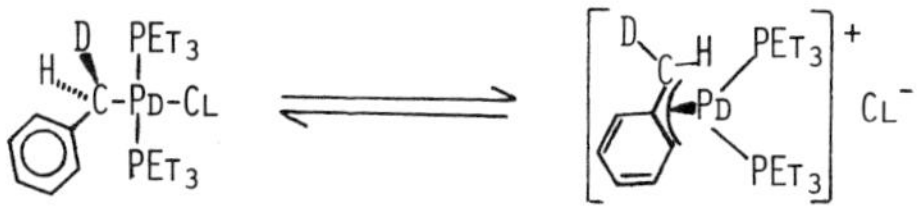

(12)

Recovered **20** had suffered partial racemization, although the source of this was also not clear. Oxidative addition to PdL_4 in the presence of CO probably does not involve Pd(CO)L$_3$ in view of the equilibrium of eq. [13] (44) and the relative reactivities of PdL$_4$ and Pd(CO)L$_3$ (45).

$$Pd(CO)(PPh_3)_3 + PPh_3 \rightleftharpoons Pd(PPh_3)_4 + CO \quad (13)$$

Results very similar to those in Scheme 7 were obtained with the secondary benzylic substrate **22**, except that in this case the alkylpalladium intermediate could not be isolated because of rapid β-elimination (42). All reactions therefore had to be performed under an atmosphere of CO.

Trialkylphosphine complex Pd(PEt$_3$)$_3$ behaved in the same way as the PPh$_3$ complexes discussed above in reactions with **20** and **21**, yielding **26** and the bromide analog of **26**, respectively. These alkyls were isolated and subsequently submitted to several atmospheres pressure of CO, followed by the usual halogen cleavage. Net inversion of configuration in the oxidative addition step amounting to 72 and 19% e.e., respectively, was inferred (39). Substantial racemization of the starting materials was also observed, particularly of **21**. Reaction of **21** also led to formation of bibenzyl, but no crossover was observed when this was carried out in the presence of Pd[CH$_2$C$_6$H$_4$=o=Me]Br(PEt$_3$)$_2$, and **26** and its bromide analog were shown to be optically stable. Thus the racemization component of these benzylic additions is apparently associated with the oxidative addition mechanism itself. The presence of a radical pair pathway in competition with an S_N2 mechanism was postulated, in complete agreement with conclusions convincingly reached by others (32,41).

One other example of an addition to platinum is shown in eq. [14]. This result is apparently further corroboration of stereospecificity in oxidative additions of secondary benzylic substrates, but the potential for prior coordination may lead to a substantial mechanistic change, and of course no configurational information may be drawn solely from specific rotations (47).

$$\text{(structure with } [\alpha]_D\ +17.3° \text{)} + Pt(PPh_3)_3 \longrightarrow \text{(structure with } [\alpha]_D\ -2.8° \text{)} \quad (14)$$

Much less work has been carried out using nickel complexes, in large part because they are not as generally useful in catalytic reactions involving alkyl halides, and because the corresponding alkylnickel intermediates are much less stable and are more prone to β-elimination than are their palladium or platinum analogs. One attempt to examine the stereochemistry of oxidative addition to Ni(PPh$_3$)$_4$, again using **20** and **21** followed by trapping of the intermediate by CO insertion, yielded racemic products (48). The source of racemization could not be determined because of the inability to observe any intermediates.

In another interesting study, the coupling of (π-allyl)nickel bromide dimer with (S)-2-iodooctane yielded racemic products (49). Neither the starting iodide

$$\underset{(S)}{\overset{Me}{\underset{C_6H_{13}}{H\cdots C-I}}} + R-(-Ni\overset{Br}{\underset{Br}{\diagdown}}Ni-)-R \xrightarrow{DMF} \underset{RACEMIC}{\overset{R\quad Me}{\diagup\!\!\!\diagdown C_6H_{13}}} \quad (15)$$

nor the product was racemized under the reaction conditions, and m-dinitrobenzene strongly inhibited the reaction. A radical chain mechanism was postulated to accommodate these observations.

3. Oxidative Addition of Alkenyl Species

Oxidative additions to both neutral and anionic metal complexes are known. These are usually attended by a high degree of retention of configuration at the trigonal carbon, although isomerization of olefinic starting material and/or products by some catalytically active species occasionally complicates the stereochemical picture. There is a large body of information available on catalytic reactions of vinyl species and stoichiometric reactions of vinyl organometallics that are structurally ill defined. No attempt is made to discuss these data exhaustively.

Configurational analysis of vinyl metal complexes is easy and reliable compared to that of metal alkyls, especially when both the Z and E geometrical isomers are available. Vicinal coupling constants in complexes of the type M–CH=CH–R are largely independent of M and R for a reasonably large number of examples, falling in the range of 5.5–10.2 Hz for cis hydrogens and being very near to 14 Hz for trans hydrogens, although values as large as 18 Hz have been reported in the latter case.

Displacement on cis and trans isomers of ethyl β-chloroacrylate (50) and β-bromostyrene (50,51) by cobaloxime anion 7 has been found to proceed with complete retention of configuration (eq. [16]). More recently, cis- and trans-1-bromooctene were observed to form octenyl(pyridine)cobaloximes, also with clean stereochemistry (52). No significant isomerization of reactants or products occurred under the reaction conditions, and no α-metallosubstituted materials could be detected. An elimination-addition mechanism could be ruled out because addition of anion 7 to phenylacetylene gave only cis-β product with no

$$[Co(DMGH)_2(PY)]^- \quad \underset{\underset{PH}{\nearrow}\overset{}{\diagdown}BR}{\overset{PH\diagdown\!\!\!\diagup BR}{\rightleftarrows}} \quad \begin{matrix} PH\diagdown\!\!\!\diagup Co(DMGH)_2(PY) \\ PH\diagup\!\!\!\diagdown Co(DMGH)_2(PY) \end{matrix} \quad (16)$$
$$\underline{7}$$

trans, and also resulted in the formation of significant amounts of α-substitution product (eq. [17]) (50,51). A free radical path could be ruled out, since electron

$$\underset{Ph}{\overset{Ph}{\diagdown}}\underset{Br}{\overset{Br}{\diagup}} \xrightarrow{7} \text{X} \xrightarrow{} Ph-C\equiv CH \xrightarrow{7} Ph\diagup\overset{}{-}\overset{}{Co}\diagdown + \diagdown\overset{Ph}{Co}\diagup \quad (17)$$

paramagnetic resonance (EPR) evidence indicated that vinyl radical, $CH_2=CH^{\cdot}$, undergoes inversion of configuration with a first-order rate constant in excess of 10^7 sec^{-1} at $-180°C$ and an E_a of ca. 2 kcal/mol (53). These data extrapolate to an inversion rate at 25°C of greater than 10^9 sec^{-1}, and for no isomerization to be observable in the products formed, unreasonably fast rates of geminate radical pair coupling and impossibly rapid rates of trapping of free vinyl radicals would be required (54). These comments regarding the extreme unlikelihood of vinyl radical intermediates in stereospecific oxidative additions of vinylic halides also apply to all the remaining discussions in Sect. II-A-3.

Nucleophilic vinyl substitution has received much attention (55), and in spite of this, no altogether satisfactory explanation for the stereochemistry observed in these reactions exists. It may be said, however, that vinyl substitution by cobalt(I) anion in some respects generally resembles nucleophilic vinyl substitution of the addition-elimination type (55). For example, vinyl halides are much less reactive than alkyl halides toward either PhS$^-$ or anion 7 (55,56), and displacement by either PhS$^-$ or 7 is highly stereospecific, as stated above. Unlike displacements by PhS$^-$, where vinyl fluorides are somewhat more reactive than the chlorides, there is no "element effect" with nucleophile 7. Vinyl iodides and bromides react readily, chlorides react sluggishly, and vinyl fluorides are unreactive (56). This implies that either a direct S_N2 displacement is occurring with 7, or the second step of an addition-elimination sequence is rate limiting. The latter possibility seems incompatible with the observed stereospecificity.

There is evidence that alkenes form π complexes with 7 (56), but there is no indication that such complexes are intermediates in vinyl halide oxidative additions. All in all, the fine mechanistic details of oxidative addition of vinyl halides to cobalt(I) anion remain obscure.

As stated above, stereochemical data concerning oxidative additions to rhodium are scarce. A single relevant example is shown in eq. [18]. Complex 29 was not isolated or otherwise observed, so its intermediacy, though reasonable, remains speculative. Assuming that reductive elimination would proceed with

$$\underset{28}{Me-RhL_3} \xrightarrow[\substack{DMF,\ \Delta \\ L\ =\ PPH_3}]{Br\quad CO_2Me} \left[\underset{29}{\overset{CO_2Me}{\underset{L\ Br}{L-Rh-Me}}}\right] \longrightarrow Me\quad CO_2Me + RhBrL_3 \quad (18)$$

retention of configuration at carbon (Sect. III-B-1), clean retention is implied in the initial oxidative addition. Whereas coupling with complete retention of configuration was observed with both *cis*- and *trans*-β-bromoacrylates, extensive isomerization was observed in the formation of 2-heptene from **28** and *cis*- or *trans*-1-iodohexene. Mechanistic interpretation of these divergent observations is unclear (57).

Iridium(I) species such as *trans*-IrCl(CO)(PPh$_3$)$_2$ and *trans*-IrCl(N$_2$)(PPh$_3$)$_2$ were initially reported to be unreactive toward vinyl halides (58). More recently these oxidative additions have been found to proceed, albeit slowly, via a free radical path (32). No data, stereochemical or otherwise, were given regarding these additions, although one would predict cis-trans isomerization in such cases.

Metals of the nickel and copper triads have extensive chemistry with vinylic species. Much of this chemistry, particularly of nickel and copper, involves ill-defined or inferred intermediates. One apparently well-defined example was afforded by the reaction of *cis*- and *trans*-β-bromostryrene with Ni(PPh$_3$)$_4$ and Ni(PEt$_3$)$_4$ (59). The corresponding products **30** and **31** were formed with retention of configuration at the vinylic carbon. In a number of other cases it may be

M = Ni, Pd, Pt
X = Cl, Br
L = PEt$_3$, PPh$_3$
Z = Ph, Cl, Br

30   **31**

reasonably inferred that stereospecific oxidative addition has occurred. For example, nickel complex **32** undergoes coupling with *cis*- and *trans*-β-bromostyrene, and upon hydrolysis this mixture yields *cis*- and *trans*-5-phenyl-4-penten-2-one, respectively (60).

MeO-⟨⟨-Ni(Br)(Br)Ni-⟩⟩-OMe $\xrightarrow{\text{1) Br-CH=CH-Ph}}{\text{2) H}_3\text{O}^+}$ [CH$_3$C(O)CH$_2$CH=CHPh]   (19)

32

It is tempting to suggest that the vinylic bromide undergoes stereospecific oxidative addition to nickel(II) in **32**, followed by reductive elimination, similarly with retention at carbon (Sect. III-B-1). The precise nature of such a stereospecific oxidative addition is enigmatic, since the reaction of eq. [19] was strongly inhibited by a few mole percent of *m*-dinitrobenzene (49). In any event, it is a significantly different process than that of the Co(I) or Ir(I) species discussed above, because in the case of complexes such as **32**, vinylic bromides are *more* reactive than are primary alkyl bromides (61).

Bis(1,5-cyclooctadiene)nickel(0) stoichiometrically coupled a variety of alkenyl halides to symmetrical 1,3-dienes. Highly activated halides, such as methyl β-bromoacrylate, afforded high yields of stereospecifically formed products, whereas unactivated halides gave low yields of partially geometrically isomerized dienes. It was not demonstrated whether the isomerization component was due to product instability (62). Another report has described the highly stereospecific Ni(0) catalyzed coupling of alkenylalanes with both (Z)- and (E)-alkenyl halides (63). Since this reaction almost certainly involves oxidative addition of vinylic halides to Ni(0), the loss of stereochemistry in the Ni(1,5-COD)$_2$-induced coupling of vinylic halides above may be a result of isomerization of the products. Nickel bromide has also been reported to catalyze the coupling of cis- and trans-β-bromostyrene with lithium enolates such as LiCMe$_2$CO$_2$Et in tetrahydrofuran (THF) with clean retention of configuration (64).

Palladium and platinum complexes have proved to be generally more mechanistically informative than their nickel congeners. As with the specific nickel reactions discussed above, the oxidative addition of vinylic halides to PtL$_3$ was faster than that of ordinary alkyl halides (32,58). For example, β-bromostyrene added to Pt(PPh$_3$)$_3$ at least 100 times more rapidly than did methyl iodide (58). The addition is highly stereospecific (58,65–68), since *trans*-1,2-dichloroethylene and the appropriate Pd(0) or Pt(0) complex afforded **31** (Z = X = Cl) where L = PPh$_3$, PPh$_2$Me and M = Pd, and L = PPh$_3$, PPh$_2$Me, and M = Pt. Similarly *cis*-1,2-dichloroethylene and Pd(PPh$_3$)$_4$ or Pd(PPh$_2$Me)$_4$ yielded *cis*-alkenyl complexes **30** (Z = X = Cl). Strangely, addition of *cis*-1,2-dichloroethylene did not readily occur with PtL$_4$ complexes (66,68). Both *cis*- and *trans*-β-bromostyrene also afforded the corresponding complexes **30** (Z = Ph, X = Br) and **31** (Z = Ph, X = Br) from reaction with Pd(PMe$_2$Ph)$_4$, Pt(PPh$_3$)$_4$, and Pt(PMe$_2$Ph)$_3$ (67).

The stereochemical outcome of all these oxidative additions could be convincingly demonstrated by ^{1}H NMR spectroscopy, as mentioned earlier. In the case of platinum complexes, additional evidence arises from the $^3J_{PtH}$ that are in the range of 31–85 Hz for **31**, and 120–160 Hz for **30** (67,68). The metal geometry can be inferred using the observation that the ^{1}H resonances of methyl-substituted phosphines are always doublets in *cis*-PtXY(PR$_3$)$_2$, but are always triplets in *trans*-PtXY(PR$_3$)$_2$ as a result of virtual coupling. An X-ray crystal structure determination has definitely established the carbon and metal configurations in *trans*-PtBr(*trans*-CH=CHPh)(PPh$_3$)$_2$ (58).

In addition to stereochemical information, it has been observed that radical inhibitors have no effect on the rates of oxidative addition of vinylic halides to Pt(PEt$_3$)$_3$ (32) and Pt(PPh$_3$)$_4$ (58). Furthermore, at $-55°$C in toluene-benzene (4:1), ^{31}P NMR data show that the equilibrium of eq. [20] is established. The relative complexation constants of 1-hexene and vinyl bromide were ca. 1:10^4 at $-55°$C, with the latter complex rearranging in a first-order reaction to the

$$\text{Pt(PEt}_3)_3 + \text{CH}_2\text{=CHR} \rightleftharpoons \text{Pt}(\pi\text{-CH}_2\text{=CHR})(\text{PEt}_3)_2 + \text{PEt}_3 \qquad (20)$$

oxidative addition product (32). A number of electron-deficient alkenes form isolable π complexes with platinum, and some of these rearrange to σ complexes under various conditions. For example, $Pt(CCl_2=CCl_2)(PPh_3)_2$ undergoes a first-order conversion to *trans*-$PtCl(CCl=CCl_2)(PPh_3)_2$ at 35°C in ethanol (69). In view of the greatly reduced reactivity of vinylic halides relative to alkyl halides in normal S_N2 substitutions, and since the opposite reactivity order is observed for oxidative addition to PdL_n and PtL_n, it is easy to believe that π-complexation of the vinylic substrate plays a definite activating role in oxidative addition, rather than just being a coincidental equilibrium not actually on the reaction coordinate to adduct formation.

If oxidative addition of vinylic halides is concerted, as implied above, then the question of metal stereochemistry becomes an interesting one. It may be that the kinetic product is cis, and that this in the presence of excess phosphine is rapidly isomerized to the more stable trans isomer (70). If the kinetic product is trans, a significant structural reorganization must be occurring in the transition state. In any event, the issue of stereochemistry at the metal is unsettled.

A wide variety of catalytic reactions of palladium complexes that almost certainly involve oxidative addition of alkenyl halides is known (71). For example, various palladium compounds catalyze the conversion of alkenyl halides with CO in alcohols to acrylic esters (72), and alkenyl halides with alkenes to dienes (73). At lower temperature, with excess phosphine, and at higher concentrations of amine bases (to consume liberated BX), these reactions become highly stereospecific. Under other conditions, significant isomerization is observed in the products, presumably subsequent to their formation. Stereospecific oxidative addition of alkenyl iodides to palladium(0) forms the basis of a recently reported catalytic preparation of butenolides (eq. [21]) (74).

$$\underset{\substack{I \\ HO}}{\overset{R}{\diagdown}} C = C \underset{R'}{\overset{H}{\diagup}} R'' \quad + \quad CO \quad \xrightarrow[\text{THF}]{PdCl_2(PPh_3)_2} \quad \underset{O}{\overset{R}{\diagdown}} \underset{O}{\diagup} \underset{R'}{\overset{R''}{\diagup}} \qquad (21)$$

A number of metal-catalyzed coupling reactions of Grignard reagents and other main group metal alkyls with vinylic halides have been reported. Again, at some point an oxidative addition of the vinylic halide is almost certainly involved.

Nickel has been found to be particularly effective in coupling alkyl, alkenyl, and aryl Grignard reagents with aryl and alkenyl halides (75). $NiCl_2(PR_3)_2$ and especially $NiCl_2[PPh_2(CH_2)_3PPh_2]$ couple Grignard reagents with vinylic halides in good yield, with high stereospecificity, and with retention of configuration. Alkenyl sulfides serve as substrates in this reaction and are coupled stereospecifically as well (eq. [22]) (76).

Lithium reagents are not coupled effectively with vinylic halides by nickel

$$\text{EtS-CH=CH-Ph} + \text{PhMgBr} \xrightarrow[\text{THF}]{\text{NiCl}_2(\text{PPh}_3)_2} \text{Ph-CH=CH-Ph} \quad (22)$$
96% YIELD
95% (Z)

complexes, but they are by Pd(PPh$_3$)$_4$, albeit stoichiometrically (77). Grignard reagents (77,78) and alkynylzinc reagents (79) are catalytically coupled to vinylic halides by palladium complexes. All these reactions are highly stereospecific.

As mentioned earlier, organocuprates, [R$_2$Cu]Li, and their derivatives have been widely used in organic synthesis. One especially useful reaction is the coupling of these reagents with vinylic halides. Approximately 40 examples of this reaction are listed in Posner's review (17b), and of these more than half involve stereochemistry and are stereospecific. Thus the reaction is a very useful way of preparing di- and trisubstituted alkenes of predetermined stereochemistry. Very little is known about the mechanism, except that vinylic halides are highly reactive, and this together with the strict retention of stereochemistry implies an oxidative addition to Cu(I) in a manner similar to that discussed for Pd(0) and Pt(0). Transient formation of Cu(III) may be the driving force for subsequent reductive elimination (see Sect. III-B-1).

Silver salts catalyze the coupling of alkyl and alkenyl bromides with Grignard reagents (80). Particularly interesting was the observation of eq. [23], which illustrates the conclusion, based on these and other data, that transmetalation between Ag(I) and Mg(II) occurred stereospecifically, whereas oxidative addition of organic halides to Ag(0) occurred by a free radical process.

$$\text{CH=CH-Br} + \text{MeMgBr} \xrightarrow{\text{Ag}^{\text{I}}} \left\{ \begin{array}{c} \text{CH=CH} \xleftarrow{\text{Ag}^{\text{I}}} \\ \cancel{\text{CH=CH}} \end{array} \right\} \text{CH=CH-MgBr} + \text{MeBr} \quad (23)$$

Oxidative addition of alkenyl bromides occurred stereospecifically to catalytic iron in a cross-coupling reaction between alkenyl halides and organomagnesium halides (81).

4. Oxidative Addition of Carbon-Carbon Bonds

Well-defined examples of unstrained and otherwise unactivated C(sp^3)–C(sp^3) bonds to soluble metal complexes (eq. [24]) are unknown. Concrete examples of even the reverse reaction, reductive elimination of two alkyl groups to form an alkane, were quite unusual until recent years.

$$\text{ML}_n + \text{>C-C<} \rightleftharpoons \text{>C-ML}_n \quad (24)$$

Metal complex catalyzed rearrangements of molecules containing strained or allylic carbon-carbon σ bonds, however, have been well-known for some time (82,83). Many of these rearrangements presumably involve reaction such as that in eq. [24]. One can imagine concerted three-centered oxidative additions to a metal that would necessarily proceed with retention of configuration at both carbon atoms, and also one can think of nonconcerted pathways that would involve carbon radicals or cations. Silver salts apparently behave in the latter way, functioning essentially as Lewis acids (82,83).

Complexes of other metals such as Fe(0), Rh(I), Ir(I), Ni(0), Pd(II), and Pt(II) probably do undergo oxidative additions to C–C bonds during catalysis of rearrangements, and in some cases strong evidence is available, since addition products can be isolated. For example, complex 33, which is apparently tetrameric, arises from reaction of cubane with $[Rh(CO)_2 Cl]_2$ with a rate that is first order in each reactant (84), and similar complexes are formed from this rhodium dimer with quadricyclane (85), and dibenzosemibullvalene (86). In the last case, the CO uninserted complex 34 could also be isolated.

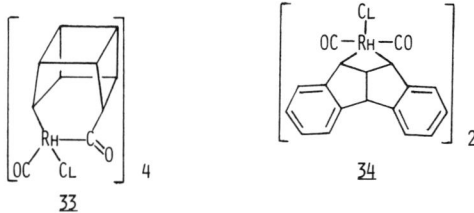

Other examples of hydrocarbon complex formation resulting from oxidative addition of strained rings are known for iron (87), platinum (88), and iridium (89), as well as rhodium (90). For all these examples, either there is no stereochemical information concerning the mode of C–C bond cleavage, or the structure of the adduct necessitates formation with retention at both carbon centers, as do structures 33 and 34. This would seem to be the most likely stereochemical path anyway, but it is nevertheless desirable to test one's prejudices on as sterically unconstrained a system as is feasible.

An elegant experiment of this kind has been reported (91). Reaction of cis-cis-n-hexycyclopropane-2,3-d_2 with $[PtCl_2(CH_2=CH_2)]_2$ followed by pyridine yielded a platinacyclobutane complex, 35, wherein NMR analysis could clearly establish the cis-cis configuration of the product (eq. [25]). trans-n-Hexylcyclopropane-*cis*-2,3-d_2 similarly yielded the product analogous to **35** except with pseudoequatorial deuterium. Hence the oxidative addition proceeded with > 90% retention at both carbon centers.

B. Transmetalation

Metal-metal interchange may proceed by a variety of mechanisms. These may be crudely classified according to oxidation state changes at the metal center of interest as being nucleophilic (R^+ transfer), homolytic ($R\cdot$ transfer), and electrophilic (R^- transfer). In principle such substitutions may be either associative or dissociative. Any dissociative mechanism would be attended by racemization of an optically active metal alkyl, whereas a concerted bimolecular mechanism should exhibit inversion at carbon for S_N2, probably inversion for S_H2, and either inversion or retention for S_E2. More complex associative mechanisms can be imagined that would be predicted to exhibit various stereochemical outcomes.

As mentioned in Sect. II.A, nucleophilic transmetalations have been implicated in the racemization of several metal alkyls such as in the exchange of $Mn(CO)_5(CHMeCO_2Et)$ with the anion $[Mn(CO)_5]^-$ (**9**), and Rh^{III}(chelate)RX, **17**·RX, with Rh^I(chelate), **17** (37).

One particularly well-documented example of this exchange was afforded by the reaction of **36** with **37** (92). The reaction was second order overall, first

36A R = *n*-OCTYL
36B R = CHDCHDPh

37 M = $[Co^I]^-$
38 M = Co^{II}(PY)

order in each **36** and **37**, and it showed a strong effect of steric hindrance as the bulk of R in **36** was increased. Most important, the rate of epimerization of **36b** in reaction with the corresponding anion, $[Co(chgH)_2py]^-$, was the same as the rate of alkyl transfer. Thus the transfer occurred with inversion of configuration.

In exactly the same kinds of exchange experiment as just outlined, it was shown that S_H2 substitution occurred between **36** and the Co(II) species **38** with inversion of configuration at carbon (92). Fortuitously, the rates of displacement on **36** by **37** and **38** were very similar, so that convincing arguments could be given that each reaction was not catalyzed by traces of the other cobalt oxidation state. Other than the work just outlined, examples of S_H2 and S_N2 transmetalations are rare, and there are no other extant well-defined stereochemical studies.

A recent study has reported the transformation shown in eq. [26] (93). Although this reaction is thought to proceed via an oxidative addition of the Hg—R bond to Pt(0), the net result is an apparent transfer of "R^+" to Pt. The reaction was purported to proceed with retention of configuration at carbon, based on small differences in specific rotations of the same sign in starting material and product diastereomers, but no further correlative reactions were carried out. Any stereochemical conclusions must therefore be regarded as questionable.

$$\text{Ph-CH-HgBr} + \text{Pt(PPh}_3)_3 \longrightarrow \underset{\underset{\text{PPh}_3}{|}}{\overset{\overset{\text{PPh}_3}{|}}{\text{Ph-CH-Pt-Br}}} + \text{Hg}° \quad (26)$$
$$\underset{\text{CO}_2\text{R}}{|} \qquad\qquad\qquad\qquad \underset{\text{RO}_2\text{C}}{|}$$

(R) or (S), R = L-MENTHYL

A mild procedure has been reported for "protonolysis" of alkenyl boranes prepared from hydroboration of internal acetylenes by catalysis with Pd(OAc)$_2$ in THF or acetone (94). This reaction proceeds with retention and probably involves the oxidative addition of the alkenyl-boron bond to Pd(0).

Examples of S_E transmetalations are numerous, but most of these are between main group alkyls and catalytic quantities of transition metal complexes, yielding species too unstable for direct structural characterization. Nevertheless, there is a large body of data dealing circumstantially and, because of its bulk, convincingly, with the stereochemistry of electrophilic transmetalation. Experiences with alkenyllithium and Grignard reagents and a few special aliphatic cases leads one to believe that all alkylations of transition metals by RLi or RMgX proceed with retention at carbon. Since configurationally stable alkyllithium and Grignard reagents cannot be prepared in the absence of special structural features, strictly speaking, one cannot say what the stereochemistry of this reaction is with certainty, although in reality there appears to be little doubt.

Several studies of transfers of alkyl groups with special structural features have appeared. Alkyllithium **39**, which is configurationally stable at $-78°$C, has been used to prepare the corresponding RCu and [R$_2$Cu]Li species that after warming to ambient temperature, were submitted to protonolysis (which almost

(S)-**39**   **40**   **41**

certainly proceeds with retention at carbon, see Sect. III-B-2), leading to recovered hydrocarbon with a high degree of retention of configuration (95). Furthermore, when the cyclopropyllithium species **40** was converted to a mixed cuprate complex, [Cu(SPh)(alkyl)]Li and this was submitted to a coupling reaction with 3-iodo-2-cyclohexenone, the product was formed with retention of configuration at the cyclopropyl ring carbon (96). Similarly, *endo*-2-norbornylmagnesium bromide, **41** (97), upon treatment with BrCuP(*n*-Bu)$_3$ yielded the corresponding organocopper complex, which in turn afforded products from several coupling reactions (Sect. III-B-1) and from reduction by DCuP(*n*-Bu)$_3$ (Sect. III-B-2), which strongly implies, all evidence taken together, that the initial alkylation of copper had occurred with retention of configuration (98).

Examples of the formation of characterizable alkenyl transition metal complexes via reaction with lithium or Grignard reagents are quite rare, and in fact, I am not aware of a single example of a transition metal alkenyl thus prepared having been stereochemically characterized either spectroscopically or crystallographically. This is surprising in view of the voluminous literature using lithium and magnesium alkenyls, but these reagents have been used almost exclusively to prepare either catalytic intermediates or unstable synthetic transition metal reagents, which are used *in situ*. Nevertheless, the extensive organic chemistry of alkenyllithium and Grignard reagents and their reactions with main group metal halides (99) both are well established to typically involve retention of configuration at carbon. Alkenyl transition metal species prepared by transmetalation have been used in catalytic coupling with organic halides (Sect. II-A), reduction by hydrides (III-B-2), electrophilic cleavage reactions (III-C), pyrolytic coupling (III-B-1), and so on, establishing that the alkenylation of transition metals occurs with a high degree of either retention or inversion of configuration. In the context of the studies just mentioned, to assume inversion would introduce insurmountable inconsistencies. Retention has been so widely satisfactory a premise that apparently no one has bothered to prove it directly.

Transmetalation of alkyl and alkenyl groups between Hg(II) and transition metals has been known for some time. Transfer from transition metals to Hg(II) is generally regarded as an "electrophilic cleavage" and so is discussed in Section III-C-3. Transfer of the organic fragment from Hg(II) to transition metals has been employed in a variety of synthetically useful reactions (71,100), and is usually highly stereospecific. For example, *endo*-2-norbornylmercuric bromide underwent transfer to CuI(PBu$_3$) in the presence of *t*-BuLi to generate a mixture that had the empirical composition of (*endo*-norbornyl)Cu·Hg·Li·*t*-butyl and possessed properties different from those of any binary mixture of the components. This mixture underwent a variety of reactions with > 95% stereospecificity which, all taken together, argued strongly for transmetalation with retention of configuration (101).

Another example of alkyl transfer as shown in eq. [27], where configuration

$$\underset{\text{HgCl}}{\text{Me}_2\text{N}\diagdown\text{C}-\text{C}\diagup\text{Me}} \xrightarrow[\text{THF}]{\text{PdCl}_2(\text{NCPh})_2} \underset{\text{Me}\diagup\text{Pd}\diagdown\text{Cl}}{\text{Me}_2\text{N}\diagdown\text{C}-\text{C}\diagup\text{Me}} \xrightarrow{\text{LiAlD}_4} \underset{\text{Me}\quad\text{D}}{\text{Me}_2\text{N}\diagdown\text{C}-\text{C}\diagup\text{Me}} \qquad (27)$$

of the amine product from the threo mecurial, and also from the corresponding erythro isomer, was shown by stereospecific deamination (102a). The general conclusion of transfer of alkyl groups from Hg(II) to Pd(II) with retention is almost certainly correct. Nevertheless, if the result were in serious doubt, one might be concerned that the potential for strong chelation by the amine in this system could render it atypical, and also that there is such a meager empirical basis for the assertion that LiAlH$_4$ reduction of the Pd—C bond proceeds with retention of configuration (Sect. III-B-2). In fact, this study is probably more significant in substantiating the last observation than the first, particularly in view of other data that bear on the stereochemistry of transfer from Hg(II) to Pd(II) (102b,114). For example, carbomethoxylation of several σ-bonded organomercurials, derived from oxymercuration of cyclohexene and norbornadiene in the presence of stoichiometric amounts of LiPdCl$_4$ and CO, was found to proceed with predominant retention of configuration at carbon, implying transmetalation with retention (102b).

A significant amount of chemistry of vinylic mercurials has been carried out catalytically (71,100). For example, a variety of Rh(I) and Rh(III) complexes catalyze the symmetrical coupling of E-vinylic mercurials quite stereospecifically with retention of configuration (103).

Aluminum alkyls have found extensive use in organometallic chemistry, principally as alkylating and reducing agents for transition metal complexes. More recently, these alkyls, and especially alkenyls, have become important reagents in organic synthesis. Often these syntheses are in conjunction with transmetalation to transition metals. General entry into a class of vinylic alanes is afforded by the regiospecific and stereospecific hydroalumination of terminal acetylenes (eq. [28]). These in turn are stoichiometrically coupled in good yield

$$\underline{i}\text{-Bu}_2\text{AlH} + \text{R-C}\equiv\text{CH} \longrightarrow \begin{array}{c} \text{R} \\ \text{H} \end{array}\!\!\!\!>\!\!\text{C=C}\!<\!\!\!\!\begin{array}{c} \text{H} \\ \text{Al}(\underline{i}\text{-Bu})_2 \end{array} \quad (28)$$

to (E,E)-dienes upon treatment with CuCl in THF (104). Transmetalation from Cp$_2$ZrClR to AlCl$_3$ has been achieved with both alkyl- and alkenylzirconium complexes (105). For example, alkyl transfer from **42** to AlCl$_3$ was shown to yield **43** with "predominant" retention of configuration (eq. [29]). The ^{1}H

NMR resonances were too broad to allow good quantitation. Vinyl transfer also proceeded with retention of configuration (105).

Coupling of alkenylalanes with vinylic halides was mentioned above to be catalytic in Ni(0) and Pd(0) (63). The stereospecificity of this reaction is consistent with retention of configuration in transmetalation between aluminum and either nickel or palladium, as well as retention in vinylic oxidative addition. Nickel(0) and Pd(0) also stereospecifically catalyze the coupling of Cp_2ZrCl-(alkenyl) with aryl and vinylic halides (106,107), and the conjugate addition of these zirconium alkenyls to α,β-unsaturated ketones is catalysed by $Ni(acac)_2$ (108).

A recent communication described the reaction of eq. [30] as involving sequential transmetalation from zirconium to zinc to either nickel or palladium

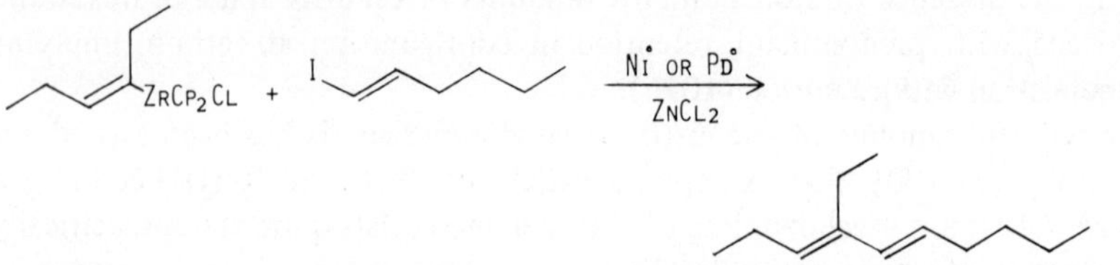

followed by reductive elimination, all of which were highly stereospecific (109). This may be correct, or it may be that $ZnCl_2$ is effectively enhancing the electrophilicity of the nickel or palladium component.

Another recent development is the use of alkenylpentafluorosilicates, **44**, as alkenylation substrates (110), as shown in Scheme 8. In view of the coordination environment of silicon in these complexes, their interactions with Pd(II) or Ag(I) are probably best regarded as stereospecific transmetalations.

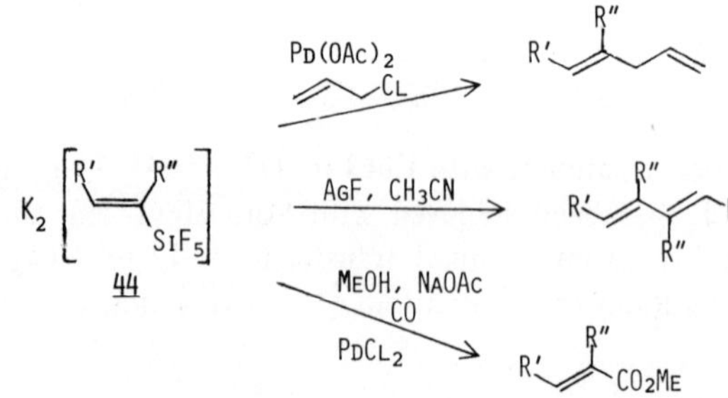

Scheme 8

Other examples, too numerous to list here, of alkenyl transmetalations exist, but those mentioned above are sufficient to illustrate the generally stereospecific nature of these reactions.

C. Additions to Complexed Alkenes and Alkynes

Virtually every important homogeneous catalytic reaction involving carbon substrates depends on an addition and/or elimination step of a complexed π system as shown in eqs. [31] and [32]. The distinction between these two paths is important, because both the nature of this addition step and the stereochemistry associated with the subsequent fate of the resulting metal alkyl determined

$$L_nM-\overset{\overset{Y}{|}}{\underset{\underset{X}{|}}{C}} + X^- \rightleftharpoons \left[L_nM-\overset{\overset{Y}{|}}{\underset{\underset{X}{|}}{C}}-X\right]^- \quad (31)$$

$$L_nM-\overset{\overset{Y}{|}}{\underset{\underset{X}{|}}{\overset{||}{C}}} \rightleftharpoons L_nM-\overset{\overset{Y}{|}}{\underset{\underset{X}{|}}{C}}\underset{}{\overset{}{\diagdown}} \quad (32)$$

the relative configurations of these two centers in the final product. Since addition-elimination reactions of more extended π-bonded systems usually result in the interconversion of different π systems rather than in formation or disruption of metal-carbon σ bonds, they are not discussed here. Other reviews may be consulted for discussions of such reaction (111). A review on optically active alkene metal complexes appeared some time ago (112).

1. Additions of Oxygen and Nitrogen Nucleophiles

Additions of water, carboxylic acids, and alcohols have all been carried out, both catalytically and stoichiometrically. The impetus for this work was the discovery of the "Wacker process", which is shown in eq. [33]. The principal step of the catalytic process is the oxidation of ethylene by Pd(II), which can also be effected stoichiometrically. Several studies of the kinetics of the latter

$$CH_2=CH_2 + \tfrac{1}{2}O_2 \xrightarrow{PdCl_4^{-2},\ CuCl_2} CH_3CHO \quad (33)$$

$$\frac{-d\left[PdCl_4^{-2}\right]}{dt} = \frac{k\left[PdCl_4^{-2}\right][CH_2=CH_2]}{[H^+][Cl^-]^2} \quad (34)$$

reaction have established the rate expression of eq. [34] (113), which was interpreted by Henry according to Scheme 9. This mechanism would predict delivery of the hydroxyl group from lead to coordinated ethylene in the cis complex **46**, that is overall syn addition to the alkene.

It has been pointed out more recently that reversible attack by water on **45** (or more likely on the trans isomer **47**, Scheme 10) would also accommodate the kinetic data and yet lead to anti addition of water (114). In spite of Henry's caveat that the kinetics do not rigorously exclude an anti addition (113), his

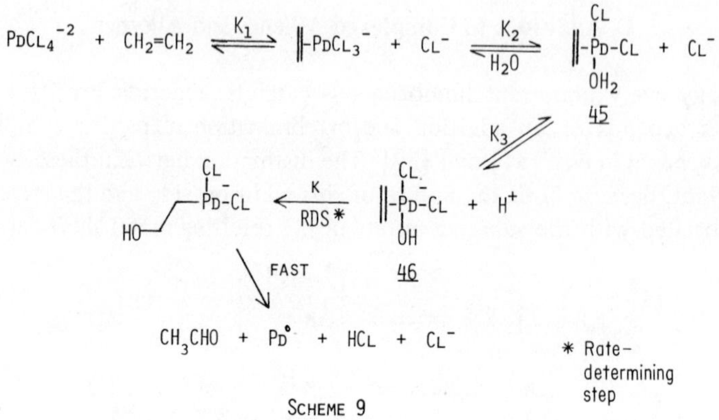

STEREOCHEMISTRY OF REACTIONS OF M—C SIGMA BONDS

SCHEME 9

* Rate-determining step

results are frequently quoted as having established syn stereochemistry for the Wacker oxidation, and other more recent kinetic studies have favored similar interpretations (115) where alternate explanations are possible.

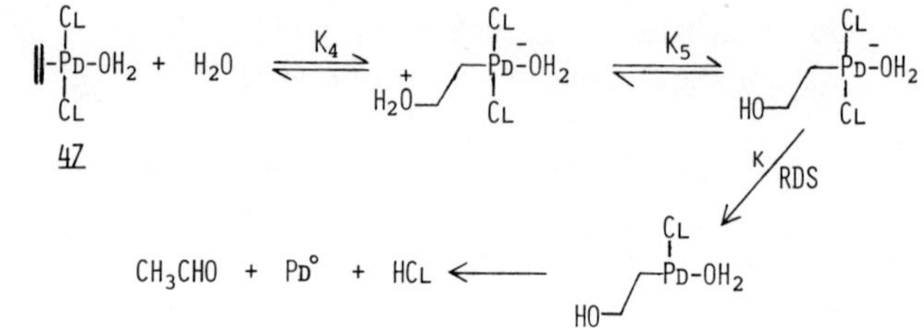

SCHEME 10

Because of the kinetic instability of β-hydroxyalkylpalladium(II) species, attempts to determine the stereochemistry of the nucleophilic addition of $^-$OH or H_2O have necessarily involved attempts either to intercept the intermediate before aldehyde or ketone can form, or to find molecules in which the alkyl-palladium moiety is unusually stable. An example of both these approaches was afforded by the hydration reaction of (1,5-cyclooctadiene)palladium dichloride, **48** (116): attack occurred anti to the metal, forming **49**, which was stable enough to isolate, but whose stereochemistry was demonstrated by subsequent CO insertion forming *trans*-lactone, **50** (eq. [35]).

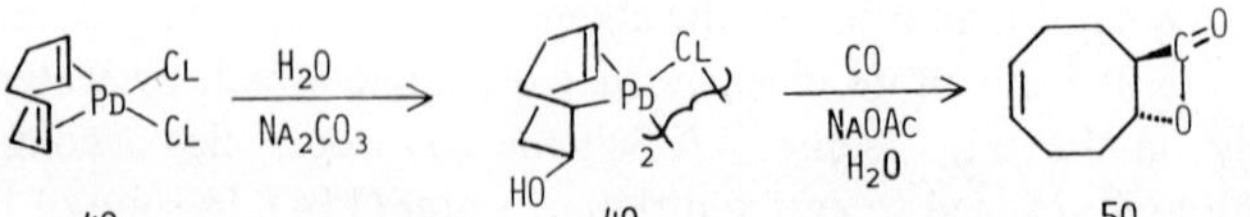

(35)

This example establishes that formal hydroxide addition can occur anti to the metal. Nevertheless, chelate complexes of this type have been considered atypical by some who claim that syn addition is hindered by the particularly rigid geometry that is maintained (see references in ref. 116). Since carbon ligands are readily transferred to the syn side of chelate ligands (Sect. II-C-2), this argument seems unconvincing.

Nevertheless, examples that are as unencumbered as possible are desirable, and two studies on the oxidation of cis- and trans-ethylene-1,2-d_2 have appeared (114,117) wherein the alkylpalladium intermediate has been trapped by different means before it could decompose to acetaldehyde. Oxidation of cis-dideuterioethylene, as in Scheme 11, in the presence of CO by Pd(II) either stoichiometrically or catalytically using stoichiometric Cu(II), resulted in the formation of trans-2,3-dideuteriopropiolactone. Assuming retention at carbon upon CO insertion, this result requires anti attack by water (117).

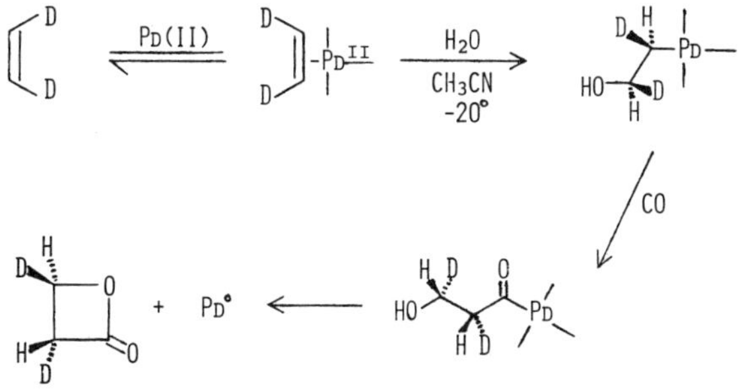

SCHEME 11

An obvious cavil can be raised here: namely, that coordination of the nucleophile may be blocked by the presence of coordinated CO. Although the affinity of Pd(II) for CO seems to be rather low (44,45), this objection is nonetheless difficult to evaluate with the data extant. Even if blocking were occurring and the preferred attack were syn, this result would nonetheless indicate that the syn and anti paths must be very close in energy.

Another interesting study concerning the Wacker oxidation was interpreted as outlined in Scheme 12 (114), except that the trans configuration of **47-d_2** is my suggestion. The stereochemical analysis was based on the observation that when the oxidation is carried out in the presence of high concentrations of Cl$^-$ and Cu(II), the main product is chlorohydrin. Analysis of this product as ethylene oxide indicated that a highly stereospecific inversion of configuration of the alkene had occurred, corresponding to syn-chlorohydrin formation. Thus either formal $^-$OH addition had occurred anti to the metal followed by inversion of configuration upon oxidative cleavage, or addition syn to the metal had occurred followed by retention of configuration upon oxidative cleavage. Taken together, pieces of evidence presented in this work (114) and elsewhere (see Sect. III-C-4)

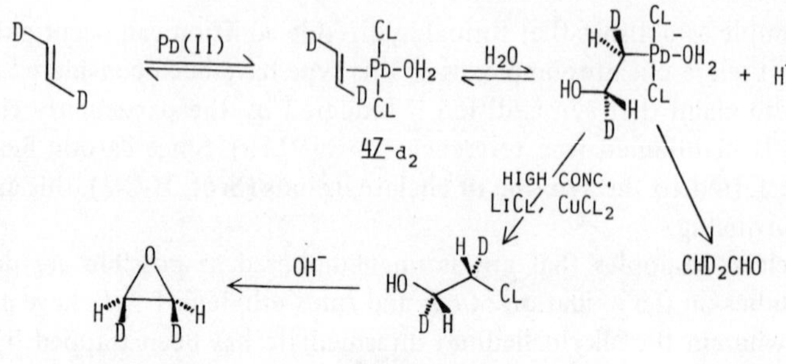

SCHEME 12

indicate that at least in the presence of excess nucleophile, the oxidative cleavage by Cu(II) does proceed with inversion. Thus attack on coordinated ethylene appears to be anti.

To be entirely rigorous, it was pointed out (114) that the conditions under which the kinetics of the Wacker process were studied involved [Cl$^-$] of up to ca. 1M (113). This stereochemical study used [Cl$^-$] of 3.3M. At such high ligand concentrations a change in mechanism is not impossible.

Very early, methanol was demonstrated to attack chelated dienes anti to palladium or platinum (118,119). For example, the dicylopentadiene complexes of PdCl$_2$ and PtCl$_2$ underwent addition of methanol as shown in eq. [36]. The stereochemistry for both **51** and **52** was deduced from ^{1}H NMR spectra, and

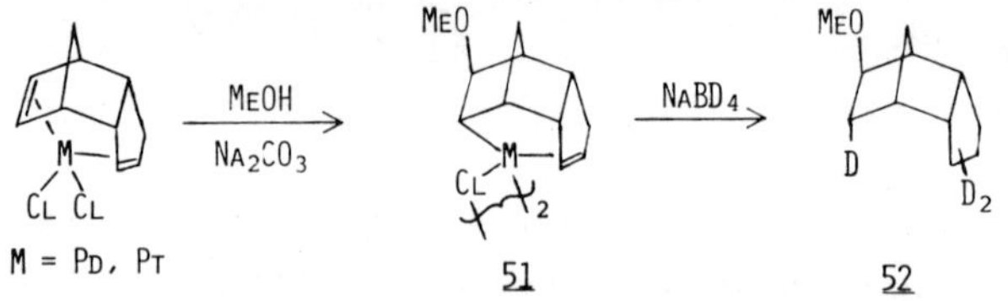

that of **51** (M = Pt) was substantiated by X-ray crystallography (120). Very similar sequences were carried out for norbornadiene and cyclooctadiene complexes, and the structure of the platinum-complexed methoxide addition product was also established by an X-ray study (121).

As mentioned above, stereochemical conclusions based on chelate complexes may not always be general. For this reason, the acyclic alkenes *cis*= and *trans*-2-butene have been methoxypalladated (122). β-Hydride elimination was avoided by intercepting the alkypalladium intermediate by CO insertion. As illustrated for *cis*-butene in eq. [37], granting that CO insertion is stereospecific with retention at carbon, the addition was clearly anti. As was the case when H$_2$O addition intermediates were trapped by CO, there is uncertainty about the effect of CO coordination on the relative availability of syn and anti nucleophilic addition pathways.

Alkylpalladium complex **54** has been isolated and characterized (123). Its ^{1}H

NMR spectrum clearly establishes the trans addition of methoxide (eq. [38]). Since the coordinate saturation of **53** would probably preclude prior coordination of the nucleophile, this example simply reemphasizes that anti addition is possible.

A kinetics investigation of the Pd(II) oxidation of ethylene in methanol has found the rate expression to be identical to that observed in water (eq. [34]) (124). In the face of mounting numbers of examples of the anti addition of methanol, these kinetics were nevertheless interpreted as before in terms of a prior coordination, cis attack mechanism.

Acetoxypalladation most probably proceeds by anti addition based on a study of the oxidation of cyclohexene-d_4, **55** (125). A mechanistic postulate such as that in Scheme 13 accounts for the large number of products formed, three of the most enlightening being **57** to **59**. Apparently acetate adds anti to palladium to form **56**, which then undergoes a series of *syn*-β-hydride elimination-addition steps (see Sect. II-C-3). Products **57** to **59** are presumably formed by

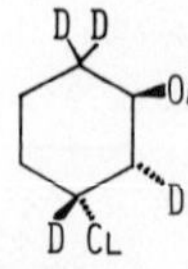

57 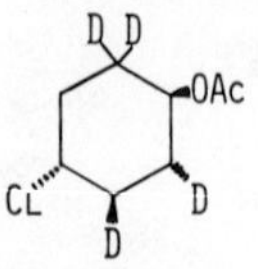 **58** 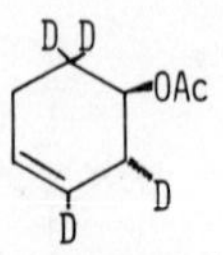 **59**

Cu(II)-induced oxidative cleavage of cyclohexylpalladium intermediates, which incidentally would be required to occur with retention at carbon, in contrast to the $CuCl_2/Cl^-$ cleavage shown in Scheme 12 (see Sect. III-C-4 for a discussion of this apparent contradiction).

Formal addition of a Pd–Cl bond to an alkene has been demonstrated in the case of complex **60** in eq. [39] (126). Similar additions to monoolefins have

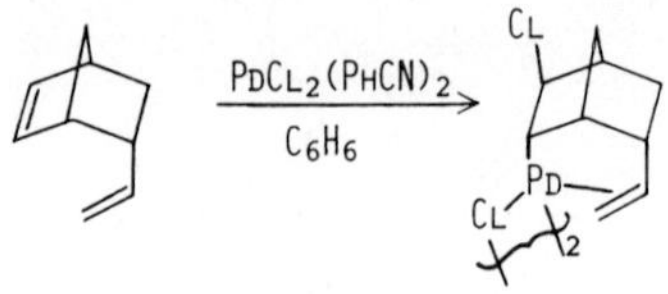

(39)

60

been investigated by observing Pd(II)-catalyzed exchange of radioisotopic Cl^- with vinyl chlorides. The picture is badly confused by rapid nonexchanging Z-E isomerization of starting materials (113). It may be observed, however, that 1-chlorocyclopentene is readily exchanged, so that if a palladium-chloride addition-elimination mechanism is operating, it cannot be stereospecific.

All extant well-defined examples of additions of amines to coordinated alkenes also exhibit anti stereochemistry. An elegant experiment using the resolved diastereomer (+)-**61**, which possesses a Pt(II)-coordinated (S)-1-butene, is outlined in Scheme 14 (127). Addition of Et_2NH to the complexed butene must

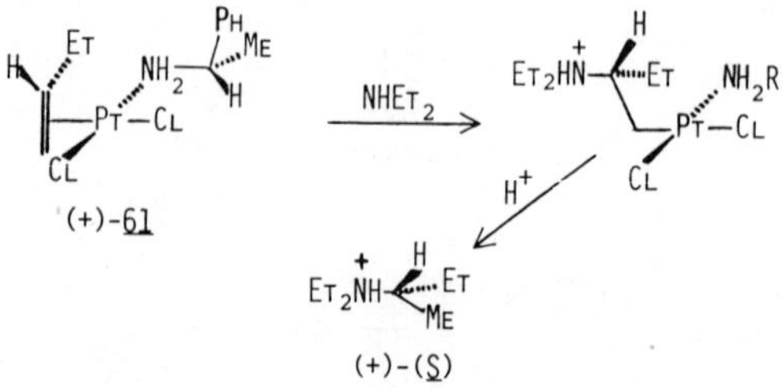

SCHEME 14

have occurred to the anti face, since the product after protonolysis was (+)-(S)-N,N-diethyl-sec-butylamine. The configuration of (+)-**61** could be demonstrated by comparison of its circular dichroism spectrum to that of a closely analogous complex whose structure had been demonstrated by X-ray crystallography.

Palladium alkene complexes probably also add amines anti. In a series of reports, several related aminopalladation reactions were examined. For example, the initially formed alkylpalladium species **62** could be reduced by $LiAlD_4$ to a tertiary amine (128); it could be oxidized, resulting in an internal displacement to yield an aziridine (when the amine originally added was primary) (129); or it could be oxidized in the presence of excess amine to form 1,2-diamines (130). All these reactions exhibited the overall stereochemistry shown in Scheme 15

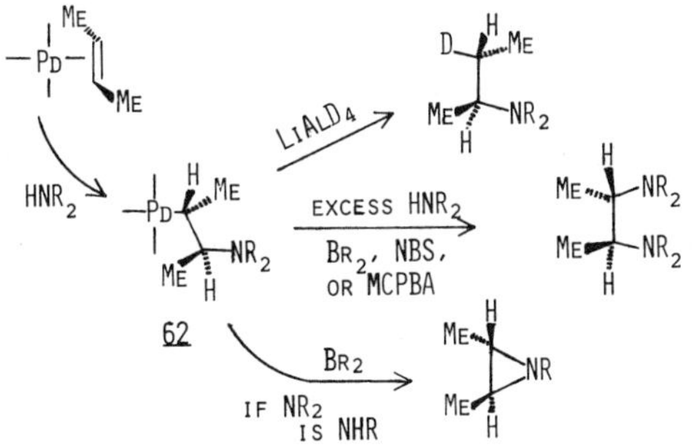

SCHEME 15

and are consistent with anti addition as represented by structure **62**. It is possible to construct a version of Scheme 15 based on initial syn addition of amine, but this would introduce inconsistencies and would necessitate new explanations for most of the chemistry in Sects. II-B and II-C.

In summary, there are many stereochemical examples of addition of oxygen and nitrogen nucleophiles to alkenes coordinated to palladium and platinum. All these are consistent with an interpretation based on anti addition, although some studies are more convincing than others. It would seem that lacking *any* example of a syn addition, the burden of proof now rests on those who would arbitrarily interpret complex kinetics data in terms of syn additions. It would seem that concrete *stereochemical* evidence would be required.

2. Additions of Carbon Species

Additions of carbon nucleophiles to complexed alkenes have been observed to proceed with both syn and anti stereochemistry with respect to the metal. The direction of addition depends on the nature of the carbanionic center and, presumably, on whether coordinate unsaturation or an appropriate leaving group is present so that the organic group can first attach itself to the metal.

With palladium and platinum diolefin complexes, the dichotomy of syn vs.

anti addition is pronounced. Delocalized carbon nucleophiles such as the anions of diethyl malonate and ethyl acetoacetate have been demonstrated to add to the exo side of the dicylopentadiene complex **63**, for example, forming the diethyl malonate adduct **64** (eq. [40]). ^{1}H NMR spectroscopy could be used to

$$\underset{\mathbf{63}}{\text{[complex with M-Cl, Cl]}} \xrightarrow[\text{NA}_2\text{CO}_3]{\text{CH}_2(\text{CO}_2\text{ET})_2} \underset{\mathbf{64}}{\text{[complex with CH(CO}_2\text{ET})_2\text{, M, Cl]}} \quad (40)$$

$$M = P_D, P_T$$

substantiate the configurations of these products (131). This reaction has been used to prepare intermediates in a prostaglandin synthesis beginning with amine **65**. In accord with the supposition that the amino group should function as a chelating ligand and should direct the complexation of Pd(II) to the same face of the ring, the carbanion was observed to attack from the opposite side. The trans nature of product **66** was demonstrated as shown in eq. [41] (132).

$$\underset{\mathbf{65}}{\text{[NMe}_2\text{ cyclopentene]}} \xrightarrow[\text{PdX}_2]{\overset{\text{Na}^+\text{ CH(CO}_2\text{ET})_2^-}{\text{THF, 0°}}} \underset{\mathbf{66}}{\text{[NMe}_2\text{ with CH(CO}_2\text{ET})_2]} \xrightarrow[\text{2) KOH, }\Delta]{\text{1) MeI}} \underset{\text{DMF}}{\text{[bicyclic lactone]}} \quad (41)$$

Acetylacetone anion added to complexed *cis*- or *trans*-ethylene-1,2-d_2 in an anti fashion in [CpPd(PPh$_3$)(C$_2$H$_2$D$_2$)]$^+$, **53**, as was clearly demonstrated from the vicinal ^{1}H coupling constants in CpPd(PPh$_3$)[CHDCHDCH(COCH$_3$)$_2$] (133). Since **53** is coordinatively saturated, however, this stereochemical outcome is not unexpected. Various other examples of these additions are known, including compounds of palladium, platinum, and iron, but in none of these was the stereochemistry clearly established.

In contrast to the anti additions above, unstabilized carbon nucleophiles generally either add in a syn manner, or lead to olefin displacement or to redox chemistry. Well-behaved additions are usually observed using carbon nucleophiles that cannot β-eliminate, and using main group organometallic derivatives that will transfer organic groups to transition metals but are otherwise of subdued reactivity. With some exceptions, lithium and Grignard reagents perform poorly, while organic derivatives of boron, mercury, and tin serve well.

Surprisingly, there are no examples of a well-behaved unimolecular insertion of the type shown in eq. [42], where the R and alkene groups are simple acyclic, unactivated groups and **67** and **68** are both characterizable. If one hopes to observe **68** directly, β-hydride elimination must be precluded in some way. Also in general it has been necessary to generate **67** indirectly, using conditions under

$$\underset{\mathbf{67}}{\overset{M-R}{\underset{C=C}{\diagup}}} + L \longrightarrow \underset{\mathbf{68}}{\overset{\underset{M-C-C-R}{|}\underset{L}{|}}{}} \qquad (42)$$

which it cannot be directly observed. If one grants that unimolecular β-hydride elimination proceeds in a syn manner (as it apparently always does; see Sect. II-C-3), then indirect but convincing evidence concerning the stereochemistry of addition is still obtainable from elimination products in appropriately designed systems.

Perhaps the earliest example of insertion involving demonstrable stereochemistry was the self-insertion of the methoxide adduct **69** induced by ligation of palladium (134). In this case, of course, the complex has no unimolecular alternative to syn addition with retention at the "migrating" center. This transformation in eq. [43] is a real one, since a crystal structure has shown that an analogous complex has structure **71** (135), rather than **72**, as some had previously conjectured.

$$\qquad (43)$$

X = Ph, Cl

Early work by Heck established that organomercurials could be used to transfer organic groups to palladium (136), and the technique has been used extensively since that time (71,100). For example, diphenylmercury arylated platinum in (norbornadiene)PtCl$_2$ either once or twice, depending on the reactant ratio, but the resulting complex **73** could not be induced to transfer the phenyl groups to norbornadiene. Conversely, (norbornadiene)PdCl$_2$ reacted with diphenylmercury forming the inserted product **71** with no intermediate analogous to **73** being observable (135). As just mentioned, the *endo*-phenyl configuration was established by X-ray crystallography, which in the light of the platinum arylation strongly implies syn addition of Pd—Ph to the complexed alkene. More recently it has been found that NaBPh$_4$ will also convert (norbornadiene)PdCl$_2$ to **71** (137), and SnMe$_3$(*p*-tolyl) forms the *p*-tolyl analog (138). None of these arylating agents are nucleophilic, and the fact that they all afford **71** or its analog argues strongly for initial arylation of the metal.

Several other very similar examples exist, such as in the preparations of **74** from (dicyclopentadiene)$PdCl_2$ and PhHgCl (139), and **75** from norbornene, Li_2PdCl_4, and PhHgCl (140). ^{1}H NMR spectroscopy was employed to elucidate the configurations of these molecules, as was reduction followed by spectral

characterization of the liberated ligands. Bis(propenyl)mercury alkenylates **76**, forming **78**, perhaps via intermediate **77** (eq. [44]). Product **78** presumably had this chelate structure, but the propenyl group was definitely endo, as was shown by analysis of its reduction product and other derivatives (141).

π-Allyl complexes will insert monoenes, affording isolable organometallic products in some instances. [(π-Methallyl)NiCl]$_2$ forms **79**, an acetate-bridged dimer, from reaction with norbornene in the presence of KOAc (142), and **80** can be isolated from a very analogous reaction (143). The structure of **79** was established by X-ray crystallography.

A number of studies have provided indirect but generally convincing evidence for syn additions of alkyl and aryl groups to complexed alkenes. As mentioned earlier, all these studies assumed that Pd–H β-elimination is strictly syn (Sect. II-C-3). As shown in Scheme 16, phenylation of Z-1-phenylpropene stoichiometrically induced by Pd(II) yielded 1,2-diphenylpropene, which is ca. 90% Z (144), and E-1-phenylpropene afforded E product. These results are consistent with syn

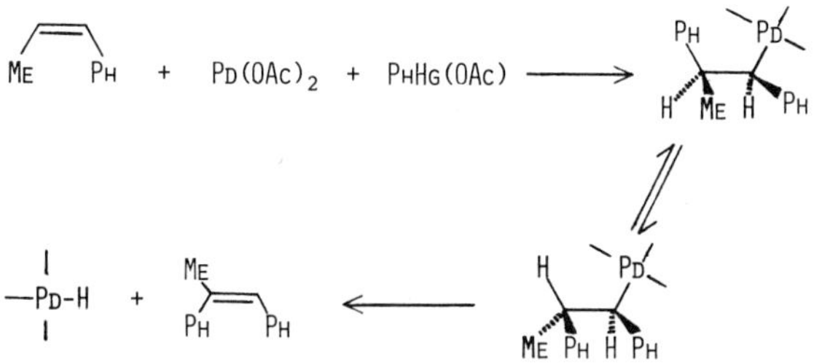

SCHEME 16

addition and syn elimination. When the isomeric phenylpropenes were phenylated using bromobenzene and catalytic palladium, essentially the same results were obtained (145). Even methyllithium has been used in an addition reaction, again with results supporting a syn addition mechanism (eq. [45]) (146).

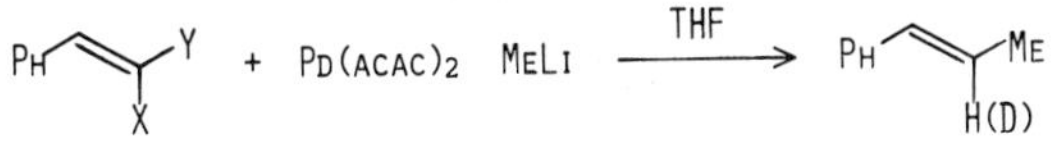

$$Ph\text{-CH=CY-X} + Pd(acac)_2 \quad MeLi \xrightarrow{THF} Ph\text{-CH=C(Me)-H(D)} \quad (45)$$

X = H, Y = D: 90%-d_1
X = D, Y = H: 8%-d_1

Cyclohexene-3,3,6,6-d_4 underwent phenylation with the results shown in eq. [46] (147). Product **81** is readily interpreted in terms of a syn phenylpalladation followed by *syn-β*-hydride elimination, addition, and finally elimination again.

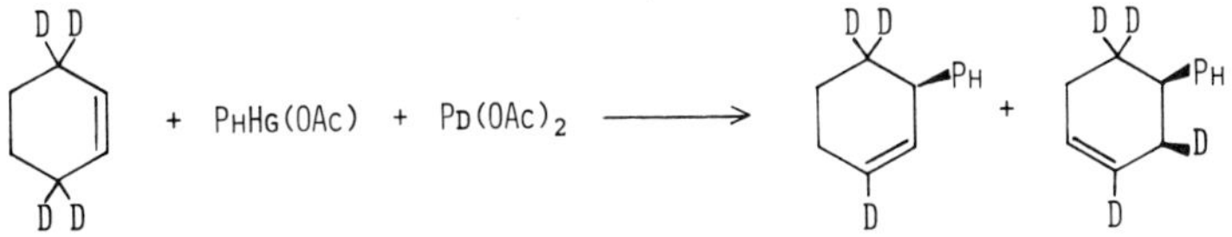

$$(46)$$

81

Additions of organic ligands to complexed aliphatic and aromatic alkynes is a widely employed, synthetically useful reaction (148). There is also a significant amount of literature concerning such insertions of highly electrophilic acetylenes, which are not discussed here. The majority of acetylene insertion reactions give polymers, oligomers, and cooligomers, many of which are of a cyclic structure (148). Naturally, to form cyclic oligomers or cooligomers, an overall (but not necessarily kinetic) syn addition to the acetylene is required, and an intramolecular reaction such as that in eq. [47] is expected yield an alkenyl product of

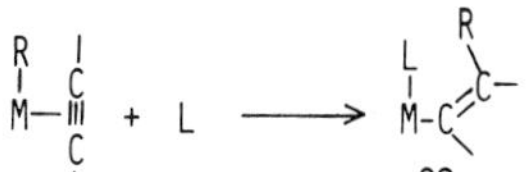

$$M-C\equiv C-R + L \longrightarrow \quad (47)$$

82

structure 82. There are a few well-defined examples of this type, as well as multitudinous indirect examples affirming that syn addition is indeed the preferred pathway. There are, however, radical paths and other mechanistic subtleties that can arise to complicate the picture, including, for example, various solvent addition paths leading to a σ-vinyl species and carbene complexes in the case of more electrophilic metals (149).

Methyl or phenylpentacarbonylmanganese(I) and phenylacetylene undergo a well-defined, 1:1 addition, presumably via the path in eq. [48]. The implications

$$Me-Mn(CO)_5 + Ph-C\equiv CH \rightleftharpoons \begin{array}{c}\text{intermediate}\end{array} \longrightarrow \begin{array}{c}\text{product}\end{array} \quad (48)$$

here are that an acetyl group will undergo transfer to the complexed alkyne, and that it does so in a syn manner (150). Closely analogous reactions are known involving the thermal reaction of $CpMo(CO)_3Me$ with 2-butyne (151), and the photochemically induced reaction of $CpW(CO)_3Me$ with acetylene (152).

One particularly clean example arises from the reaction of nickel complex 83 with 2-butyne (eq. [49]). Configurational assignment of Z-84 was readily made by CO insertion. Relatively small changes in the acetylene had an apparently

$$\underset{\underline{83}}{\begin{array}{c}PPh_3\\|\\Br-Ni-Ph\\|\\PPh_3\end{array}} \xrightarrow{Me-C\equiv C-Me} \underset{\underline{84}}{\begin{array}{c}PPh_3\,Ph\\|\quad\diagup\\Br-Ni\!-\!\diagdown\!Me\\|\quad\quad Me\\PPh_3\end{array}} \xrightarrow[MeOH]{CO} \begin{array}{c}Me\quad\quad Me\\\diagdown\!=\!\diagup\\Ph\quad\quad CO_2Me\end{array} \quad (49)$$

profound effect on the overall reaction, since 3-hexyne yielded Z- and E-3-phenyl-3-hexene. Added O_2 enhanced the yield of the phenylhexenes, which were monodeuterated when CD_3OH was used as solvent. Apparently the phenylhexenyl analog of 84 is significantly less stable to oxidatively induced homolysis (153).

An unexpected answer was obtained from an insertion quite similar to the one just discussed. When various acetylenes reacted with 85, kinetic product ratios were obtained that usually did not correspond to a syn addition. The data in eq. [50] illustrate this point. A phosphine-dissociated intermediate that rapidly isomerized was involved to explain these results. The interesting point was made from this work that even if a complex is found to be the kinetic product of an insertion reaction, the observed stereochemistry is not necessarily that of the crucial insertion step (154).

In 1971 it was reported (155) that alkylcopper reagents would add to terminal acetylenes to form alkenylcopper species, 86, (eq. [51]), which have since been

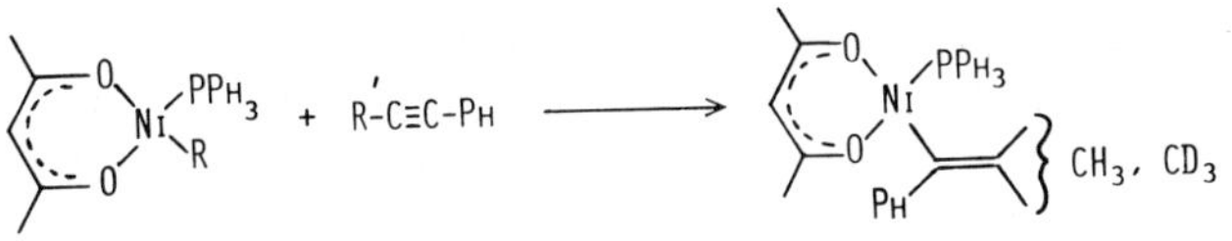

[Equation (50) scheme showing complex 85 reacting with R'-C≡C-Ph]

R	R'	Z / E
CH₃	CD₃	35/65
CD₃	CH₃	61/39

found to be highly useful in organic synthesis (Scheme 17). The stereochemistry of this reaction has not been determined by direct observation of **86**; however

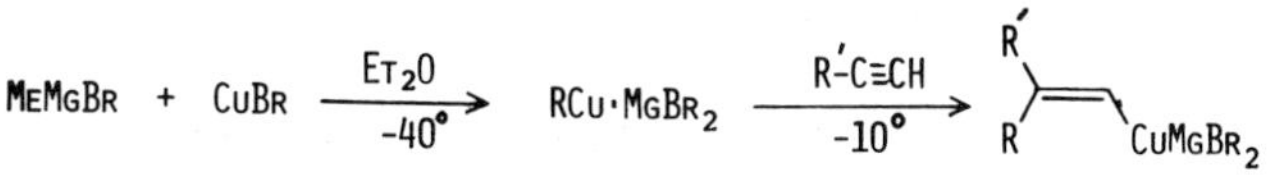

$$\text{MeMgBr} + \text{CuBr} \xrightarrow[-40°]{\text{Et}_2\text{O}} \text{RCu}\cdot\text{MgBr}_2 \xrightarrow[-10°]{\text{R'-C≡CH}} \underset{\mathbf{86}}{\overset{R'}{\underset{R}{\diagup}}\!\!\!=\!\!\!\overset{}{\underset{\text{CuMgBr}_2}{\diagdown}}} \quad (51)$$

the wide variety of reactions that have been carried out all point to a highly stereoselective syn addition. Although addition to a terminal acetylene normally

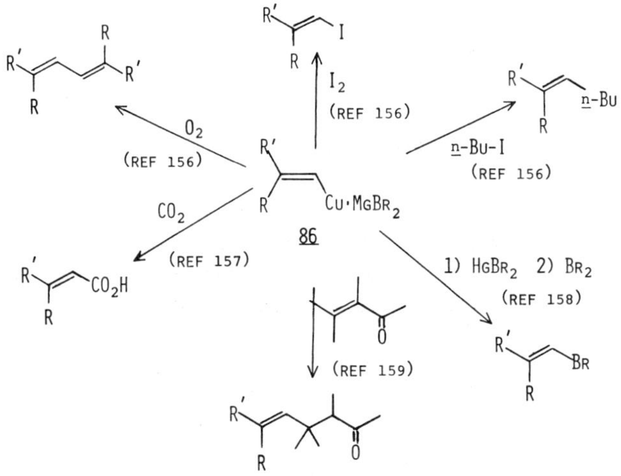

[Scheme 17: cyclic diagram showing transformations of alkenylcopper **86** (R'C=CR-CuMgBr₂) with various reagents: O₂ → dimer (REF 156); I₂ → vinyl iodide (REF 156); n-Bu-I → n-Bu vinyl (REF 156); 1) HgBr₂ 2) Br₂ → vinyl bromide (REF 158); (REF 159) → ketone product; CO₂ → carboxylic acid R'CR=CHCO₂H (REF 157)]

SCHEME 17

gives the primary alkenyl structure **86**, the presence of potential chelating groups, such as in HC≡C–(CH₂)$_n$X, where X = NEt₂, SEt, OR, or halogen, can induce addition of the opposite regiochemistry to form the secondary alkenylcopper reagent (160).

And finally, one mechanistically rather ill-defined acetylene addition reaction is particularly noteworthy because it involves a transformation that is quite unusual. Whereas copper reagents will add in a useful way to terminal acetylenes, they generally will not add to internal acetylenes. In one case this limitation has been overcome by the use of a two-component system, one example of which is shown in eq. [52] (161). The resulting vinylmetal species is useful in standard coupling and cleavage reactions and affords a highly stereoselective synthesis of trisubstituted alkenyl groups.

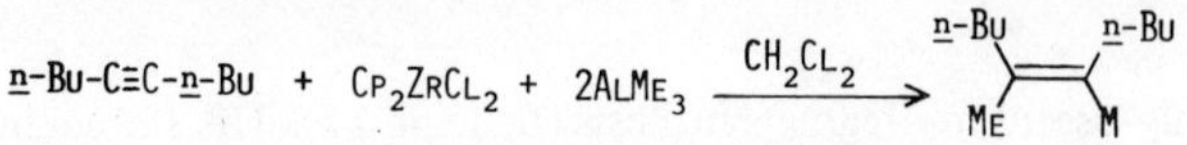

$$\underline{n}\text{-Bu-C}\equiv\text{C-}\underline{n}\text{-Bu} + \text{Cp}_2\text{ZrCl}_2 + 2\text{AlMe}_3 \xrightarrow{\text{CH}_2\text{Cl}_2} \underset{\text{Me}}{\overset{\underline{n}\text{-Bu}}{}}\text{C}=\text{C}\underset{\text{M}}{\overset{\underline{n}\text{-Bu}}{}} \qquad (52)$$

3. Additions of Hydride, β-Additions, and β-Eliminations

Insertions of alkenes and alkynes into metal hydride bonds (e.g., eq. [53]) and the reverse, β-hydride elimination, are a cornerstone of organometallic reactions. They are important in catalytic hydrogenation, hydroformylation, isomerization, dimerization, hydrosilylation, and so on, of alkenes, and comprise a major route for the thermal decomposition of metal alkyls.

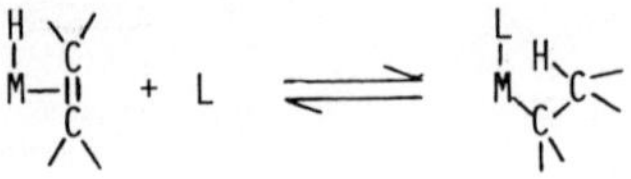

$$\text{M}-\overset{\text{H}}{\underset{\text{C}}{\text{C}}}\text{=C} + \text{L} \rightleftharpoons \overset{\text{L}}{\underset{\text{M}}{}}\overset{\text{H}}{\underset{}{\text{C}}}-\text{C} \qquad (53)$$

In homogeneous hydrogenation, for example (162), there were originally questions of whether alkene complexation by the catalyst was required, and whether alkane formation occurred by simultaneous transfer of two hydrogens from the metal (path A, Scheme 18), or by a stepwise transfer (steps B and C). Careful kinetics studies have established the presence of an alkene metal complex on the reaction coordinate for hydrogenations by Wilkinson's catalyst,

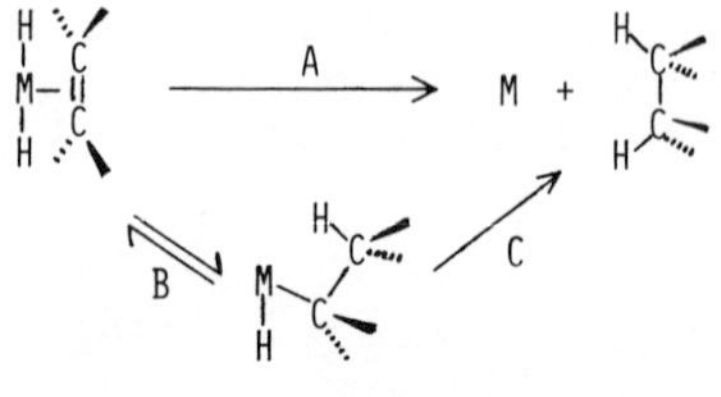

SCHEME 18

RhCl(PPh$_3$)$_3$ (163), and an alkylmetal hydride has now been observed directly in a catalytic system by NMR spectroscopy at low temperatures in the case of [Rh(diphos)H(alkyl)]$^+$ (164).

Since deuteration catalyzed by RhCl(PPh$_3$)$_3$ has been demonstrated in a variety of cases to yield 1,2-d_2 products resulting clearly from syn addition

(162), then either each of steps B and C in Scheme 18 proceeds with retention, or each with inversion. For a unimolecular decomposition of MH_2(alkene), two inversion steps are very difficult to imagine; nonetheless it is important to establish the answer to the question of whether the reaction of eq. [53] is syn.

In addition to syn hydrogenation, other early indications of the nature of M–H additions arose from hydroformylation and related studies. Nickel carbonyl

87

88

and CO in D_2O upon reaction with norbornene yielded **87** resulting from syn-exo addition (165), and **88** arose from cobalt-catalyzed hydroformylation of the corresponding alkene with D_2 and CO (166). Granting that it is well established that the CO insertion proceeds with retention, these both represent likely syn M–H additions.

Convincing data are now available for a variety of metal hydride systems. For example, unequivocal NMR evidence indicated that Cp_2ZrClH (or D) added to the appropriately labeled t-butylethylenes as shown in eq. [54] to form products

89 Cp_2ZrClR **90**   (54)

of syn addition (167). In a similar way, Cp_2MoD_2 was shown to undergo syn addition to form *threo-* and *erythro-*$Cp_2MoD[CH(CO_2Me)CHD(CO_2Me)]$ from dimethyl fumarate and maleate, respectively. In this case, however, the kinetics were complex and included an induction period; this and the coordinatively saturated nature of the metal hydride made mechanistic interpretation of the stereochemical results difficult (168).

Observation of the stereochemistry of β-elimination from rhodium and iridium alkyls has been difficult owing to the propensity of the resulting hydridometal complexes to re-add to the olefinic products, eventually resulting in randomization of any stereochemical label (169–171). Nevertheless, in one instance success was attained. Oxidative addition of *threo-* or *erythro-*2,3-diphenylbutanoyl chloride to $PhCl(PPh_3)_3$ formed the corresponding $RhCl_2(PPh_3)_2$(acyl) species, which upon heating decarbonylated to the rhodium alkyl. This then underwent β-elimination (169). As illustrated in eq. [55] for the threo isomer, if one assumes that the decarbonylation proceeded with retention, then *cis*-1,2-

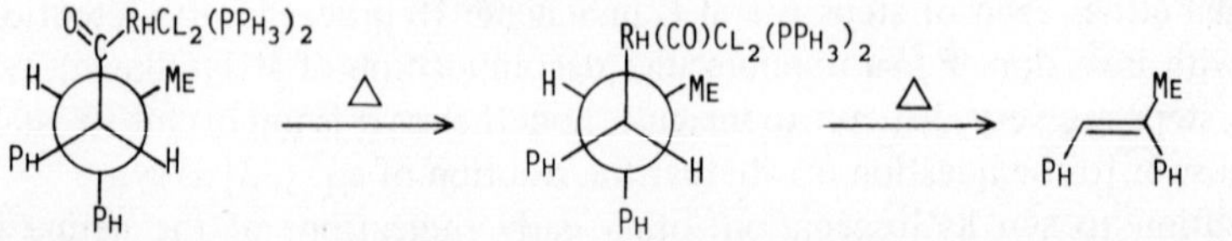

diphenylpropene must have resulted from syn elimination. To isomerize this product, rhodium hydride would presumably have to add such that rhodium would be bonded to a tertiary benzylic carbon — something that apparently is difficult, since only 10% of the trans isomer was observed. Only trans product was detected from *erythro*-acylrhodium starting material.

Alkoxypalladium complex **54**, which was discussed in Sect. II-C-1, underwent pyrolysis affording an 85% yield of the vinyl ethers shown in eq. [56], and only 5% of other isomers, which is consistent with a stereospecific syn elimination. Presumably prior dissociation of triphenylphosphine to provide a vacant coordi-

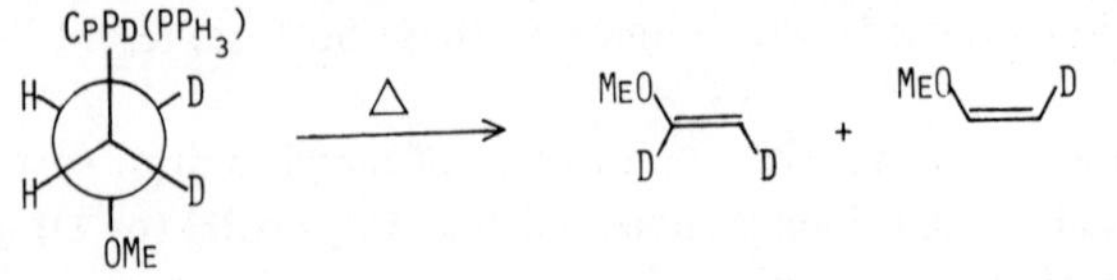

nation site is involved (123). For reasons that are not clear, mixtures of *cis*- and *trans*-CHD=CDCH(COMe)$_2$ were obtained from pyrolysis of both *threo*- and *erythro*-CpPd(PPh$_3$)[CHDCHDCH(COMe)$_2$] (133).

Additional indirect evidence for the syn nature of M—H additions to alkenes arises from a number of cases of the occurrence of multiple β-addition-elimination sequences that are more rapid than olefin decomplexation. Examples of this have already been discussed in Sect. II-C-1 (Scheme 13) and Sect. II-C-2 (eq. [46]). In another instance, when 1-methylcyclohexene was hydroformylated using Co$_2$(CO)$_8$, then as can be seen in Scheme 19, hydride addition-elimination generated a series of cis and a series of trans products, with the trans predominant

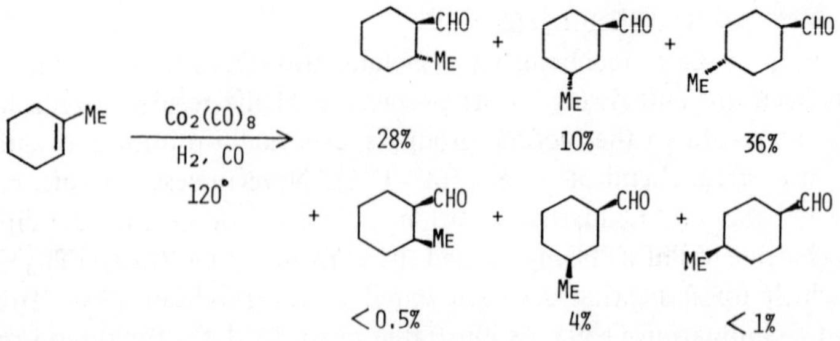

SCHEME 19

by far (172). Presumably complex **91** β-adds to form *trans*-**92** (eq. [57]), whose subsequent rearrangement intermediates interconvert more rapidly than they exchange olefins.

$$\text{cyclohexene-Me} \xrightarrow[-CO]{HCo(CO)_4} \mathbf{91} \rightleftharpoons \mathbf{92} \quad (57)$$

A particularly elegant example arose from work in two different laboratories (173,174). Hydroformylation of (S)-3-methyl-1-hexene yielded a small amount of (R)-3-ethylhexanal in addition to the normal products. Carrying out the same reaction on 3-methyl-1-hexene-3-d_1, **95**, gave the deuterium-migrated material **96**,

$$(S) \xrightarrow[H_2,CO]{Co_2(CO)_8} \text{USUAL PRODUCTS} + (R)\text{-}\mathbf{93} \quad 3.6\% \quad (58)$$

94, **95**, **96**

demonstrating that **93** had not formed via some intramolecular insertion such as in **94** and that hydride addition-elimination rearrangements were more rapid than olefin exchange on the cobalt catalyst. A similar example has been reported (175).

Though alkene formation from decomposition of a metal alkyl is consistent with a β-hydride elimination pathway, it does not logically require it. Radical disproportionation, or, in the presence of strong electrophiles, intermolecular hydride transfer, may occur. In the case of iron alkyls, at least, hydride abstraction has been shown to be β (176) (eq. [59]) and anti (177) (eq. [60]).

$$CpFe(CO)_2\text{-}CD(CH_3)_2 + Ph_3C^+ \longrightarrow [Cp\text{-}Fe(CO)_2(CD(CH_3)=CH_2)]^+ \quad (59)$$

$$\xrightarrow{1)\ Ph_3C^+,\ 2)\ PPh_3} \quad (60)$$

Acetylenes do not show as consistent stereochemical behavior in their reactions with metal hydrides as do alkenes, since both syn and anti additions occur. The former is the case with zirconium hydrides of the types Cp_2ZrH_2 and Cp_2ZrHCl, which undergo insertions of terminal (178) and internal (178,179) alkynes to form alkenyl complexes with completely E stereochemistry. An interesting regioselectivity has been found with internal alkynes (eq. [61]) (179).

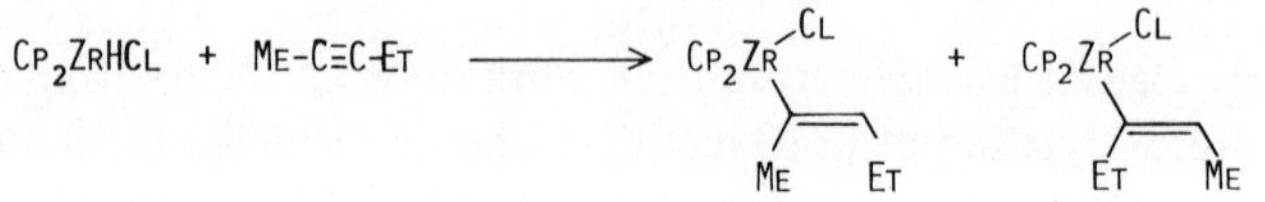

$$Cp_2ZrHCl + Me-C\equiv C-Et \longrightarrow Cp_2Zr\begin{smallmatrix}Cl\\ \diagdown\end{smallmatrix} + Cp_2Zr\begin{smallmatrix}Cl\\ \diagdown\end{smallmatrix} \qquad (61)$$
$$\qquad\qquad\qquad\qquad\qquad\qquad Me\quad Et \qquad\qquad Et\quad Me$$

INITIAL RATIO – 55:45
AFTER TREATMENT
WITH Cp_2ZrHCl – 91:9

The kinetic ratio of products from reaction with 2-hexyne was nearly 50:50, but upon treatment with catalytic amounts of Cp_2ZrHCl, equilibration resulted in a 10:1 ratio favoring the less sterically hindered product. The isomerization was explained in terms of a diaddition intermediate, **97** (179), which is reasonable because **98** has been isolated (178).

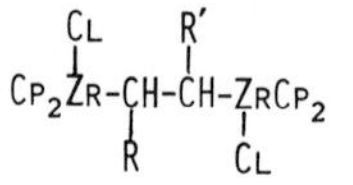

$$Cp_2Zr\text{-}CH\text{-}CH\text{-}ZrCp_2 \qquad\qquad Cp_2Zr\text{-}CH\text{-}CH\text{-}ZrCp_2$$
with Cl, R' on top and R, Cl on bottom for **97**; Cl, Ph on top and Ph, Cl on bottom for **98**.

97 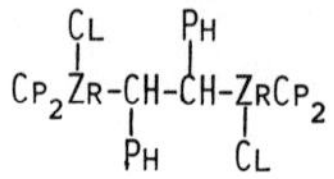   **98**

Platinum hydrides also undergo syn additions with acetylenes. For example, *trans*-PtHX(PEt$_3$)$_2$ reacted with several disubstituted acetylenes, typically 1-phenylpropyne, to afford the *trans*-alkenyl product. Complexes that were either cis or trans at the metal center could be isolated depending on the reaction conditions. The relative stabilities of these isomers led to postulation of the reaction sequence shown in Scheme 20 (180).

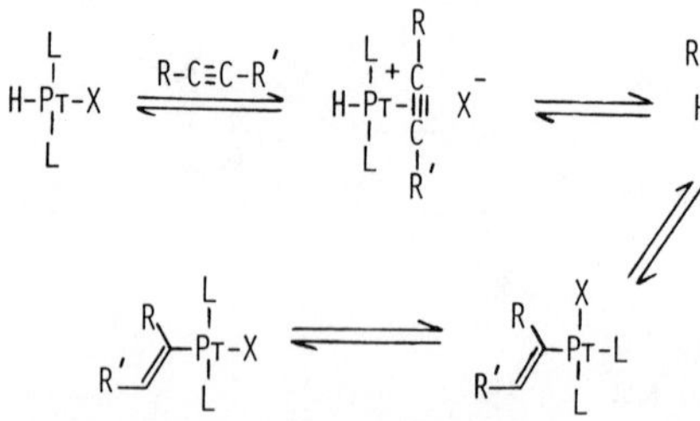

SCHEME 20

Pyrolysis of $(MeO_2CC{\equiv}CCO_2Me)Pt(PPh_3)_2$ at 130°C in toluene led to the formation of **99**, which was at first thought to be cis (181), but was later demonstrated by X-ray diffraction to be the trans isomer shown (182). Presumably ortho metalation followed by a syn addition to the alkyne was involved. Treatment with strong acids of the same acetylene complex, $Pt(alkyne)(PPh_3)_2$, has

[Structure **99**: orthometalated Pt complex with PH_2P, PH_3P, CO_2Me groups]

[Structure **100**: $PH_3P-Pt-PPh_3$ vinyl complex with X, R, R' substituents; X = Cl, OCOCF$_3$; R = R' = Ph, tolyl; R' = H; R = Me, Et, Ph]

also led to formation of vinyl complexes, most likely in this case via oxidative addition of HX to the metal followed by syn addition to the alkyne, and has afforded complexes **100** (67).

Dimethyl acetylenedicarboxylate was reported to react with $RhH(CO)(PPh_3)_3$ to form *trans*-$Rh(CO)(PPh_3)_2$ [Z-$C(CO_2Me){=}CH(CO_2Me)$], presumably via an anti addition (183). It was subsequently found however, that the addition product was actually E (184) (see Sect. III-C-1).

The alkyne $HC{\equiv}C-R$, where R is 1,2-carborane, $C_2B_{10}H_{11}$, underwent initial oxidative addition to *trans*-$IrCl(CO)(PPh_3)_2$ forming a hydridoacetylide complex which then underwent a syn addition to a second molecule of the same acetylene (185). The structure was determined by X-ray crystallography.

Syn addition also affords $Cp_2Re(E-CH{=}CHCO_2Me)$ from reaction of $HC{\equiv}CCO_2Me$ with Cp_2ReH (186). Since the rhenium starting material is coordinatively saturated and the product should be the thermodynamic one, it is difficult to speculate regarding mechanisms.

An interesting dichotomy arises with additions of $HMn(CO)_5$ and *cis*-$HMn(CO)_4(PPh_3)$ to activated alkynes such as $HC{\equiv}CCO_2Me$, $HC{\equiv}CCHO$, and $MeO_2CC{\equiv}CCO_2Me$ (187). With $HMn(CO)_5$, the additions all yield Z products (i.e., the additions are anti); but with the phosphine-substituted metal hydride, syn additions are obtained. These observations are consistent with the first reaction being a conjugate addition to the α,β-unsaturated organic group by $[Mn(CO)_5]^-$. This should yield an anion trans to the metal, which is rapidly protonated by another $HMn(CO)_5$. Unfortunately, no attempt to observe base catalysis was reported. Conversely, $HMn(CO)_4(PPh_3)$ would be several orders of magnitude less acidic, and thus would not be prone to additions catalyzed by adventitious base. In the latter case, prior coordination of the alkyne followed by a syn insertion is more likely. Other very similar insertions have been observed with manganese hydrides (150).

As with $HMn(CO)_5$, $[Co(dmgH)_2py]^-$, **7**, adds to a variety of acetylenes and, in each case, by an anti addition (50,51,56).

III. REACTIONS THAT DISRUPT METAL–CARBON SIGMA BONDS

Transition metal–carbon σ bonds suffer a variety of fates including thermal decomposition, insertion, reductive elimination, and electrophilic cleavage. Thermal decomposition occasionally proceeds by homolytic cleavage, but the clear outcome of this, except for very efficient solvent cage reactions, is invariably stereorandomization, so this phenomenon is not explicitly discussed. By far the predominant path for decomposition of metal alkyls bearing β-hydrogens is the β-hydride elimination. Since this is also a bond-forming reaction and is often observed to proceed in both directions, it has already been discussed (Sect. II-C-3).

All the other reactions of M–C bonds are of course in principle, as most of them are in fact, the reverse of bond-forming reactions; thus in effect many of them have already been considered in Sect. II. Nevertheless, since the organizational philosophy of this chapter is to discuss reactions in the context of the direction in which they are observed to proceed, there are a number of interesting transformations that have not yet been covered.

A. Insertion Reactions

Insertion reactions, as illustrated in Scheme 21, comprise a class that formally involves the breaking of the M–C bond with concomitant formation of two new

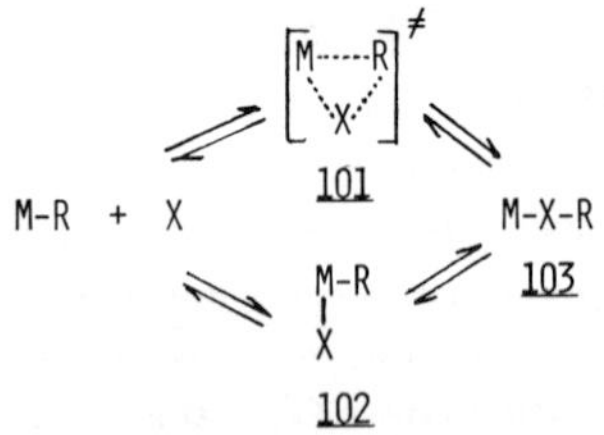

SCHEME 21

bonds to some intervening species X. As indicated, one can envisage processes wherein species X somehow affixes itself to the metal center as in **102**, and then by a concerted or nonconcerted bonding rearrangement, forms **103**. Many insertion reactions do appear to proceed via species like **102**, and their rearrangements to **103** usually seem to be concerted. These migratory insertions, as they are sometimes called, have interesting questions associated with the stereochemical features at the metal (Does R move to X? Does X move to R? Can one tell the difference?) and at the carbon terminus (Is there inversion or retention of configuration?).

There does not appear to be a single, well-defined example of an overall concerted insertion, presumably proceeding via a transition state such as **101**, in which X has not interacted previously with the metal center. Any insertions that appear not to involve prior coordination of X with the metal seem to involve ionic or radical intermediates.

1. Carbon Monoxide Insertion

Carbon monoxide insertion has been the subject of several good reviews (188), and only stereochemistry is stressed here. The reaction is frequently observed in the forward and reverse (decarbonylation) directions, and both are discussed. Judging from a wide variety of data, it is evident that the general reaction sequence is as in eq. [62], although with good nucleophiles and low-polarity solvents an apparently second-order reaction can dominate (188). In every case that has been tested, the inserting CO is first coordinated to the metal.

$$\text{M-R} \rightleftharpoons \underset{\underset{\text{CO}}{|}}{\text{M}}\text{-R} \xrightarrow{\text{L}} \underset{\underset{\text{L}}{|}}{\text{M}}\text{-}\overset{\overset{\text{O}}{\|}}{\text{C}}\text{-R} \quad (62)$$

The earliest unequivocal demonstration of carbon stereochemistry in the CO insertion reaction was accomplished by the PPh_3-induced conversion of *threo*-$CpFe(CO)_2(CHDCHD$-*t*-Bu$)$, **2**, to *threo*-$CpFe(CO)(PPh_3)(COCHDCHD$-*t*-Bu$)$, indicating that a high degree of retention of configuration at carbon was involved (6). This result was duplicated by the preparation of *threo*-$CpFe(CO)(PPh_3)$-$(COCHDCHDPh)$ from the corresponding *threo*-alkyliron starting material (13). Secondary alkyliron complex **4** and its trans isomer upon oxidation in methanol yielded **104** and its trans isomer, respectively, again demonstrating retention at carbon (10).

$$\underset{\underset{\text{Me} \quad \textbf{4}}{}}{\text{[cyclohexyl-FeCp(CO)}_2\text{]}} \xrightarrow[\text{MeOH}]{CuCl_2} \underset{\underset{\text{Me} \quad \underline{104}}{}}{\text{[cyclohexyl-CO}_2\text{Me]}} \quad (63)$$

Palladium complex **69**, upon treatment with CO in methanol, afforded ester **106**, which could be assigned the endo configuration based on NMR analysis (189).

$$\underset{\underline{69}}{\text{[MeO-norbornyl-PdCl]}_2} \xrightarrow[\text{MeOH}]{CO} \left[\underset{\underline{105}}{\text{[MeO-norbornyl-Pd-CO]}} \right] \longrightarrow \underset{\underline{106}}{\text{[MeO-norbornyl-CO}_2\text{Me]}} \quad (64)$$

By analogy to a reaction discussed above (Sect. II-C-2, eq. [43]), an internally inserted intermediate such as **105** is a reasonable assumption and implies that retention of configuration at carbon accompanies the CO insertion. The same arguments applied in using the corresponding diene palladium methanol addition products to produce esters **107** and **108** from CO insertion with retention at carbon (190). The preceding three examples all utilized chelated metal alkyl

107

108

complexes and so demonstrate that insertion with retention is available, but not necessarily that it is preferred in the general case. Various CO-inserted chelate species have been isolated from interactions of strained rings with CO-containing metal complexes as described above (Sect. II-A-4), further substantiating this observation. Nevertheless, there are several additional examples of CO insertions into nonchelated complexes, indicating that retention of configuration at carbon is quite general. For example, *threo*-Cp_2ZrCl(CHDCHD-*t*-Bu), **90**, underwent CO insertion to form *threo*-Cp_2Zr-Cl(COCHDCHD-*t*-Bu) (167), and *threo*-$IrCl_2$-$(PPh_3)_2$(COCHDCHDPh) was thermally decarbonylated to afford *threo*-Ir(CO)-$Cl_2(PPh_3)_2$(CHDCHDPh) (170).

Thus there are an unusual number of well-defined stereochemical examples that give weight to the generalization that CO insertion always involves retention at carbon. Numerous examples have been discussed elsewhere in this chapter where CO insertion or decarbonylation was one step in a sequence, several steps of which might have stereochemical implications. Taken individually, these studies would not be convincing; but taken together and considering the far-reaching inconsistencies that would arise if the contrary were assumed, all these data lend strong additional corroboration to the stereochemical generalization of CO insertion.

One curious set of data pertains to the stoichiometric decarbonylation of aldehydes using $RhCl(PPh_3)_3$ (191). For example, (−)-(*R*)-2-phenyl-2-methyl-butanal was decarbonylated during prolonged heating at 160°C, yielding (+)-(*S*)-2-phenylbutane, but with only 81% net retention of configuration. Similarly, as shown in eq. [65], almost complete racemization was found in some cases for

X	OPTICAL % YIELD
Me	94
Cl	83
F	73
OMe	6

(65)

109 (OPTICALLY ACTIVE) → **110**

$RhCl(PPh_3)_3$ / XYLENE / Δ

the decarbonylation of the series of aldehydes **109**. This loss of stereochemistry was attributed to the intrusion of a radical pair mechanism for the decarbonylation step (191). A mechanistic rationale similar to that in Scheme 22 was suggested (the structural hypotheses are this author's) wherein an intervening radical pair might resemble **112**. However it would seem that other possibilities cannot be ruled out. For example, the decarbonylation intermediate might have

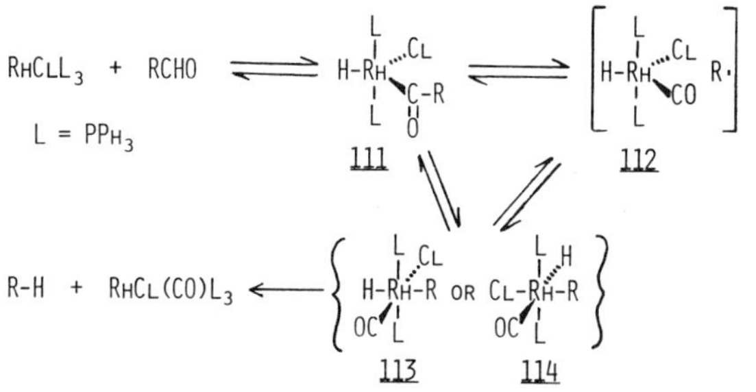

SCHEME 22

structure **113**, for which reductive elimination of alkane would be difficult, and under the vigorous reaction conditions reported, homolyses or intermolecular reactions might become competitive with the rate of isomerization of **113** to **114**. This kind of question exemplifies one very practical reason for being concerned about the stereochemistry at the metal in these reactions. One additional question regarding the decarbonylation of **109** is that X—C—M species are known to quite labile in a number of organometallic systems, and it is difficult to predict whether this structural feature of the reactive intermediates for reactions of **109** might be important here as well.

The CO insertion of alkenylmetal complexes and the reverse reaction are both highly stereospecific, as has been found in the $RhCl(PPh_3)_3$-induced decarbonylation of E-PhCH=C(Et)CHO (191), the $RhCl(CO)[C_6H_4\text{-}p\text{-}OCH_3)_3]_2$-catalyzed decarbonylation of E-PhCH=C(R)CHO (R = Me, Et) (192), and the CO insertion into $trans$-NiBr(cis- or $trans$-CH=CHPh) $(PPh_3)_2$ (59).

A complete understanding of the mechanism of the CO insertion also requires detailed knowledge of the stereochemistry at metal centers. The best known example of this has dealt with octahedral alkylmanganese species. $RMn(CO)_5$ kinetically forms cis-(acyl)$Mn(CO)_3L$ with incoming ligands, L = ^{13}CO, phosphine, or phosphite; but with larger ligands the product rapidly establishes a cis-trans equilibrium (188). When the kinetic product is stable, in principle one can ascertain whether the alkyl group has moved to CO or whether the CO has moved to the alkyl group. In an elegant experiment using **115** where specifically one cis CO site was ^{13}C-labeled, ^{12}CO was used to induce the CO insertion in

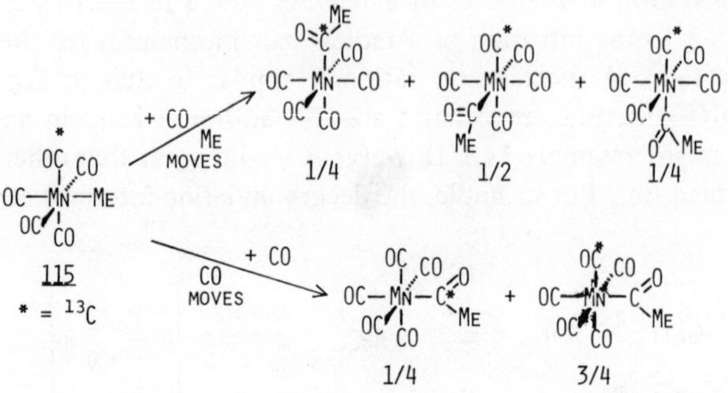

SCHEME 23

ether solvent. As shown in Scheme 23, methyl migration would result in two terminally CO-labeled products in a 2:1 ratio, and CO migration would result in one product. Infrared spectra revealed two ^{13}CO stretching bonds in the appropriate positions, consistent with methyl migration. Several permutations of this experiment were run with internally consistent results (193).

Since publication of the foregoing experiments, many pentacoordinate species have been found to be highly nonrigid ("fluxional"). Table I contains lists of possible geometries for a pentacoordinate intermediate and the resulting label distribution in the products. Ratios of cis to trans-labeled products range from only one product to ratios of 2:1, 3:1, and 5:1. Infrared intensity ratios can be accurately measured, but the inherent intensities of cis and trans CO stretching bonds may be significantly different. Hence the experiment above is somewhat ambiguous.

Table I
Positions of ^{13}CO Label in Products of Eq. [66]

Intermediate	L = CO[a]		L = P(OCH$_2$)CCH$_3$[b]		
Alkyl migration	$(b+c):d$		b	c	d
Square pyramid	2	1	2	0	1
Trigonal bipyramid, axial acyl	2	1	2	1	1.5
Trigonal bipyramid, radial acyl	5	1	2	0,5	0.5
CO migration					
Square pyramid	All	0	2	1	0
Trigonal bipyramid, axial acyl	All	0	2	1	0
Trigonal bipyramid, radial acyl	5	1	2	0.5	0.5
Stereorandom intermediate	3	1	2	1	1

[a] Normalized to $d = 1$.
[b] Normalized to $b = 2$.

$$\underset{\underset{\bullet\,=\,^{13}C}{\underline{115}}}{\overset{\overset{\bullet}{C}O}{OC-\underset{OC}{\overset{CO}{Mn}}-Me}} \xrightarrow{L} \underset{\underline{a}}{\overset{\overset{L}{|}\overset{CO}{}}{OC-\underset{OC}{\overset{|}{Mn}}-\overset{\bullet}{\underset{\parallel}{C}}-Me}} + \underset{\underline{b}}{\overset{\overset{L}{|}\overset{CO}{}}{OC-\underset{\overset{\bullet}{O}C}{\overset{|}{Mn}}-\overset{}{\underset{\parallel}{C}}-Me}} + \underset{\underline{c}}{\overset{\overset{L}{|}\overset{CO}{}}{OC-\underset{OC}{\overset{|}{Mn}}-\overset{}{\underset{\parallel}{C}}-Me}} + \underset{\underline{d}}{\overset{\overset{L}{|}\overset{CO}{}}{O\overset{\bullet}{C}-\underset{OC}{\overset{|}{Mn}}-\overset{}{\underset{\parallel}{C}}-Me}} \quad (66)$$

More recently it has been found (194) that reliable intensity measurements can be made by ^{13}C NMR at $-110°$C, and using this method the isomer ratio was confirmed to be 2.0 : 1 in the experiment of Scheme 23 when carried out in the THF-acetone. In HMPA, the product ratio was 3.0 : 1. Presumably coordination of the intermediate by HMPA increases its lifetime and yet allows fluxional behaviour, leading to randomly labeled product. As seen from Table I, even a ratio of 2 : 1 is ambiguous. Again, using ^{13}C NMR it could be reliably shown that when the insertion was induced by P(OCH$_2$)$_3$CCH$_3$, which is known to produce only cis product (195), the product ratio (as lettered in eq. [66]) of b : c : d was 2 : 0 : 1 in either THF-acetone or HMPA. This gives the unambiguous answers to some long-standing questions: (a) the alkyl group does migrate, and (b) the intermediate is a rigid square pyramid unless (c) a good coordinating solvent complexes the intermediates. Solvent intervention can be bypassed with a ligand that is sufficiently nucleophile (194).

A phosphine-induced insertion of CpMo(CO)$_3$(CH$_2$CH$_2$CH$_2$Br) has been shown to afford a cis kinetic product (196), but no information is available on which ligand moves to the other.

Some information on the stereochemistry of decarbonylation at rhodium is available. Acyl chlorides oxidatively add to RhCl(PPh$_3$)$_3$, yielding RhCl$_2$(COR)(PPh$_3$)$_2$, probably **116**, but of uncertain structure. This rapidly rearranges to **117**, which decarbonylates to **118** (197).

$$\text{RhCl(PPh}_3)_3 + \text{RCOCl} \longrightarrow \left[\begin{array}{c}\overset{O}{\underset{}{}}\diagdown_{C}\diagup^{R}\\\text{Cl}-\overset{|}{\underset{\text{Cl}}{\text{Rh}}}-\text{PPh}_3\\\text{PPh}_3\end{array}\right] \longrightarrow \underset{\underline{117}}{\text{Cl}-\underset{\text{PH}_3\text{P}}{\overset{\overset{O}{\diagdown_{C}\diagup^{R}}\,\text{PPh}_3}{\text{Rh}}}-\text{Cl}} \longrightarrow \underset{\underline{118}}{\text{Cl}-\underset{\text{PH}_3\text{P}}{\overset{\overset{CO}{}\text{PPh}_3}{\text{Rh}}}-\text{R}} \quad (67)$$

$$\underline{116}$$

The structure of **117** (R = CH$_2$CH$_2$Ph) has been determined by X-ray crystallography to be square pyramidal (198), although another report of preliminary X-ray results determined to trigonal bipyramidal structure (axial phosphines) for this same species **117** (R = CH$_2$CH$_2$Ph) (171). It is not impossible, of course, that two isomeric forms do exist (for examples, see ref. 199). One radioisotopic Cl$^-$ labeling experiment demonstrated that the chlorides in PhCH$_2$COCl and RhCl(PPh$_3$)$_3$ were totally randomized (171); unfortunately, therefore, the study gave essentially no structural information. None of the work just cited allowed comment on whether R moved to CO, CO moved to R, or some intermediate behavior occurred.

An interesting study dealt with the stereochemistry of the CO insertion into the Ir–Et bond in **119** (200). Complex **120** slowly rearranged to a more stable isomer, and so probably was the kinetic product. The stereochemistry in this

$$\underset{\mathbf{119}}{\text{Et}-\underset{\underset{\text{Cl}}{|}}{\overset{\overset{\text{COCl}}{|}}{\text{Ir}}}-\text{AsMe}_2\text{Ph}} \xrightarrow{\text{AsMePh}_2} \underset{\mathbf{120}}{\text{Et}-\overset{\overset{\text{O}}{\|}}{\text{C}}-\underset{\underset{\text{Cl}}{|}}{\overset{\overset{\text{COCl}}{|}}{\text{Ir}}}-\text{AsMePh}_2} \qquad (68)$$

case cannot be explained in terms of CO or R moving, one to the other, because some other ligand migration must be occurring either during or instantly after the insertion. For example, one speculative path might be as in eq. [69], wherein a strong trans effect of the acyl group might cause rapid rearrangement of **121** to **122**. Details of this kind cannot presently be elucidated.

$$\mathbf{119} \longrightarrow \underset{\mathbf{121}}{\text{Et}-\overset{\overset{\text{O}}{\|}}{\text{C}}-\underset{\underset{\text{Cl}}{|}}{\overset{\overset{\text{CO}}{|}}{\text{Ir}}}-\text{Cl}} \longrightarrow \underset{\mathbf{122}}{\text{Et}-\overset{\overset{\text{O}}{\|}}{\text{C}}-\underset{\underset{\text{Cl}}{|}}{\overset{\overset{\text{COCl}}{|}}{\text{Ir}}}---} \xrightarrow{\text{AsMePh}_2} \mathbf{120} \qquad (69)$$

Substantiation of the idea that trans effects may be important in insertion reactions comes from the observation that whereas **123** exhibits a very facile insertion equilibrium with **124**, neither **125** nor **126** can be induced to insert, even though **126** has a cis arrangement of the two potentially interacting ligands (201).

$$\underset{\mathbf{123}}{\overset{\text{Ph}_2\text{MeP}}{\underset{\text{OC}}{\diagdown}}\overset{\text{Cl}}{\underset{\text{Ph}}{\text{Pt}}\diagup}} \qquad \underset{\mathbf{124}}{\overset{\text{Ph}_2\text{MeP}}{\underset{\text{O=C}}{\diagdown}}\overset{\text{Cl}}{\underset{\text{Ph}}{\text{Pt}}\diagup}\overset{\overset{\text{Ph}}{\diagdown}\text{C=O}}{\underset{\text{PMePh}_2}{\diagup}}} \qquad \underset{\mathbf{125}}{\overset{\text{Ph}_2\text{MeP}}{\underset{\text{OC}}{\diagdown}}\overset{\text{Ph}}{\underset{\text{Cl}}{\text{Pt}}\diagup}} \qquad \underset{\mathbf{126}}{\overset{\text{Ph}_2\text{MeP}}{\underset{\text{Cl}}{\diagdown}}\overset{\text{Ph}}{\underset{\text{CO}}{\text{Pt}}\diagup}}$$

Pseudotetrahedral iron systems of the type CpFe(CO)LR are well suited to addressing the question of the stereochemistry of CO-induced CO insertion (eq. [70]), since the outcome should be conceptually easy to interpret – inversion means R has moved, retention means CO has moved – and there are no competing trans effects. The first studies of this type were reported using pure but racemic diastereomers such as **127** (202) and **128** (203) in photochemical de-

$$\underset{\text{OC}}{\overset{\text{Cp}}{\underset{}{\text{Fe}}}}\overset{\text{R}}{\underset{\text{O}}{\diagdown}} \underset{\underset{\text{MOVED}}{\text{CO}}}{\overset{+\text{CO}}{\rightleftharpoons}} \underset{\text{OC}}{\overset{\text{Cp}}{\underset{}{\text{Fe}}}}\text{R} \underset{\underset{\text{MOVED}}{\text{R}}}{\overset{+\text{CO}}{\rightleftharpoons}} \underset{\text{R}}{\overset{\text{Cp}}{\underset{\text{C}}{\text{Fe}}}}\overset{\text{CO}}{\underset{\text{O}}{\diagdown}} \qquad (70)$$

carbonylations. Photolysis of these species in benzene afforded the corresponding decarbonylated alkyliron complexes. Since the products are photolabile, the lower the conversions were, the higher the diasteromeric excess of the alkyl products, with limiting stereospecificities above 85%. The reaction was shown to proceed with inversion at iron by the elegant set of experiments outlined in Scheme 24 (204). Beginning with an enantiomerically pure diastereomer of **129**

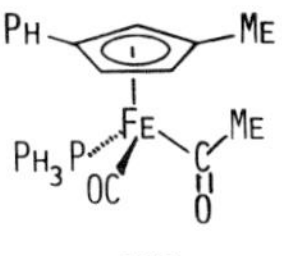

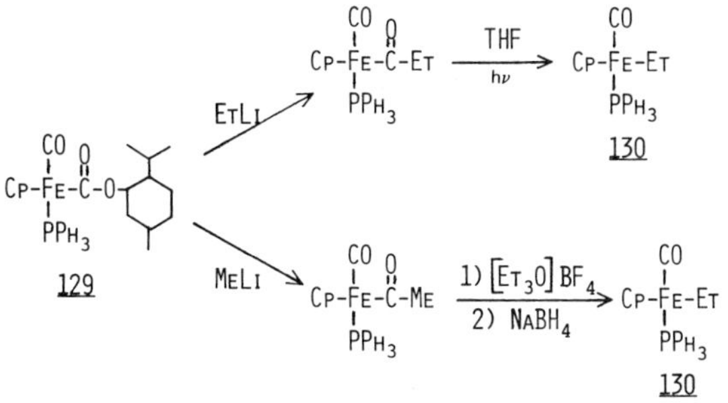

SCHEME 24

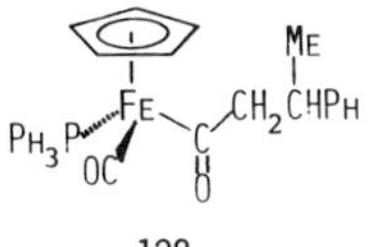

(205), the sequences led to opposite enantiomers of **130**, and since the only uncertain reaction is the photocarbonylation, it must proceed with inversion, that is, by alkyl migration. The stereochemistry of the thermal process is also of interest, and recently it has been found (206) that CO insertion will proceed at reasonable rates under mild conditions with enantiomerically pure complexes **131** (207) and **132** (206) affording the corresponding acyl complexes **133** and

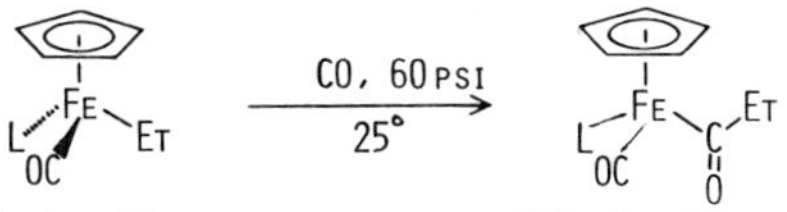

131, L = PPH_3
132 L = $P(OCH_2)_3CME$

133, L = PPH_3
134, L = $P(OCH_2)_3CME$

(71)

134. The reaction is highly stereospecific in nitroethane, the acyls being of >90% e.e. and forming with inversion of configuration. However, a study of solvent dependence using **131** revealed a continum of stereochemical behavior, the opposite extreme being in HMPA where retention was found with ca. 75% e.e. (206). When the insertion was induced with cyclohexylisonitrile, the resulting

$$\underset{\textbf{131}}{\text{Ph}_3\text{P}\cdots\overset{\text{Cp}}{\underset{\text{OC}}{\text{Fe}}}-\text{Et}} \xrightarrow[25°]{\text{C}\equiv\text{N}-\bigcirc} \underset{\textbf{135}}{\text{Ph}_3\text{P}\cdots\overset{\text{Cp}}{\underset{\text{O}}{\text{Fe}}}-\text{C}\underset{\text{Et}}{\overset{}{\equiv}}\text{N}}\bigcirc \qquad (72)$$

product **135** was probably of net inverted configuration (from circular dichroism data) either in EtNO$_2$ (ca. 40% e.e.) or in HMPA (ca. 20% e.e.). These results clearly establish for the thermal reaction that (1) in the absence of solvent intervention, a process occurs that is equivalent to alkyl migration to CO, (2) solvent intervention can profoundly alter the stereochemistry, but (3) solvent intervention must occur after the transition state, otherwise isonitrile would not obviate its effect (kinetics show the formation of an intermediate in the isonitrile reaction).

The discussion above raises the question of the structure of the intermediates in CO insertion reactions. Recently several otherwise coordinatively unsaturated acyls of titanium, zirconium, and tungsten have been shown to satisfy the metal by η^2-complexation as in **136** (152,208). Iminoacyl complexes also often exhibit this coordination as shown in **137** (209).

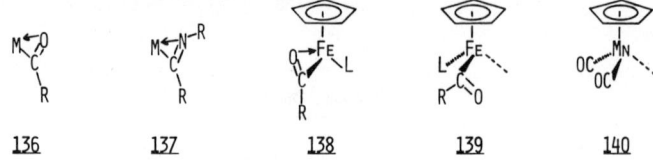

Conversely, there are no substantiated η^2-acyls other than these, and there are a number of coordinatively unsaturated rhodium, iridium, and platinum acyls known. In any event, the stereochemical results for the CO insertion into **131** discussed above are best rationalized in terms of intermediate **138** (206), although structure **139** is not unreasonable for a low-spin, d^6, pseudopentacoordinate complex, as recent calculations attest for CpMn(CO)$_2$, **140** (210).

2. Alkene, Alkyne, Sulfur Dioxide, and Other Insertions

The stereochemistry of insertions of alkenes and alkynes was discussed with regard to the unsaturated carbon group in Sect. II-C-2. Data explicitly dealing with the carbon center that is moving from the metal are scarce. One very facile insertion that was discussed previously (eq. [43]) is an intramolecular insertion of a palladium alkyl bond in a chelate complex wherein only retention at the migrating center was possible. Similarly, an alkenylpalladium group has been observed to add across an alkene with retention at the migrating sp^2 center (eq. [44], ref. 141). In a sense, the numerous examples of conjugate additions of vinylic cuprates to α,β-unsaturated carbonyl compounds represent insertions of unsaturated organic groups into the vinyl-copper bond with retention at the migrating center (e.g., Scheme 17, ref. 159).

Several examples are available for the reverse reaction where ring strain is the driving force for elimination of the alkene. These involve metal-induced ring opening of Feist's esters, *cis*- and *trans*-**141**. Reaction of *cis*-**141** with platinum

$$\text{cis-141} + \text{142} \xrightarrow{\text{MeOH}} \text{143} \tag{73}$$

hydride **142** resulted in formation of **143**, whose configuration corresponded to C—C bond cleavage with retention of configuration. The corresponding result was also obtained with *trans*-**141** (211). In the same way, $Pd(NCMe)_2Cl_2$ afforded a ring-opened product in reaction with **141**, again with retention (212). Both these reactions are likely to proceed via an intermediate similar to **144**.

FROM **142**, X = H, M = PT
FROM $PD(NCME)_2CL_2$, X = CL, M = PD

144

Systems of considerable practical importance are the Ziegler-Natta polymerization catalysts, which convert vinyl monomers to highly stereoregular polymers, either isotacticly or syndiotacticly (213). If *trans*-propylene-d_1 is employed with an isotactic catalyst, then *threo*-diisotactic polypropylene, **145**, is formed, and similarly *cis*-propylene-d_1 affords *erythro*-diisotactic polypropylene **146** (214). If one accepts the Cossee or any closely related mechanism for Ziegler polymerization (213), which in essence is simply the insertion reaction of eq. [42], where L is the monomer and R is the growing polymer chain, these tacticity data say that chain growth occurs either by addition of M and R to the same side of the monomer with retention at the migrating R carbon center, or by addition of

M and R to opposite sides of the monomer with inversion of configuration at R. Although the former seems much more likely on the basis of known solution chemistry (Sect. II-C-2), the latter cannot be ruled out in an electrophilic multimetallic system.

Insertion of $MeO_2C-C{\equiv}C-CO_2Me$ into *threo-* or *erythro*-CpFe(CO)$_2$CHD-CHD-*t*-Bu yielded CpFe(CO)$_2$[C(CO$_2$Me)=C(CO$_2$Me)CHDCHD-*t*-Bu] for which NMR spectra of only poor resolution could be obtained (6a). Analysis of the vicinal coupling pattern showed that the insertion had occurred with net retention at the migrating carbon, but the exact extent of stereospecificity could not be determined. In view of the electrophilic nature of the acetylene, the inability to determine whether the product alkene was cis or trans, the inexact diastereomer analysis, and the coordinative saturation of the iron center, the mechanistic implications of this reaction remain obscure.

Metal–carbon bonds undergo insertions by isonitriles (209). Several cases are known involving chelated organic complexes of palladium and platinum wherein the insertion must proceed with retention of configuration at the migrating carbon and does so (215). Treatment of *trans*-MBr(PPh$_3$)$_2$(Z-CH=CHCO$_2$Me), where M is palladium or platinum, with *p*-MeC$_6$H$_4$N≡C, resulted in insertion with retention of configuration to form **147** (216). In contrast to this, oxidative addition of Z-BrCH=CHCO$_2$Me to Pt(CN-*t*-Bu)$_3$ gave **148** (and its cis isomer) wherein the olefinic geometry had been inverted (216). The inversion was rationalized as arising from a special Michael type of interaction of the metal center with the α,β-unsaturated ligand in **148** with C–C bond rotation in the resultant dipolar form, and was not thought to arise from the insertion itself.

The insertion of sulfur dioxide into transition metal–carbon σ bonds was thought at first to be a reaction very similar to CO insertion. Intensive study (217) has revealed that it is indeed very different. The stereochemistry of the insertion differs depending on the complex examined. For iron, formation of distinct diastereomers of CpFe(CO)$_2$S(O)$_2$CHDCHDR (R = *t*-Bu, ref. 6; R = Ph, ref. 13) from the corresponding alkyl diastereomers indicated that inversion of configuration at carbon had occurred. Similarly, *cis*-Mn(CO)$_4$(PEt$_3$)(*threo*-CHDCHDPh) and *trans*-CpW(CO)$_2$(PEt$_3$)(*threo*-CHDCHDPh) underwent SO$_2$ insertion with inversion at carbon, and, incidentally, retention at the metal, although the latter is probably just a thermodynamic effect (13). Isoelectronic analogs of SO_2 such as $OSN-S(O)_2Ph$ and $S(=N-C_6H_4\text{-}p\text{-Cl})_2$ also insert into $CpFe(CO)_2CHDCHDPh$ with inversion of configuration at carbon, affording **149** and **150** respectively (218).

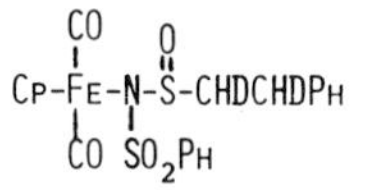

$$\begin{array}{cc}
\text{CO} & \text{O} \\
| & || \\
\text{Cp-Fe-N-S-CHDCHDPh} \\
| & | \\
\text{CO} & \text{SO}_2\text{Ph}
\end{array}
\qquad
\begin{array}{cc}
\text{CO} & \text{N-SO}_2\text{C}_6\text{H}_4\text{-}p\text{-Cl} \\
| & || \\
\text{Cp-Fe-N-S-CHDCHDPh} \\
| & | \\
\text{CO} & \text{SO}_2\text{C}_6\text{H}_4\text{-}p\text{-Cl}
\end{array}$$

<u>**149**</u> 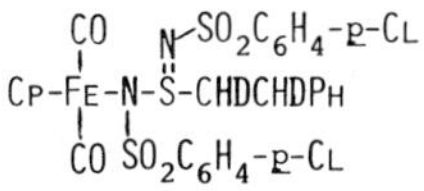   <u>**150**</u>

In contrast, the formation of a structurally somewhat ill-defined Cp_2ZrClS-$(O)_2CHDCHD$-t-Bu from its corresponding alkylzirconium precursor occurred with retention of configuration at carbon (167).

A related result is outlined in Scheme 25, wherein the configuration of **153** could be established via CO insertion to form **154**, which could be correlated with a known acyl chloride. The implication of this sequence is that the extrusion of SO_2 in going from presumed intermediate **152** to **153** occurs with inversion (43). Of course it is not clear that **152** is involved. It might be that this is a nucleophilic type of oxidative addition with attack by Pd(0) directly at the carbon of **151**, as was discussed in Sect. II-A-2.

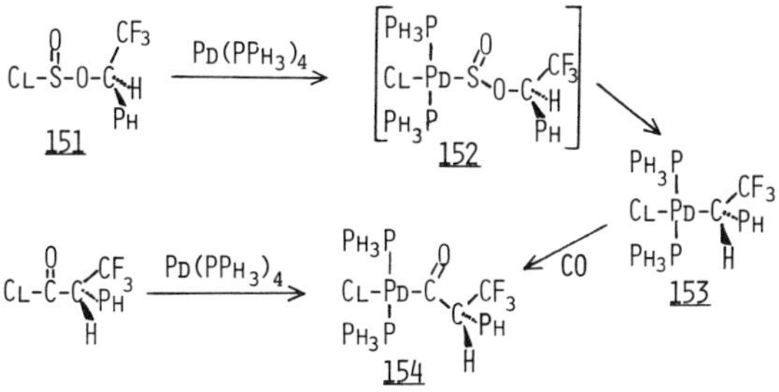

SCHEME 25

The metal stereochemistry of the SO_2 insertion has been extensively studied for iron and has also been examined for titanium. Molecular systems related to those used for studies of CO insertions (Sect. III-A-1) were again employed, including those iron alkyls that would yield the corresponding sulfinates, **155** as a racemic diastereomer (219), **156** as a racemic diastereomer where $R = CH_2CH\text{-}(Me)Ph$ (203), **156** as resolved enantiomers and diastereomers, where $R = Me$, Et, Bz, CH_2CO_2R, and so on, (207), and **157** as a resolved enantiomer (206). Also, the titanium sulfinate **158** has been prepared stereospecifically by SO_2 insertion into the corresponding titanium alkyl (220). All these SO_2 insertions are highly stereospecific at the metal, and in the case of **156** where $R = i$-Bu, X-ray crystallography has established the absolute configuration (221). This, in conjunction

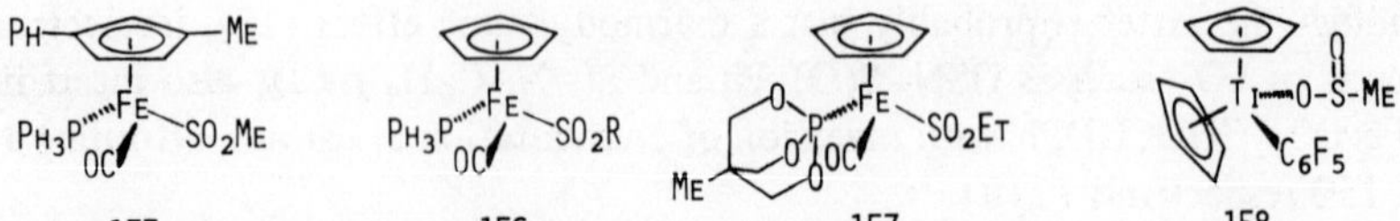

155 **156** **157** **158**

with an X-ray determination of the configuration of the alkyl precursor to **156** (222), establishes that the SO_2 insertion proceeds with retention of configuration at the metal, and it probably does so for all the complexes **155** to **158**.

The SO_2 insertion exhibits first-order kinetics with respect to $CpFe(CO)_2R$, is severely retarded by steric bulk in R, and is very responsive to increases in electron density at R (217); it now is known that this insertion proceeds with retention of configuration at the metal and inversion at carbon (except in the case of Cp_2ZrClR). These data have generally been interpreted in terms of an S_E2 (inversion) mechanism, which must therefore involve an ion pair as in Scheme 26.

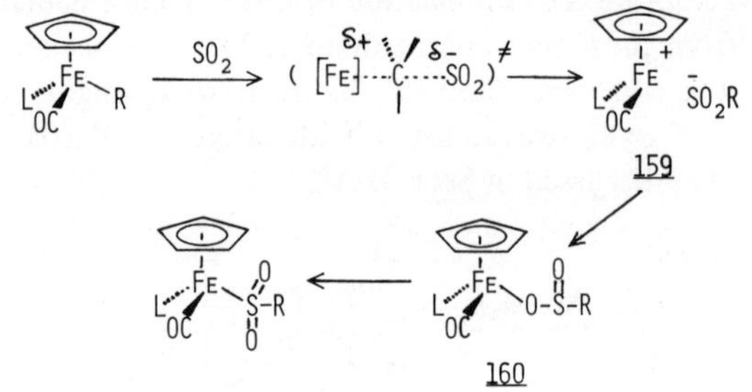

159

160

SCHEME 26

O-Sulfinates (e.g., **160**) appear generally to be involved, and in the case of titanium complex **158**, and also probably $Cp_2ZrCl(SO_2R)$, is the stable product. The implication of this Scheme is that ion pairs such as **159** may be of significant configurational stability; thus the interpretation of stereochemical data of pseudotetrahedral organometallic species must be interpreted with care, lest ionic or dissociative reactions be mistaken for concerted processes.

B. Reductive Elimination

1. Forming Carbon-Carbon Bonds

Reductive elimination is a principal path for carbon-carbon bond forming reactions in organo–transition metal chemistry. It occurs with metal alkyls, alkenyls, alkynyls, aryls, and acyls in virtually all the possible binary combinations.

Several of these reactions were considered in the reverse direction in Sect. II-A-4, where it was observed that oxidative additions of metal complexes to saturated, strained C–C bonds proceeded with retention of configuration at both centers. Otherwise, about the closest thing to a well-defined stereochemical example of a reductive elimination at a saturated carbon center that is available is the oxidative coupling of certain organocuprates. For example, eq. [74] illus-

$$\underset{\underset{\text{MgBr}}{}}{\triangle}\text{-H} \xrightarrow[\text{Et}_2\text{O}]{\text{CuBr(PBu}_3\text{)}} \underset{\underset{\text{CuPBu}_3}{\underline{161}}}{\triangle}\text{-H} \xrightarrow[\text{2) PhNO}_2]{\text{1) MeLi}} \underset{\underset{\text{Me}}{\underline{162}}}{\triangle}\text{-H} \qquad (74)$$

trates the retention of configuration obtained in oxidatively induced reductive elimination of the "-ate" complex derived from copper reagent **161**. Formation of **161** from the corresponding Grignard reagent almost certainly occurs with retention at carbon (Sect. II-B) (98).

A number of reasonably well defined examples of reductive elimination of alkenylmetal complexes are known. Either as solids or as their soluble phosphine complexes, *cis*- and *trans*-propenylcopper(I) or silver(I) yielded the corresponding 2,4-hexadienes stereospecifically with retention upon pyrolysis (54), or upon oxidation by a variety of reagents in the case of the copper species (223). Also, RhMeI(CO)(PPh$_3$)$_2$ [E-C(CO$_2$Me)=CH(CO$_2$Me)] afforded Z-MeO$_2$CC(Me)=CH-CO$_2$Me upon pyrolysis (184). The geometry of the alkenyl ligand was determined by protonolysis (Sect. III-C-2).

Just as was the case for generalizing that organolithium and -magnesium reagents alkylate transition metals with retention at carbon (Sect. II-B), the most compelling reason to generalize that reductive elimination to form carbon-carbon bonds occurs with retention of configuration is that it is so universally a satisfactory and consistent hypothesis. Numerous examples were quoted in Sects. II-A-3 and II-B, such as coupling of alkenyl halides with (π-allyl)nickel complexes (60), nickel coupling of alkenyl halides (62), coupling of alkenyl mercurials by various metals (100,103), and coupling of organometallic species with organic halides catalyzed by nickel (63,75), palladium (63,77–79,106,107,109,110), copper (17b,101,104), silver (80), and iron (81). Many of these reactions are not only related to oxidative addition and reductive elimination, but also are correlated with electrophilic cleavage, transmetalation, insertion reactions, and so on. The entire scheme is most consistent with retention of configuration. Reductive elimination of R–CHDPh from palladium (4) has recently been shown to proceed with net retention at the benzylic center (263).

Some interesting problems remain in this area: why are alkyl-alkyl couplings so much less common than other types of C–C bond forming reactions (even methyl-methyl coupling where β-elimination cannot compete is rare), and why is coupling of R$_2$CuLi with R'X so specific for formation of R–R'^{2}. The former question is outside of the scope of this review, but the answer to the latter may reside in metal stereochemistry.

It has been proposed that cuprate coupling may reasonably proceed by a nucleophilic oxidative addition of R'X of the type discussed for palladium and platinum complexes in Sects. II-A-2 and II-A-3, which would form a Cu(III) trialkyl complex with inversion at the aliphatic electrophilic carbon of R' (eq. [75]) (20). Copper(III) should be square planar, and if R' were trans to the

$$R_2CuLi + R'-X \longrightarrow \underset{\underset{163}{(L)}}{R-\overset{R'}{\underset{|}{Cu}}-R} \longrightarrow R-R' \qquad (75)$$

fourth ligand (phosphine, halide, solvent, etc.) as shown in **163**, and if reductive elimination occurred from cis positions on the metal as expected, and with retention at carbon, the entire reaction could be understood.

Analogous chemistry of gold is known and is consistent with eq. [75]. Lithium dimethylaurate(I) is active toward oxidative addition and has been shown to yield only trans product in reaction with ethyl iodide (eq. [76]) (224). Although cis-trans isomerization complicated the study, nevertheless under appropriate

$$[AuMe_2(PPh_3)]^- Li^+ \xrightarrow{EtI} Et-\underset{\underset{Me}{|}}{\overset{\overset{Me}{|}}{Au}}-PPh_3 \longrightarrow PROPANE \qquad (76)$$

conditions, *trans*-AuMe$_2$Et(PPh$_3$) reductively eliminated only propane, whereas the cis isomer afforded both ethane and propane, consistent with a cis reductive elimination from Au(III) (225). A similar study of deuterium-labeled PtMe$_3$I-(PMe$_2$Ph)$_2$ supports a concerted cis reductive elimination pathway (226).

2. Forming Carbon-Hydrogen Bonds

Section II-C-3 presented the argument that (*a*) homogeneous catalytic hydrogenation that is first order in a mononuclear metal catalyst effects cis transfer of H$_2$ to an alkene, (*b*) β-hydride addition-elimination reactions that are unimolecular are syn, and (*c*) an R–M–H intermediate is involved in alkene reductions. The inescapable conclusion is that reductive elimination of R–M–H to form alkane occurs with retention of configuration at carbon.

Given the data above, the recent observation (164) of an alkyl intermediate [Rh(diphos)H(alkyl)]$^+$ in a catalytically active system must be taken as a specific example of a reductive elimination with retention of configuration. The only other stereochemical example of which I am aware is the report of *threo*- and *erythro*-Cp$_2$MoD[CH(CO$_2$Me)CHD(CO$_2$Me)] undergoing first-order reductive elimination to form *dl*- and *meso*-dimethyl succinate-1,2-d_2, respectively (168). The configurations of the starting molybdenum alkyls were established directly by NMR spectroscopy in this case. Some label scrambling did occur in this alkane elimination, but this was attributed to a competing β-hydride addition-elimination process.

A closely related question is of the stereochemistry of the reduction of metal-carbon bonds by complex metal hydrides (NaBH$_4$ and LiAlH$_4$) and by other transition metal hydrides. Probably the earliest example was the NaBD$_4$ reduction of **51** to **52** (eq. [36]) (118), which was mentioned in Sect. II-C-1. The stereochemistry of the reduction of **164** to **165**, and **76** to **166** by NaBD$_4$ was

164   **165**   **76**   **166**

shown by a variety of spectroscopic and reaction techniques to be retention at carbon. Reduction of the remaining double bond in **165** was avoided by carrying out the reaction in the presence of norbornene. In this way it was shown that mixtures of both **165** and **166** were not obtained in either reaction, as would be expected for a radical process (227). Corroboration of the result from NaBD$_4$ reduction of **51** has been obtained in a very analogous system (139). Also, an example of a nonchelated but nonetheless constrained system involves the NaBD$_4$ reduction of **75** affording **167**. Since the structural assignment of **75** is almost certainly correct, this represents reduction with retention of configuration (140).

75 → NaBD$_4$ → **167**   (77)

Lithium aluminum deuteride was used to reduce palladium-alkyl bonds in several related studies reported above. These involved three different reactions: (1) transfer of a β-aminoalkyl group from mercury to palladium (Sect. II-B, eq. [27], ref. 102a), (2) addition of amines to alkenes complexed to palladium (Sect. II-C-1, Scheme 15, ref. 128), and (3) the LiAlD$_4$ reductions in both the above. Since the stated intent of these publications was to demonstrate the stereochemistry of the first two reactions, and yet they assumed the stereochemistry of the third, the arguments in this work must be taken as somewhat circular. Nevertheless, if one makes the reasonable assumptions on the basis of other transmetalations (q.v.), and other amine additions to platinum alkene complexes (q.v.) that reactions (1) and (2) above are established, this work may be taken as evidence that LiAlH$_4$ reductions proceed with retention at carbon.

$$\underset{\text{M-R}}{\overset{X}{|}} + \text{LiAlH}_4 \longrightarrow \underset{\underset{\textbf{168}}{\text{M-R}}}{\overset{H}{|}} \longrightarrow \text{R-H} \qquad (78)$$

Although the mechanisms of these reductions are not certain, a reasonable hypothesis would be as represented in eq. [78], where in each case the metal bears a leaving group that may be converted to an M—H bond. Reductive elimination from **168** would then be expected to proceed with retention.

Nevertheless, transition metal hydrides have been demonstrated in some cases to reduce M—C bonds when an obvious leaving group is not present. Various alkenyl complexes of $Mn(CO)_5$ may undergo reduction upon treatment with $HMn(CO)_5$ with retention at carbon, but the results are somewhat ambiguous, since the better-defined examples yield thermodynamically favored products, and other examples, which are reactions of acetylenes with excess $HMn(CO)_5$, are subject to variation in the stereochemistry of the initial addition of MnH to the acetylene (228).

endo-2-Norbornyl(tri-*n*-butylphosphine)copper(I), **161**, formed as in eq. [74], underwent reduction with $DCuPBu_3$ yielding *endo*-norbornane-2-d_1 and so proceeded by some undefined concerted mechanism rather than by a radical path (98). Mechanisms of such reactions remain to be elucidated.

3. *Forming Carbon-Heteroatom Bonds*

Reductive elimination to form carbon-heteroatom bonds is the reverse of oxidative addition, which was discussed in detail in Sects. II-A-1 to II-A-3. As seen in Sect. II-A, oxidative additions may proceed by a variety of mechanisms; but the stereochemical results have always been either inversion or racemization in additions of aliphatic halides or tosylates, and retention or isomerization with alkenyl species. In view of this, the following study is intriguing (171). Stoichiometric decarbonylation of acid chloride **169** by $RhCl(PPh_3)_3$ led to formation of labeled benzyl chloride (*S*)-**20** with only ca. 25% e.e. and with net retention of configuration. A control experiment ruled out product racemization, although a check of optical stability of the starting acyl halide was not mentioned. A conventional mechanism would proceed as in Scheme 27 via oxidative addition of **169** to Rh(I) and decarbonylation of **170** to **171** with retention at carbon; therefore the reductive elimination must proceed with net retention of configuration at carbon. A radical pair, presumably resembling **172**, was suggested as a possible intermediate intervening in the **170**–**171** interconversion, which might account for the racemization component. In the event that **172** does contribute to the reaction path, there is also the question of whether it might just as well disproportionate by chlorine atom transfer, resulting in formation of **20** with retention. In any event, this is a result that has interesting implications whatever the

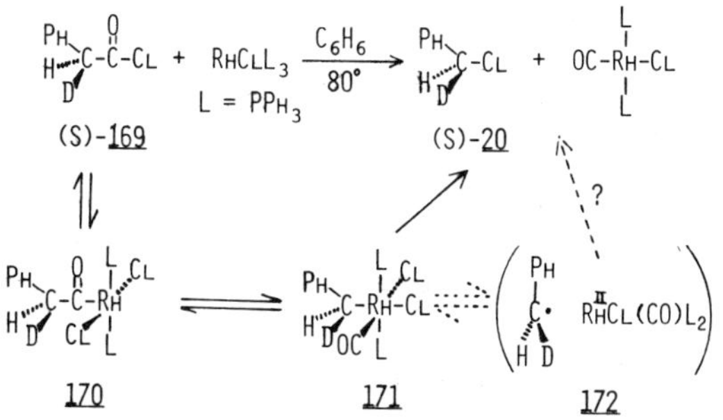

SCHEME 27

actual path may turn out to be. Several other examples of apparent reductive elimination of RX with retention of configuration at carbon are discussed in Sects. III-C-2 and III-C-4.

C. Electrophilic Cleavage

Interactions of electrophiles with organo–transition metal species is an extensive and complex area, some mechanistic aspects of which have been the subject of a recent excellent review (229). Electrophilic cleavage reactions of metal-carbon bonds have found significant use in the removal of organic groups from a variety of organometallic systems. One of the principal applications of such cleavage has been as one step in a series of reactions in stereochemical studies, with the result that an understanding of the stereochemistry associated with the cleavage reactions themselves becomes of considerable importance.

As a framework for discussions in the sections that follow, Scheme 28 outlines some of the basic mechanisms that are believed to operate in electrophilic cleavage. These are fundamentally of two types – bimolecular attack directly at the M–C bond (S_E2 reactions, paths A and B), and those that involve some kind of prior oxidation of the metal, either electron transfer or oxidative addition (paths C, D, and E). Each of these paths is expected to have particular stereochemical consequences. The accompanying implication is that the stereochemistry may vary widely as a function of subtle factors.

1. Protonolysis

Although a number of stereochemical studies have employed secondary alkyl groups, very few data are available on the stereochemistry of protonolysis. One example is the D_2O hydrolysis of the species RCu and [R$_2$Cu] Li, where R was derived from the cyclopropyllithium reagent (S)-39 (95). The overall sequence of conversion of 39 to copper reagents to deuteriocyclopropane occurred with a high degree of retention of configuration.

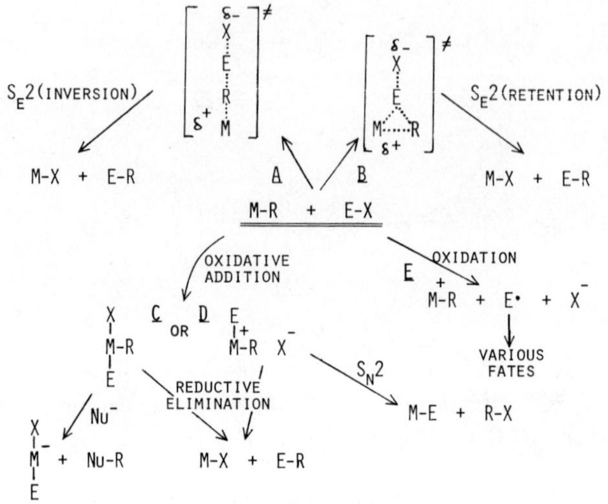

SCHEME 28

(S)-**39**

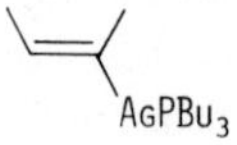

173

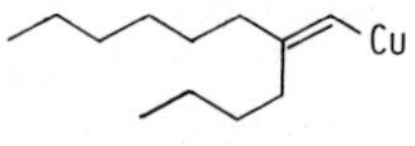

174

There are a number of examples of protonolyses of alkenylmetal complexes. For example, hydrolysis of the (E-2-butenyl)silver species **173** gave only cis-2-butene (54), alkenylcuprate **174** underwent deuteration in $D_2O/DOAc$ with complete retention of configuration (156), and trans-NiBr(PPh$_3$)$_2$(Z-CH=CHPh) was cleaved by D_2SO_4 to give cis-deuterated styrene (56).

There are many examples of the protonation of low valent transition metal complexes (230), and oxidative addition paths for protonolysis have been demonstrated (**C** and **D** in Scheme 28) (229). The resulting reductive elimination would be expected to proceed with retention (Sect. III-B-2). Nevertheless, the direct S_E2 path (**B** in Scheme 28) probably does occur in some cases because, for example, Cp$_2$ZrCl(E-CH=CH-t-Bu) is cleaved stereospecifically in D_2SO_4/D_2O with retention at carbon, and Zr(IV) in this complex is d^0 (167).

Care is necessary is assigning configuration based on electrophilic cleavage. For example, Sect. II.C.3 mentioned that trans-RH(CO)(PPh$_3$)$_2$ [C(CO$_2$Me)CH=(CO$_2$Me)] was originally thought to contain a Z-alkenyl ligand (183), but it was later found that the alkenyl group was actually E (184). The stereochemistry was not determined directly, but rather by cleavage of the Rh–C bond by anhydrous HCl. Initially dimethyl fumarate was found, but more careful analysis indicated that the kinetic product was dimethyl maleate, which was rapidly isomerized to the trans isomer by the acid.

The paucity of data on protonolysis is probably due at least in part to the lesser utility of protonolysis, as opposed to other cleavages (e.g., conformational or configurational analysis of RBr is easier than RD), and it is widely assumed to always proceed with retention of configuration, since there are no counterexamples in main group or transition metal chemistry.

2. Cleavage by Halogens

Cleavage of transition metal–alkyl bonds by halogens, as implied by Scheme 28, has proved to be a very diverse reaction, routinely exhibiting retention or inversion, with radical-related racemization pathways often being difficult to subdue.

The standard stereochemical test for iron, *threo*-, or *erythro*-CpFe(CO)$_2$-CHDCHD-*t*-Bu (6), has been applied, and it was found that I$_2$ and Br$_2$ in nonpolar solvents yielded *erythro*- or *threo*-XCHDCHD-*t*-Bu, respectively, with clean inversion of configuration at carbon. Chlorine in any solvent or Br$_2$ in MeOH yielded acyl products resulting from oxidatively induced CO insertion (6).

In contrast, cleavage of CpFe(CO)$_2$CHDCHDPh by halogens was observed to proceed with retention of configuration (177).

These observations are explicable on the basis of the mechanism in Scheme 29. Oxidative addition of the halogen to the iron center generates the good

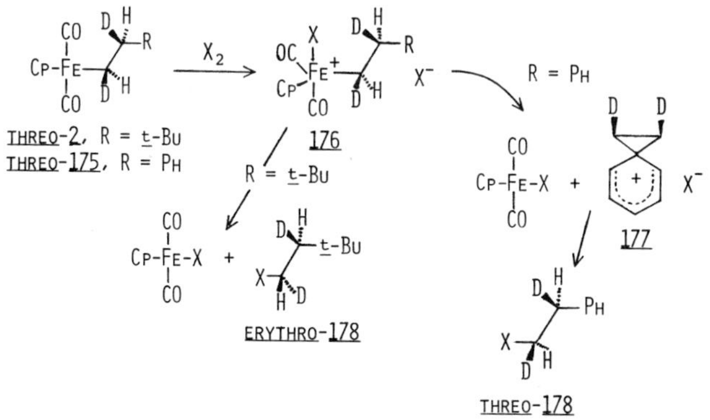

SCHEME 29

leaving group, CpFe(CO)$_2$X, which is displaced from carbon by the counterion X$^-$ in ion pair 176 if R = *t*-Bu, affording *erythro*-178 via an inversion path. If R = Ph, then phenyl participation leads to phenonium ion 177 and thence to *threo*-178 with overall retention.

Evidence for intermediate 177 was provided by the observation that halogen cleavage of CpFe(CO)$_2$CD$_2$CH$_2$Ph led to formation of a 1 : 1 mixture of XCD$_2$-CH$_2$Ph and XCH$_2$CD$_2$Ph (231). This result was confirmed using ^{13}C labeling (232).

Although electron transfer also could account for the chemistry of Scheme 29, it is possible that the one-electron path is actually responsible for the CO insertion chemistry instead (e.g., see ref. 10). Thus the reactions of $CpFe(CO)_2R$ are probably variations on path **D** in Scheme 28.

Alkylation of $[Mn(CO)_5]^-$ by a secondary alkyl halide was shown to proceed with inversion of configuration (Scheme 3, Sect. II-A-1) (9), as was the alkylation of $[Mn(CO)_4(PEt_3)]^-$ by a primary halide (13). Thus cleavage of the Mn—C bond in $Mn(CO)_5$-(CHMeCO$_2$Et) by Br$_2$ could be shown to proceed with net retention, albeit only with about 20% e.e. (9). More recently, use of the substrate cis-$Mn(CO)_4(PEt_3)$(threo-CHDCHDPh) has revealed that the carbon stereochemistry varies considerably from 76% net inversion to 54% net retention, depending on the solvent and halogen used (233).

There is now general recognition that the path taken by an electrophilic attack depends on a variety of factors, the major ones being whether a complex is coordinatively saturated, how sterically encumbered it is, and perhaps most important, the nature of the highest occupied molecular orbital (HOMO) of the complex (229,232,234). These factors and others, such as solvent polarity and polarizability properties of the electrophiles, apparently all meet in a fine balance in the manganese cleavage reaction above (233).

The halogen cleavage reactions of octahedral cobalt(III) alkyls have likewise posed an interesting puzzle. Virtually all the stereochemical work has been carried out using (pyridine)cobaloxime derivatives, $RCo(dmgH)_2(py)$, **179**, which were shown in Sect. II-A-1 to be formed with inversion of configuration at carbon upon alkylation of $[Co(dmgH)_2(py)]^-$, **7**. Cleavage of **179a** to **179c** by Br$_2$ (235–237) and I$_2$ (235) proceeded with net inversion of configuration to give the corresponding alkyl halides.

179A, R = (cyclohexyl with Br)

179B, R = (cyclohexyl with t-Bu)

(+)-**179C**, R = $-\underset{\text{Me}}{\text{CH}}-C_6H_{13}$

179D, R = THREO-CHDCHD-t-Bu

Oxidation of (+)-**179c** $IrCl_6^{2-}$ in the presence of bromide ion afforded (−)-2-bromooctane, from which (+)-**179c** had been prepared, suggesting that the actual cleavage path by halogens might involve the prior one-electron oxidation of **179** followed by backside attack at carbon by the halide counterion displacing Co(II) (path E, Scheme 28) (238). In fact, the one-electron oxidized cations of **179** have been observed at low temperature by ESR (239). However Co(IV) might simply oxidize Br$^-$ back to Br$_2$, which could then cleave the Co(III)—C bond by an S_E2(inversion) mechanism. For this reason, (−)-**179c** was oxidized by one equivalent of Ce(IV) at −60°C, and then Cl⁻, which would not be oxidized by Co(IV), was added. (+)-2-Chlorooctane was formed with ca. 90% inversion of configuration (240). The mechanism of Scheme 30 is therefore a viable possibility for this reaction.

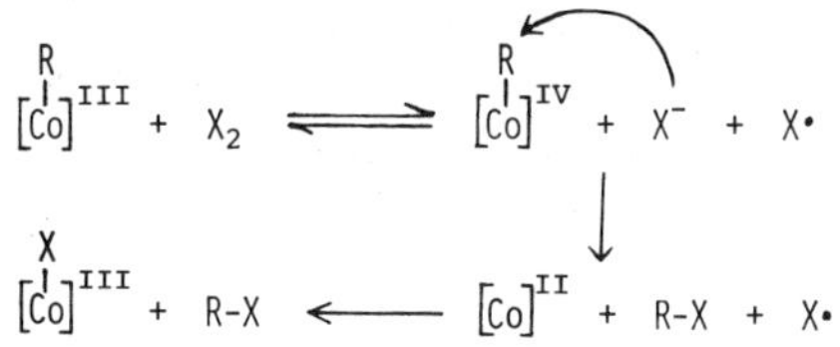

SCHEME 30

Varied stereochemical behavior has been observed in investigations of halogen cleavage of palladium-carbon bonds. Complexes **69** (241) and **70** (134), shown in eq. [43], have been used in these studies, but the results with **180** are representative (241). From comparison of proposed intermediate **184** with the three products (Scheme 31, Table 2), it can be seen that **181** would have been formed

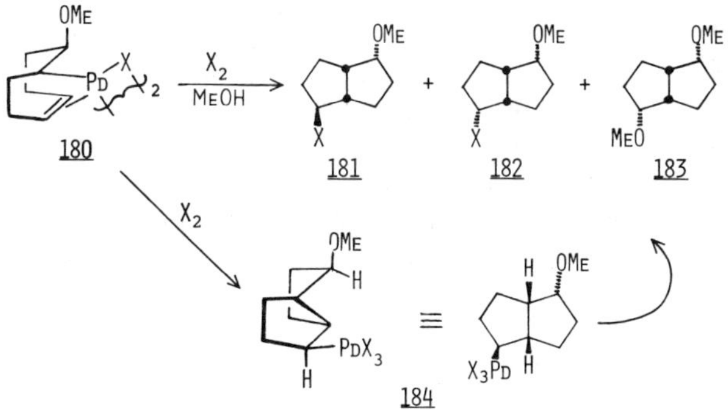

SCHEME 31

Table 2 Products of Reaction in Scheme 31

Halogen	Percentage from Reaction		
	181	182	183
Cl_2	13	64	23
Br_2	26	32	42
Cl_2/Cl^-	14	84	2
Br_2/Br^-	10	90	—

with retention, and **182** and **183** with inversion of configuration. In the absence of added halide ion, apparently there is a competition between reductive elimination and nucleophilic displacement, the latter by a combination of adventitious or dissociated halide ion and solvent. With added halide ion, nucleophilic attack at carbon with displacement of Pd(II) predominates.

The bromine cleavage of *threo*- or *erythro*-$Cp_2ZrCl(CHDCHD$-t-$Bu)$ proceeds with retention of configuration and is the only extant stereochemical example of the cleavage of a d^0 complex. Presumably the unavailability of a prior oxidation pathway and the presence of vacant metal orbitals to interact with the developing anion favor an S_E2(retention) pathway (167).

Halogen cleavage of alkenylmetal species also shows significant stereochemical variation. Alkenylzirconium complexes, $Cp_2ZrCl(E$-$CR{=}CHR)$ underwent high-yield conversions to the corresponding alkenyl bromides or chlorides with complete retention of configuration using *N*-bromo- or *N*-chlorosuccinimide, respectively (179). In contrast to this, bromination of vinyl(pyridine)cobaloximes, **179**, where R = alkenyl, may proceed with either retention or inversion (52,242,243). For example, brominolysis of **179**, where R = Z-CH=CHPh, afforded Z-β-bromostyrene (242,243), whereas the same reaction of **179**, where R = E-1-octenyl, when carried out at $-50°C$ in CS_2, yielded 1-bromooctene that was 72% Z. Under the same conditions, the Z cobalt complex yielded E-1-bromooctene (52).

In a similar dichotomy, bromine cleavage of $Mn(CO)_5 [Z$-$C(CF_3){=}CH(CF_3)]$ proceeded with clean retention at carbon, but $Mn(CO)_5 [Z$-$C(CO_2Me){=}CH(CO_2$-$Me)]$ gave products with ca. a 70 : 30 ratio of inversion to retention. The reasons for these changes in stereochemistry are uncertain (228).

Vinylic copper reagents are cleaved cleanly with retention of configuration by iodine (156). Bromine or chlorine causes oxidative alkenyl coupling rather than halogen cleavage, but this can be readily circumvented by the use of *N*-bromo- or *N*-chlorosuccinimide, which affords vinyl halides stereospecifically (244).

3. Cleavage by Mercuric Salts

Transfers of organic groups between transition metals and mercury(II) are well known in both directions and are formally transmetalations, but it appears that they are often not simply S_E2 reactions at carbon. When these transfers occur from transition metals to mercury, they have been conventionally viewed as electrophilic cleavage reactions and so are discussed here.

Cleavage of iron alkyls $CpFe(CO)_2R$, for a variety of R groups, has been found to proceed with third-order kinetics, rate = $k[FeR][HgX_2]^2$. Furthermore, when R is primary, the products are $CpFe(CO)_2X$ and $RHgX$, and when R is tertiary or benzylic, $CpFe(CO)_2HgX$ and RX are formed (245). The stereochemistry at carbon accompanying the cleavage of primary substrates such as

CpFe(CO)$_2$CHDCHD-t-Bu (6) and CpFe(CO)$_2$CHDCHDPh (177) is retention of configuration. These data are consistent with mechanism of Scheme 32, where for primary substrates the reductive elimination step should proceed with retention at carbon.

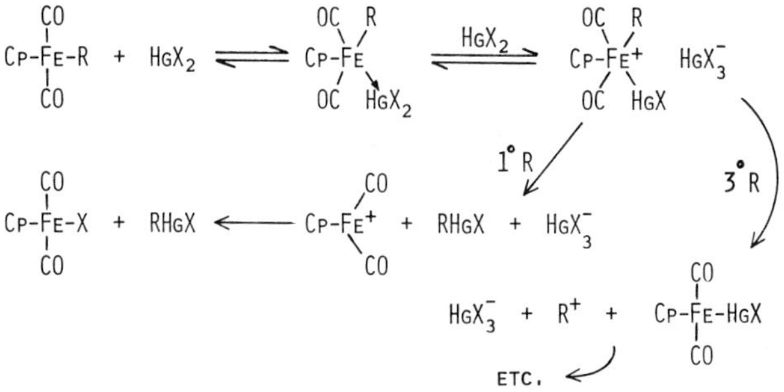

$$\begin{array}{c}
\text{Cp-Fe-R} + \text{HgX}_2 \rightleftharpoons \text{Cp-Fe} \overset{\text{OC}\ \text{R}}{\underset{\text{OC}\ \text{HgX}_2}{\diagup}} \xrightleftharpoons{\text{HgX}_2} \text{Cp-Fe}^+ \overset{\text{OC}\ \text{R}}{\underset{\text{OC}\ \text{HgX}}{\diagup}} \text{HgX}_3^-
\end{array}$$

$$\begin{array}{c}
\text{Cp-Fe-X} + \text{RHgX} \longleftarrow \text{Cp-Fe}^+ \diagup \text{CO} + \text{RHgX} + \text{HgX}_3^- \\
\qquad\qquad\qquad \text{HgX}_3^- + \text{R}^+ + \text{Cp-Fe-HgX} \\
\qquad\qquad\qquad \text{ETC.}
\end{array}$$

SCHEME 32

Retention of configuration has also been noted for the HgCl$_2$ cleavage of *trans*-CpW(CO)$_2$(PEt$_3$)(*threo*-CHDCHDPh). A mechanism similar to that in Scheme 32 was suggested here as well (246).

In contrast to the iron and tungsten systems, cleavage of *cis*-Mn(CO)$_4$(PEt$_3$)-(*threo*-CHDCHDPh) (246) and Co(py)(dmgH)$_2$(*threo*-CHDCHD-t-Bu) (247) by HgCl$_2$ and Hg(ClO$_4$)$_2$, respectively, proceeded with inversion of configuration. The latter result was corroborated using the secondary substrates *cis*- and *trans*-**179b** (236). The difference between the inversion and the retention systems has been attributed to steric hindrance to frontside approach in the latter (247) and to the likelihood that a metal lone pair is the HOMO in the former systems (246).

Section II-B presented data suggesting that saturated carbon is transferred to Pd(II) with retention of configuration (102). This reaction has been effected in the reverse direction, also with retention of configuration, as shown in eq. [79] (262).

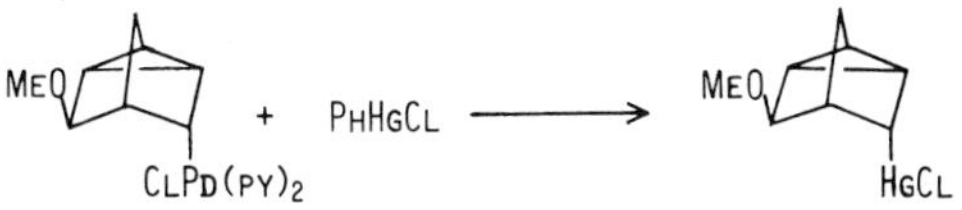

$$\text{MeO-}\underset{\text{ClPd(py)}_2}{\bigtriangleup} + \text{PhHgCl} \longrightarrow \text{MeO-}\underset{\text{HgCl}}{\bigtriangleup} \qquad (79)$$

Vinyl(pyridine)cobaloximes, **179** where R = *cis*- or *trans*-β-styryl, were cleaved by Hg(OAc)$_2$ in DMF with complete retention (242), but **179**, where R = *cis*- or *trans*-1-octenyl, afforded mixtures of isomeric vinylmercury products (52). Recovered starting material was not isomerized in the latter case.

A number of research groups have sought to examine the stereochemistry of a variety of the electrophilic cleavage reactions discussed herein using pseudotetrahedral metal complexes of the types already described: **185** (219), **186** (248), and **187** (249).

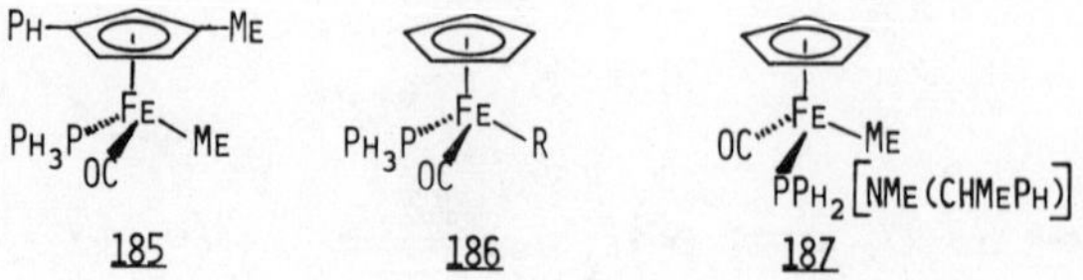

185   **186**   **187**

Presumably in the early stages of this work it was believed that an S_E2(retention) path might be common and should be differentiable from the other paths in Scheme 28 by retention of configuration at the metal as opposed to racemization. It now appears that S_E mechanisms are uncommon, and the metal stereochemistry is consistent with this observation. For example, I_2 cleavage of **185** to **187** proceeds, depending on conditions, with from ca. 4 to 67% stereospecificity. Recent X-ray crystal structure determinations of the acetyl precursor of **185** and the I_2-cleavage product have established that cleavage by HgI_2, I_2, and HI all procced with net retention of configuration at iron (219). A cyclic set of cleavage and alkylation reactions involving **187** independently led to the same conclusions (249). It is likely that the retention observed in these reactions is a function of the bulk of the phosphine ligand, and perhaps a type of trans effect, such that the electrophile preferentially forms **188** (Scheme 33). The stereochemical fate of this would be directly determined by nucleophilic attack for path *H* or by ion pairing and possible nonplanarity of the cation for path *J*.

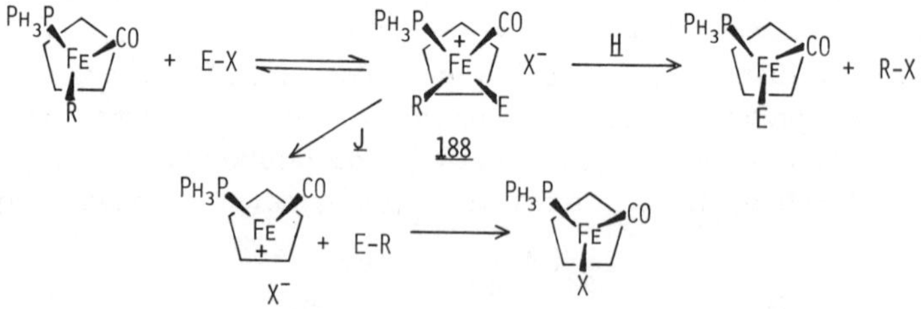

SCHEME 33

Electrophilic cleavage of a presumably equilibrated cis-trans mixture of $CpMo(CO)_2 LMe$ (L = PPh_3 or PBu_3) tends to be stereospecific, the outcome being dependent on the electrophile used (250). Mercuric halide cleavage afforded entirely *cis*-$CpMo(CO)_2 LX$, whereas halogen cleavage gave mixtures that were rich in trans isomer. Control experiments established that the product ratios were kinetically determined. Mechanistic explanations for these results were not obvious.

4. Oxidative Cleavage and Autoxidation

If one views concerted S_E2 reactions (Scheme 28) to be "normal" electrophilic cleavages (the oxidation state of the transition metal never changes), the other mechanistic types represented in Scheme 28 are oxidatively induced cleavages, since the oxidation state of the metal increases by 1 or 2 in various intermediates. Most of the reactions discussed in Sect. III-C thus far are probably oxidative reactions. Nevertheless, there are additional cases of a reagent that has been added expressly for the purpose of inducing oxidation of the metal or for oxygen transfer to the organic ligand.

For example, in Sect. II-C-1 several aminopalladation reactions were outlined in Scheme 15 (129,130). In these cases, a (β-aminoalkyl)palladium intermediate, initially formed by external attack by an amine on an (alkene)palladium complex, was decomposed by oxidation of the metal. Oxidation of Pd(II) presumably to Pd(IV) by Br_2, N-bromosuccinimide, or m-chloroperbenzoic acid labilized the Pd—C bond toward nucleophilic attack by either the β-amino group (to form aziridines) or excess amine in solution (to form diamines). Each of these reactions was found to be highly stereospecific.

The same β-aminoalkylpalladium species could be oxidized by $Pb(OAc)_4$ to afford β-aminoacetates with inversion of configuration (251). In a related reaction, (1,5-cyclooctadiene)$PdCl_2$, **189**, could be oxidized by $Pb(OAc)_4$ in

acetic acid to yield **190** and **191** (252,253), which correspond to reductive elimination of RCl from **192** with retention of configuration, or displacement by acetate on **192** with inversion, in complete analogy with Scheme 31 in Sect. III-C-2.

The presumed driving force for either of these would be reduction of Pd(IV) back to Pd(II). Other interpretations of these results are possible, however (251), and further elucidation is desirable.

The oxidation of Pd–C bonds by $CuCl_2$ was a key step in several of the investigations concerning the stereochemistry of nucleophilic additions to complexed alkenes (Sect. II-C-1). In the case of oxidation of CHD=CHD by $PdCl_2/H_2O/CuCl_2$ to form ClCHDCHDOH, it was argued that the $CuCl_2$ cleavage proceeded with inversion of configuration (Scheme 12) (114). Corroborative evidence came from several sources. First, the overall addition of chloride and acetate in eq. [81] proceeded with syn stereochemistry (254). Since the aim of

$$C_8H_{17}\diagup\!\!\!\diagdown D \xrightarrow[\text{HOAc}]{PdCl_2 \atop CuCl_2/LiCl} \left[\begin{array}{c}AcO\quad D\\ \diagdown\!\!\!\diagup H \\ R'' \diagup\!\!\!\diagdown Cl \\ H\quad Pd \end{array}\right] \longrightarrow \begin{array}{c}AcO\quad Cl\\ \diagdown\!\!\!\diagup \\ R'' \diagup\!\!\!\diagdown D \\ H\quad H\end{array} \quad (81)$$

193

the former study was to establish the direction of addition of an oxygen nucleophile to a complexed alkene, the assumption in the latter study that acetate adds anti to the metal borders on being tautological. Second, and more convincingly, it was shown that HOCHDCHDHgCl was transferred to $PdCl_2$ and thence cleaved by $CuCl_2$ with overall inversion of configuration (114). Since the transfer from Hg(II) to Pd(II) very probably proceeds with retention (Sect. II-B), the cleavage probably causes inversion.

Ironically, part of the argument presented to support the idea that intermediate **193** formed via anti acetate addition was a paper discussed earlier (Sect. II-C-1) (125) that required as part of its mechanistic rationale that the cyclohexylpalladium bond be cleaved by $CuCl_2$ with retention of configuration (Scheme 13 and structure **58**).

And finally, to further confuse the issue, $CuCl_2$ cleavage of Cp_2ZrCl(CHD-CHD-t-Bu), ClHg(CHDCHD-t-Bu), **194a**, and **194b** were all reported to proceed in a completely stereorandom manner (255).

194A, M = Pd(diphos)Cl
194B, M = HgCl

One may suggest that $CuCl_2$ cleavage of a primary Pd–C bond in the presence of excess Cl^- could proceed with inversion, while the secondary Pd–C bond of an oxidized complex might homolyze in the absence of added nucleophile. This leaves unexplained the retention result mentioned above, however. Needless to say, the $CuCl_2$ cleavage issue is far from settled.

Oxidation of $PtCl(PPh_3)_2$ [(R)-CHDPh] by m-chloroperbenzoic acid yielded (R)-benzyl-α-d_1 m-chlorobenzoate, consistent with an oxidative addition–reductive elimination mechanism (256).

Turning to autoxidations, treatment of *threo-* or *erythro*-Cp$_2$ZrCl(CHDCHD-*t*-Bu) with dry O$_2$ followed by acidic hydrolysis afforded HOCHDCHD-*t*-Bu with 50% diastereomeric excess (167). This was believed to arise from half the sample undergoing an epimerizing radical autoxidation to form a zirconium alkylperoxide (eq. [82]), which in turn rapidly oxidized a second equivalent of zirconium alkyl stereospecifically with retention (eq. [83], where R* is alkyl of retained configuration).

$$R^*-(Z_R) + O_2 \longrightarrow R-O-O-(Z_R) \qquad (82)$$

$$RO_2^-(Z_R) + R^*-(Z_R) \longrightarrow R-O-(Z_R) + R^*-O-(Z_R) \qquad (83)$$

The only other data dealing with the stereochemistry of autoxidation of transition metal—carbon bonds pertain to the cobaloximes, structures **179** (see Sect. III-C-2 also). Initially it was reported that either thermal or photochemical oxygenation of optically active **179e** yielded optically active alkylperoxy complex **195**, although the relative configurations of starting material and product were unknown (257). Similarly, **179f** afforded the analogous **195** that, it was

$$R-Co(DMGH)_2(PY) \xrightarrow{O_2} R-O-O-Co(DMGH)_2(PY) \xrightarrow[\text{OR}]{\text{NaBH}_4} ROH \qquad (84)$$
$$\underline{179} \qquad \qquad \underline{195} \qquad \qquad \text{LiAlH}_4$$

179B, R = (cyclohexyl-methyl)   **179F**, R = (HO-indanyl)

179E, R = $-\underset{\underset{\text{Ph}}{|}}{\text{C}}\text{HCH}_2\text{OH}$ **179G**, R = $-\underset{\underset{\text{Me}}{|}}{\text{C}}\text{H-Et}$

argued, had formed with retention of configuration based on NMR analysis (258). Later two different groups prepared alkylperoxy species **195** from optically active **179e** and **179g** (259), and the cis and trans isomers of **179b** (260). These intermediates were reduced by LiAlH$_4$ or NaBH$_4$ and yielded inactive or epimeric alcohols. Unfortunately, the latter results were uninformative because it was shown that ketones are intermediates on the reduction path of the alkylperoxycobaloximes (261). This stereochemical issue thus remains unresolved.

IV. CONCLUSIONS

Investigations of the stereochemistry of reactions of σ bonds between organo—transition metals and carbon has progressed to the point that a few generalizations have emerged and others are taking shape. The carbon monoxide insertion reaction is apparently almost always highly stereospecific with retention of configuration at carbon. Racemization has intruded in a few cases that have arisen but these examples are multistep processes and the source of racemization cannot be definitely located. The alkylation of metalate anions by organic halides and, particularly, organic sulfonates is another reaction that has dependably exhibited stereospecific behavior, in this case inversion at carbon.

In general, there is surprisingly little definitive work that clearly establishes the stereochemistry with which single-step reactions proceed. Nevertheless, there is a great deal of information on multistep reactions, which leads one to formulate what appear to be reliable generalizations, although some of these generalizations do verge on purely being assumptions. Generalizations based on indirect but substantial data include the following: insertion of $\diagdown$C=C$\diagup$ into M–H is syn; insertion of $\diagdown$C=C$\diagup$ into M–R is with retention in R; transmetalations occur with retention at carbon; formation of R–H, R–R, R–(alkenyl), R–(aryl), R–(acyl), and so on, by reductive elimination from metal complexes proceeds with retention; and nucleophilic additions of oxygen, nitrogen, and resonance-stabilized carbanions to metal-complexed alkenes occur anti, while additions of nonstabilized carbanions occur syn. Some generalizations exist in the literature that are based on almost no corroboration, including the idea that protonolysis of M–R will always occur with retention.

Often in organometallic reactions, various amounts of stereorandom behavior are encountered, which may be due to a radical component of the mechanism, or to the stereochemical instability of one or another intermediate or product species. Mechanistic interpretation of even apparently stereospecific reactions is sometimes not straightforward, since as a recent study has pointed out (154), in a multistep reaction there is no guarantee that the kinetic product of the overall reaction is also the kinetic product resulting from the key step about which one wishes to infer information.

In the introduction it was mentioned that many reactions exhibit highly variable mechanistic, therefore stereochemical, behavior. Clearly among this class of reactions are oxidative additions of organic halides to neutral metal complexes, as well as halogen and other electrophilic and oxidative cleavages of metal-carbon bonds. The use of those reactions to infer stereochemical information about other reactions in a multistep sequence is hazardous unless one has definite stereochemical evidence bearing on the exact system under consideration.

ACKNOWLEDGMENTS

I thank Mrs. Sarah Lowery and Mr. John Statler for their assistance in the preparation of the manuscript for this chapter.

REFERENCES

1. F. J. McQuillin, *Tetrahedron*, **30**, 1661 (1974).
2. F. A. Cotton and G. Wilkinson, *Advanced Inorganic Chemistry*, 4th ed., Interscience, New York, 1980, Chapters 27, 29, 30.
3. J. E. Huheey, *Inorganic Chemistry*, 2nd ed., Harper & Row, New York, 1978, Chapter 13.
4. J. P. Collman, *Acc. Chem. Res.*, **1**, 136 (1968).
5. J. K. Stille and K. S. Y. Lau, *Acc. Chem. Res.*, **10**, 434 (1977).
6. (a) P. L. Bock, D. J. Boschetto, J. R. Rasmussen, J. P. Demers, and G. M. Whitesides, *J. Am. Chem. Soc.*, **96**, 2814 (1974); (b) G. M. Whitesides and D. J. Boschetto, *ibid.*, **93**, 1529 (1971); (c) G. M. Whitesides and D. J. Boschetto, *ibid.*, **91**, 4313 (1969).
7. (a) G. S. Koermer, M. L. Hall, and T. G. Traylor, *J. Am. Chem. Soc.*, **94**, 7205 (1972); (b) H. G. Kuivila, J. L. Considine and J. D. Kennedy, *J. Am. Chem. Soc.*, **94**, 7206 (1972).
8. D. A. Slack and M. C. Baird, *J. Chem. Soc., Chem. Commun.*, **1974**, 701; *J. Am. Chem. Soc.*, **98**, 5539 (1976).
9. R. W. Johnson and R. G. Pearson, *J. Chem. Soc., Chem. Commun.*, **1970**, 986.
10. K. M. Nicholas and M. Rosenblum, *J. Am. Chem. Soc.*, **95**, 4449 (1973).
11. J. P. Collman, S. R. Winter, and D. R. Clark, *J. Am. Chem. Soc.*, **94**, 1788 (1972).
12. J. P. Collman, R. G. Finke, J. N. Cawse, and J. I. Brauman, *J. Am. Chem. Soc.*, **99**, 2515 (1977).
13. D. Dong, D. A. Slack, and M. C. Baird, *J. Organomet. Chem.*, **153**, 219 (1978).
14. R. F. Heck, *J. Am. Chem. Soc.*, **85**, 1460 (1963).
15. F. R. Jensen, V. Madan, and D. H. Buchanan, *J. Am. Chem. Soc.*, **92**, 1414 (1970).
16. P. L. Bock and G. M. Whitesides, *J. Am. Chem. Soc.*, **96**, 2826 (1974).
17. (a) G. H. Posner, *Org. Reactions*, **19**, 1 (1972); (b) G. H. Posner, *ibid.*, **22**, 253 (1975).
18. G. M. Whitesides, W. F. Fischer, J. San Filippo, Jr., R. W. Bashe, and H. O. House, *J. Am. Chem. Soc.*, **91**, 4871 (1969).
19. E. J. Corey and G. H. Posner, *J. Am. Chem. Soc.*, **89**, 3911 (1967).
20. C. R. Johnson and G. A. Dutra, *J. Am. Chem. Soc.*, **95**, 7783 (1973).
21. J. Schaffler and J. Retey, *Angew. Chem., Int. Ed. Engl.*, **17**, 845 (1978).
22. H. Eckert, D. Lenoir, and I. Ugi, *J. Organomet. Chem.*, **141**, C23 (1977).
23. F. R. Jensen and D. H. Buchanan, *J. Chem. Soc., Chem. Commun.*, **1973**, 153.
24. L. Walder, G. Rytz, K. Meier, and R. Scheffold, *Helv. Chim. Acta*, **61**, 3013 (1978).
25. P. J. Krusic, P. J. Fagan, and J. San Filippo, Jr., *J. Am. Chem. Soc.*, **99**, 250 (1977).
26. T. Bauch, A. Sanders, C. V. Magatti, P. Waterman, D. Judelson, and W. P. Giering, *J. Organomet. Chem.*, **99**, 269 (1975).
27. J. A. Labinger, R. J. Braus, D. Dophin, and J. A. Osborn, *J. Chem. Soc., Chem. Commun.*, **1970**, 612.
28. R. G. Pearson and W. R. Muir, *J. Am. Chem. Soc.*, **92**, 5519 (1970).
29. F. R. Jensen, and B. Knickel, *J. Am. Chem. Soc.*, **93**, 6339 (1971).
30. J. S. Bradley, D. E. Connor, D. Dolphin, J. A. Labinger and J. A. Osborn, *J. Am. Chem. Soc.*, **94**, 4043 (1972).
31. J. A. Labinger, A. V. Kramer, and J. A. Osborn, *J. Am. Chem. Soc.*, **95**, 7908 (1973).
32. J. A. Osborn, in *Organotransition-Metal Chemistry*, Y. Ishii and M. Tsutsui, Eds., Plenum Press, New York, 1975, p. 65.
33. P. B. Chock and J. Halpern, *J. Am. Chem. Soc.*, **88**, 3511 (1966).
34. P. B. Chock and J. Halpern, *Proc. 10th Int. Conf. Coord. Chem.*, **1967**, 135.

35. J. P. Collman and C. T. Sears, *Inorg. Chem.*, **7**, 27 (1968).
36. A. R. Rossi and R. Hoffmann, *Inorg. Chem.*, **14**, 365 (1975).
37. J. P. Collman and M. R. MacLaury, *J. Am. Chem. Soc.*, **96**, 3019 (1974).
38. S. Otsuka, A. Nakamura, T. Yoshida, M. Naruto, and K. Ataka, *J. Am. Chem. Soc.*, **95**, 3180 (1973).
39. Y. Becker and J. K. Stille, *J. Am. Chem. Soc.*, **100**, 838 (1978).
40. A. V. Kramer, J. A. Labinger, J. S. Bradley, and J. A. Osborn, *J. Am. Chem. Soc.*, **96**, 7145 (1974).
41. A. V. Kramer and J. A. Osborn, *J. Am. Chem. Soc.*, **96**, 7832 (1974).
42. K. S. Y. Lau, R. W. Fries, and J. K. Stille, *J. Am. Chem. Soc.*, **96**, 4983 (1974); P. K. Wong, K. S. Y. Lau, and J. K. Stille, *ibid.*, **96**, 5956 (1974); K. S. Y. Lau, P. K. Wong, and J. K. Stille, *ibid.*, **98**, 5832 (1976).
43. J. K. Stille, L. F. Hines, R. W. Fries, P. K. Wong, D. E. James, and K. S. Y. Lau, *Adv. Chem. Ser.*, **132**, 90 (1974).
44. K. Kudo, M. Hidai, and Y. Uchida, *J. Organomet. Chem.* **33**, 393 (1971).
45. K. Kudo, M. Sato, M. Hidai, and Y. Uchida, *Bull. Chem. Soc. Jap.*, **46**, 2820 (1973).
46. Y. Becker and J. K. Stille, *J. Am. Chem. Soc.*, **100**, 845 (1978).
47. V. I. Sokolov, *Inorg. Chim. Acta*, **18**, L9 (1976).
48. J. K. Stille and A. B. Cowell, *J. Organomet. Chem.*, **124**, 253 (1977).
49. L. S. Hegedus and L. L. Miller, *J. Am. Chem. Soc.*, **97**, 459 (1975).
50. K. N. V. Duong and A. Gaudemer, *J. Organomet. Chem.*, **22**, 473 (1970).
51. M. D. Johnson and B. S. Meeks, *J. Chem. Soc. B*, **1971**, 185.
52. M. Tada, M. Kubota, and H. Shinozaki, *Bull. Chem. Soc. Jap.*, **49**, 1097 (1976).
53. R. W. Fessenden and R. H. Schuler, *J. Chem. Phys.*, **39**, 2147 (1963).
54. G. M. Whitesides, C. P. Casey, and J. K. Krieger, *J. Am. Chem. Soc.*, **93**, 1379 (1971).
55. Z. Rappoport, *Adv. Phys. Org. Chem.*, **7**, 1 (1969).
56. D. Dodd and M. D. Johnson, *J. Organomet. Chem.*, **52**, 1 (1973).
57. M. F. Semmelhack and L. Ryono, *Tetrahedron Lett.*, **1973**, 2967.
58. J. Rajaram, R. G. Pearson, and J. A. Ibers, *J. Am. Chem. Soc.*, **96**, 2103 (1974).
59. L. Cassar and A. Giarrusso, *Gazz. Chim. Ital.*, **103**, 793 (1973).
60. L. S. Hegedus and R. K. Stiverson, *J. Am. Chem. Soc.*, **96**, 3250 (1974).
61. M. F. Semmelhack, *Org. React.*, **19**, 115 (1972).
62. M. F. Semmelhack, P. M. Helquist, and J. D. Gorzynski, *J. Am. Chem. Soc.*, **94**, 9234 (1972).
63. S. Baba and E. Negishi, *J. Am. Chem. Soc.*, **98**, 6729 (1976).
64. A. A. Millard and M. W. Rathke, *J. Am. Chem. Soc.*, **99**, 4833 (1977).
65. P. Fitton and J. E. McKeon, *J. Chem. Soc., Chem. Commun.*, **1968**, 4.
66. W. J. Bland and R. D. W. Kemmitt, *J. Chem. Soc. A*, **1968**, 1278.
67. B. E. Mann, B. L. Shaw, and N. I. Tucker, *J. Chem. Soc. A*, **1971**, 2667; *J. Chem. Soc., Chem. Commun.*, **1970**, 1333.
68. B. F. G. Johnson, J. Lewis, J. D. Jones, and K. A. Taylor, *J. Chem. Soc., Dalton Trans.*, **1974**, 34; *J. Organomet. Chem.*, **32**, C62 (1971).
69. W. J. Bland, J. Burgess, and R. D. W. Kemmitt, *J. Organomet. Chem.*, **14**, 201 (1968).
70. J. Chatt and B. L. Shaw, *J. Chem. Soc.* **1959**, 4020; R. Romeo, D. Minniti, and M. Trozzi, *Inorg. Chem.*, **15**, 1134 (1976); and references therein.
71. R. F. Heck, *Adv. Cataly.*, **26**, 323 (1977).
72. A. Schoenberg, I. Bartoletti, and R. Heck, *J. Org. Chem.*, **39**, 3318 (1974).
73. H. A. Dieck and R. F. Heck, *J. Org. Chem.*, **40**, 1083 (1975).
74. A. Cowell and J. K. Stille, *Tetrahedron Lett.*, **1979**, 133.
75. K. Tamao, K. Sumitani, Y. Kiso, M. Zembayashi, A. Fujioka, S. Kodama, I. Nakajima, A. Minato, and M. Kumada, *Bull. Chem. Soc. Jap.*, **49**, 1958 (1976), and references

therein: in particular, K. Tamao, M. Zembayashi, Y. Kiso, and M. Kumada, *J. Organomet. Chem.*, **55**, C91 (1973).
76. H. Okamura, M. Miura, and H. Takei, *Tetrahedron Lett.*, **1979**, 43.
77. M. Yamamura, I. Moritani, and S. Murahashi, *J. Organomet. Chem.*, **91**, C39 (1975).
78. H. P. Dang and G. Linstrumelle, *Tetrahedron Lett.*, **1978**, 191.
79. A. O. King, N. Okukado, and E. Negishi, *J. Chem. Soc., Chem. Commun.*, **1977**, 683.
80. M. Tamura and J. Kochi, *J. Am. Chem. Soc.*, **93**, 1483 (1971).
81. R. S. Smith and J. K. Kochi, *J. Org. Chem.*, **41**, 502 (1976), and references therein.
82. K. C. Bishop III, *Chem. Rev.*, **76**, 461 (1976).
83. J. Halpern, in *Organic Synthesis via Metal Carbonyls*, Vol. 2, I. Wender and P. Pino, Eds., Wiley, New York, 1977, p. 705.
84. L. Cassar, P. E. Eaton, and J. Halpern, *J. Am. Chem. Soc.*, **92**, 3515 (1970).
85. L. Cassar and J. Halpern, *J. Chem. Soc., Chem. Commun.*, **1970**, 1082.
86. B. F. G. Johnson, J. Lewis, and S. W. Tam, *J. Organomet. Chem.*, **105**, 271 (1976).
87. R. M. Moriarty, K. N. Chen, C. L. Yeh, J. L. Flippen, and J. Karle, *J. Am. Chem. Soc.*, **94**, 8944 (1972).
88. P. W. Hall, R. J. Puddephatt, and C. F. H. Tipper, *J. Organomet. Chem.*, **71**, 145 (1974), and references therein.
89. T. H. Tulip and J. A. Ibers, *J. Am. Chem. Soc.*, **100**, 3252 (1978).
90. M. Sohn, J. Blum, and J. Halpern, *J. Am. Chem. Soc.*, **101**, 2694 (1979), and references therein.
91. N. Dominelli and A. C. Oehlschlager, *Can. J. Chem.*, **55**, 364 (1977).
92. D. Dodd, M. D. Johnson, and B. L. Lockman, *J. Am. Chem. Soc.*, **99**, 3664 (1977).
93. O. A. Reutov, V. I. Sokolov, G. Z. Suleimanov, and V. V. Bashilov, *J. Organomet. Chem.*, **160**, 7 (1978).
94. H. Yatagai, Y. Yamamoto, and K. Maruyama, *J. Chem. Soc., Chem. Commun.*, **1978**, 702.
95. M. P. Periasamy and H. M. Walborsky, *J. Am. Chem. Soc.*, **97**, 5930 (1975).
96. E. Piers and E. H. Ruediger, *J. Chem. Soc., Chem. Commun.*, **1979**, 166.
97. F. R. Jensen and K. L. Nakamaye, *J. Am. Chem. Soc.*, **88**, 3437 (1966).
98. G. M. Whitesides, J. San Filippo, Jr., E. R. Stedronsky, and C. P. Casey, *J. Am. Chem. Soc.*, **91**, 6542 (1969).
99. D. Seyferth, *Prog. Inorg. Chem.*, **3**, 129 (1962).
100. R. C. Larock, *Angew. Chem., Int. Ed. Engl.*, **17**, 27 (1978).
101. D. E. Bergbreiter and G. M. Whitesides, *J. Am. Chem. Soc.*, **96**, 4937 (1974).
102. (a) J. -E. Bäckvall and B. Åkermark, *J. Chem. Soc., Chem. Commun.*, **1975**, 82; (b) J. K. Stille and P. K. Wong, *J. Org. Chem.*, **40**, 335 (1975).
103. R. C. Larock and J. C. Bernhardt, *J. Org. Chem.*, **42**, 1680 (1977).
104. G. Zweifel and R. L. Miller, *J. Am. Chem. Soc.*, **92**, 6678 (1970).
105. D. B. Carr and J. Schwartz, *J. Am. Chem. Soc.*, **101**, 3521 (1979).
106. E. Negishi and D. E. Van Horn, *J. Am. Chem. Soc.*, **99**, 3168 (1977).
107. N. Okukado, D. E. Van Horn, W. L. Klima, and E. Negishi, *Tetrahedron Lett.*, **1978**, 1027.
108. M. J. Loots and J. Schwartz, *J. Am. Chem. Soc.*, **99**, 8046 (1977).
109. E. Negishi, N. Okukado, A. O. King, D. E. Van Horn, and B. I. Spiegel, *J. Am. Chem. Soc.*, **100**, 2254 (1978).
110. K. Tamao, H. Matsumoto, T. Kakui, and M. Kumada, *Tetrahedron Lett.*, **1979**, 1137; K. Tamao, T. Kakui, and M. Kumada, *ibid.*, **1979**, 619, and references therein.
111. (a) G. Marr and B. W. Rockett, *J. Organomet. Chem.*, **163**, 325 (1978), and previous annual surveys on π-complexes referenced therein; (b) S. G. Davies, M. L. H. Green, and D. M. P. Mingos, *Tetrahedron*, **34**, 3047 (1978).

112. G. Paiaro, *Organomet. Chem. Rev. A.*, **6**, 319 (1970).
113. P. M. Henry, *Adv. Organomet. Chem.*, **13**, 363 (1975), and references therein.
114. J. -E. Bäckvall, B. Åkermark, and S. O. Ljunggren, *J. Am. Chem. Soc.*, **101**, 2411 (1979); *J. Chem. Soc., Chem. Commun.*, **1977**, 264.
115. C. Burgess, F. R. Hartley, and G. W. Searle, *J. Organomet. Chem.*, **76**, 247 (1974).
116. J. K. Stille and D. E. James, *J. Organomet. Chem.*, **108**, 401 (1976); *J. Am. Chem. Soc.*, **97**, 674 (1975).
117. J. K. Stille and R. Divakaruni, *J. Organomet. Chem.*, **169**, 239 (1979). *J. Am. Chem. Soc.*, **100**, 1303 (1978).
118. J. K. Stille and R. A. Morgan, *J. Am. Chem. Soc.*, **88**, 5135 (1966), and references therein.
119. M. Green and R. I. Hancock, *J. Chem. Soc. A.*, **1967**, 2054.
120. W. A. Whitla, H. M. Powell, and L. M. Venanzi, *J. Chem. Soc., Chem. Commun.*, **1966**, 310.
121. C. Panattoni, G. Bombieri, E. Forsellini, and B. Crociani, *J. Chem. Soc., Chem. Commun.*, **1969**, 187.
122. D. E. James, L. F. Hines, and J. K. Stille, *J. Am. Chem. Soc.*, **98**, 1806 (1976), and references therein.
123. T. Majima and H. Kurosawa, *J. Chem. Soc., Chem. Commun.*, **1977**, 610.
124. H. -B. Lee and P. M. Henry, *Can. J. Chem.*, **54**, 1726 (1976).
125. P. M. Henry and G. A. Ward, *J. Am. Chem. Soc.*, **94**, 7305 (1972).
126. W. T. Wipke and G. L. Goeke, *J. Am. Chem. Soc.*, **96**, 4244 (1974).
127. A. Panunzi, A. De Renzi, G. Paiaro, *J. Am. Chem. Soc.*, **92**; 3488 (1970).
128. B. Åkermark, J. -E. Bäckvall, K. Siirala-Hansén, K. Sjöberg, and K. Zetterberg, *Tetrahedron Lett.*, **1974**, 1363; B. Åkermark and J. -E. Bäckvall, *ibid.*, **1975**, 819.
129. J. -E. Bäckvall, *J. Chem. Soc., Chem. Commun.*, **1977**, 413.
130. J. -E. Bäckvall, *Tetrahedron Lett.*, **1978**, 163.
131. J. K. Stille and D. B. Fox, *J. Am. Chem. Soc.*, **92**, 1274 (1970).
132. R. A. Holton, *J. Am. Chem. Soc.*, **99**, 8083 (1977).
133. H. Kurosawa and N. Asada, *Tetrahedron Lett.*, **1979**, 255.
134. D. R. Coulson, *J. Am. Chem. Soc.*, **91**, 200 (1969).
135. A. Segnitz, P. M. Bailey, and P. M. Maitlis, *J. Chem. Soc., Chem. Commun.*, **1973**, 698.
136. R. F. Heck, *J. Am. Chem. Soc.*, **90**, 5518 (1968).
137. A. Segnitz, E. Kelly, S. H. Taylor, and P. M. Maitlis, *J. Organomet. Chem.*, **124**, 113 (1977).
138. C. Eaborn, K. J. Odell, and A. Pidcock, *J. Chem. Soc. Dalton Trans.*, **1978**, 357.
139. A. Kasahara, T. Izumi, K. Endo, T. Takeda, and M. Ookita, *Bull. Chem. Soc. Jap.*, **47** 1967 (1974).
140. H. Horino, M. Arai, and N. Inoue, *Tetrahedron Lett.*, **1974**, 647.
141. E. Vedejs and P. D. Weeks, *J. Chem. Soc., Chem. Commun.*, **1974**, 223.
142. M. C. Gallazzi, L. Porri, and G. Vitulli, *J. Organomet. Chem.*, **97**, 131 (1975).
143. R. P. Hughes and J. Powell, *J. Organomet. Chem.*, **30**, C45 (1971).
144. R. F. Heck, *J. Am. Chem. Soc.*, **91**, 6707 (1969).
145. H. A. Dieck and R. F. Heck, *J. Am. Chem. Soc.*, **96**, 1133 (1974).
146. S. -I. Murahashi, M. Yamamura, and N. Mita, *J. Org. Chem.*, **42**, 2870 (1977).
147. P. M. Henry and G. A. Ward, *J. Am. Chem. Soc.*, **94**, 673 (1972).
148. R. F. Heck, *Organotransition Metal Chemistry*, Academic Press, New York, 1974, Chapters 7 and 8.
149. M. H. Chisholm and H. C. Clark, *Acc. Chem. Res.*, **6**, 202 (1973).
150. B. L. Booth and R. G. Hargreaves, *J. Chem. Soc. A.* **1970**, 308.

151. J. L. Davidson, M. Green, J. Z. Nyathi, C. Scott, F. G. A. Stone, and A. J. Welch, *J. Chem. Soc., Chem. Commun.*, **1976**, 714.
152. H. G. Alt, *J. Organomet. Chem.*, **127**, 349 (1977).
153. S. J. Tremont and R. G. Berman, *J. Organomet. Chem.*, **140**, C12 (1977).
154. J. M. Huggins and R. G. Bergman, *J. Am. Chem. Soc.*, **101**, 4410 (1979).
155. J. F. Normant and M. Bourgain, *Tetrahedron Lett.*, **1971**, 2583.
156. J. F. Normant, G. Cahiez, C. Chuit, and J. Villieras, *J. Organomet. Chem.*, **77**, 269 (1974).
157. J. F. Normant, G. Caheiz, C. Chuit, and J. Villieras, *J. Organomet. Chem.*, **77**, 281 (1974).
158. J. F. Normant, C. Chuit, G. Cahiez, and J. Villieras, *Synthesis*, **1974**, 803.
159. P. R. McGuirk, A. Marfat, and P. Helquist, *Tetrahedron Lett.*, **1978**, 2973, and references therein.
160. A. Alexakis, J. Normant, and J. Villieras, *J. Organomet. Chem.*, **96**, 471 (1975).
161. D. E. Van Horn and E. Negishi, *J. Am. Chem. Soc.*, **100**, 2252 (1978).
162. (a) A. J. Birch and D. H. Williamson, *Org. React.*, **24**, 1 (1976); (b) B. R. James, *Adv. Organomet. Chem.*, **17**, 319 (1979).
163. J. Halpern, in *Organotransition-Metal Chemistry*, Y. Ishii and M. Tsutsui, Eds., Plenum Press, New York, 1975, p. 109.
164. A. S. C. Chan and J. Halpern, *J. Am. Chem. Soc.*, **102**, 838 (1980).
165. C. W. Bird, R. C. Cookson, J. Hudec, and R. O. Williams, *J. Chem. Soc.*, **1963**, 410.
166. A. Rosenthal, and D. Abson, *Can. J. Chem.* **42**, 1811 (1964).
167. J. A. Labinger, D. W. Hart, W. E. Seibert III, and J. Schwartz, *J. Am. Chem. Soc.*, **97**, 3851 (1975).
168. A. Nakamura and S. Otsuka, *J. Am. Chem. Soc.*, **95**, 7262 (1973).
169. J. K. Stille, F. Huang, and M. T. Regan, *J. Am. Chem. Soc.*, **96**, 1518 (1974).
170. N. A. Dunham and M. C. Baird, *J. Chem. Soc., Dalton Trans.*, **1975**, 774.
171. K. S. Y. Lau, Y. Becker, F. Huang, N. Baenziger, and J. K. Stille, *J. Am. Chem. Soc.*, **99**, 5664 (1977).
172. A. Stefani, G. Consiglio, C. Botteghi, and P. Pino, *J. Am. Chem. Soc.*, **99**, 1058 (1977).
173. F. Piacenti, S. Pucci, M. Bianchi, and P. Pino, *J. Am. Chem. Soc.*, **90**, 6847 (1968).
174. C. P. Casey and C. R. Cyr, *J. Am. Chem. Soc.*, **95**, 2240 (1973).
175. C. P. Casey, C. R. Cyr, and J. A. Grant, *Inorg. Chem.*, **13**, 910 (1974).
176. M. L. H. Green and P. L. I. Nagy, *J. Organomet. Chem.*, **1**, 58 (1963).
177. D. Slack and M. C. Baird, *J. Chem. Soc., Chem. Commun.*, **1974**, 701.
178. P. C. Wailes, H. Weigold and A. P. Bell, *J. Organomet. Chem.*, **27**, 373 (1971).
179. D. W. Hart, T. F. Blackburn, and J. Schwartz, *J. Am. Chem. Soc.*, **97**, 679 (1975).
180. H. C. Clark and C. S. Wong, *J. Organomet. Chem.*, **92**, C31 (1975).
181. H. C. Clark and K. E. Hine, *J. Organomet. Chem.*, **105**, C32 (1976).
182. N. C. Rice and J. D. Oliver, *J. Organomet. Chem.*, **145**, 121 (1978).
183. B. L. Booth and A. D. Lloyd, *J. Organomet. Chem.*, **35**, 195 (1972).
184. J. Schwartz, D. W. Hart, and J. L. Holden, *J. Am. Chem. Soc.*, **94**, 9269 (1972); D. W. Hart and J. Schwartz, *J. Organomet. Chem.*, **87**, C11 (1975).
185. K. P. Callahan and M. F. Hawthorne, *J. Am. Chem. Soc.*, **95**, 4574 (1973).
186. M. Dubeck and R. A. Schell, *Inorg. Chem.*, **3**, 1757 (1964).
187. B. L. Booth and R. G. Hargreaves, *J. Chem. Soc. A*, **1969**, 2766.
188. (a) A. Wojcicki, *Adv. Organomet. Chem.*, **11**, 87 (1973); (b) F. Calderazzo, *Angew. Chem., Int. Ed. Engl.*, **16**, 299 (1977).
189. J. K. Stille and L. F. Hines, *J. Am. Chem. Soc.*, **92**, 1798 (1970).
190. L. F. Hines and J. K. Stille, *J. Am. Chem. Soc.*, **94**, 485 (1972).

191. H. M. Walborsky and L. E. Allen, *J. Am. Chem. Soc.*, **93**, 5465 (1971).
192. K. Ohno and J. Tsuji, *J. Am. Chem. Soc.*, **90**, 99 (1968).
193. K. Noack and F. Calderazzo, *J. Organomet. Chem.*, **10**, 101 (1967).
194. (a) T. C. Flood and J. E. Jensen, *Abstr. IXth Int. Conf. Organomet. Chem.*, Dijon, France, 1979; (b) J. A. Statler and T. C. Flood, unpublished results.
195. M. Green and D. C. Wood, *J. Am. Chem. Soc.*, **88**, 4106 (1966).
196. F. A. Cotton and C. M. Lukehart, *J. Am. Chem. Soc.*, **93**, 2672 (1971).
197. D. A. Slack, D. L. Egglestone, and M. C. Baird, *J. Organomet. Chem.*, **146**, 71 (1978).
198. D. L. Egglestone, M. C. Baird, J. C. Lock, and G. Turner, *J. Chem. Soc., Dalton Trans.*, **1977**, 1576.
199. (a) K. N. Raymond, P. W. R. Corfield, and J. A. Ibers, *Inorg. Chem.*, 7, 1362 (1968); (b) J. K. Stalick, P. W. R. Corfield, and D. W. Meek, *Inorg. Chem.*, **12**, 1668 (1973).
200. R. W. Glyde and R. J. Mawby, *Inorg. Chim. Acta.*, **5**, 317 (1971).
201. G. K. Anderson and R. J. Cross, *J. Chem. Soc., Chem. Commun.*, **1978**, 819.
202. T. G. Attig and A. Wojcicki, *J. Organomet. Chem.*, **82**, 397 (1974).
203. P. Reich-Rohrwig and A. Wojcicki, *Inorg. Chem.*, **13**, 2457 (1974).
204. A. Davison and N. Martinez, *J. Organomet. Chem.*, **74**, C17 (1974).
205. H. Brunner and E. Schmidt, *J. Organomet. Chem.*, **36**, C18 (1972).
206. T. C. Flood, K. D. Campbell, and H. H. Downs, unpublished results.
207. T. C. Flood, F. J. DiSanti, and D. L. Miles, *Inorg. Chem.*, **15**, 1910 (1976), and references therein.
208. G. Fachinetti and C. Floriani, *J. Chem. Soc., Dalton Trans.*, **1977**, 2297, and references therein.
209. R. D. Adams and D. F. Chodosh, *Inorg. Chem.*, **17**, 41 (1978), and references therein.
210. P. Hofmann, *Angew. Chem., Int. Ed. Engl.*, **16**, 536 (1977).
211. T. G. Attig, *Inorg. Chem.*, **17**, 3097 (1978).
212. M. Green and R. P. Hughes, *J. Chem. Soc., Dalton Trans.*, **1976**, 1880.
213. A. D. Caunt, in *Catalysis*, Vol. 1, C. Kemball, Ed., Chemical Society, London, 1977, p. 234.
214. G. Natta, *Science*, **147**, 261 (1965).
215. (a) R. Zanella, G. Carturan, M. Graziani, and U. Belluco, *J. Organomet. Chem.*, **65**, 417 (1974); (b) G. Carturan, R. Zanella, M. Graziani, and U. Belluco, *ibid.*, **82**, 421 (1974).
216. S. Otsuka and K. Ataka, *J. Chem. Soc., Dalton Trans.*, **1976**, 327.
217. (a) A. Wojcicki, *Acc. Chem. Res.*, **4**, 344 (1971); (b) A. Wojcicki, *Adv. Organomet. Chem.*, **12**, 32 (1974).
218. R. G. Severson, T. W. Leung, and A. Wokcicki, *Inorg. Chem.*, **19**, 915 (1980).
219. T. G. Attig, R. G. Teller, S. -M. Wu, R. Bau, and A. Wojcicki, *J. Am. Chem. Soc.*, **101**, 619 (1979), and references therein.
220. A. Dormond, C. Moise, A. Dahchour, and J. Tirouflet, *J. Organomet. Chem.*, **168**, C53 (1979).
221. S. L. Miles, D. L. Miles, R. Bau, and T. C. Flood, *J. Am. Chem. Soc.*, **100**, 7278 (1978).
222. C. -K. Chou, D. L. Miles, R. Bau, and T. C. Flood, *J. Am. Chem. Soc.*, **100**, 7271 (1978).
223. G. M. Whitesides, J. San Filippo, Jr., C. P. Casey, and E. J. Panek, *J. Am. Chem. Soc.*, **89**, 5302 (1967).
224. A. Tamaki and J. K. Kochi, *J. Organomet. Chem.*, **51**, C39 (1973).
225. S. Komiya, T. A. Albright, R. Hoffmann, and J. K. Kochi, *J. Am. Chem. Soc.*, **98**, 7255 (1976).
226. M. P. Brown, R. J. Puddephatt, and C. E. E. Upton, *J. Chem. Soc., Dalton Trans.*, **1974**, 2457.

227. E. Vedejs and M. F. Salomon, *J. Am. Chem. Soc.*, **92**, 6965 (1970).
228. B. L. Booth and R. G. Hargreaves, *J. Organomet. Chem.*, **33**, 365 (1971).
229. M. D. Johnson, *Acc. Chem. Res.*, **11**, 57 (1978).
230. T. C. Flood, E. Rosenberg, and A. Sarhangi, *J. Am. Chem. Soc.*, **99**, 4334 (1977), and references therein.
231. T. C. Flood and F. J. DiSanti, *J. Chem. Soc., Chem. Commun.*, **1975**, 18.
232. D. A. Slack and M. C. Baird, *J. Am. Chem. Soc.*, **98**, 5539 (1976).
233. D. Dong, B. K. Hunter, and M. C. Baird, *J. Chem. Soc., Chem. Commun.*, **1978**, 11.
234. W. A. Nugent and J. K. Kocki, *J. Am. Chem. Soc.*, **98**, 5979 (1976).
235. F. R. Jensen, V. Madan, and D. H. Buchanan, *J. Am. Chem. Soc.*, **93**, 5283 (1971).
236. H. Shinozaki, H. Ogawa, and M. Tada, *Bull. Chem. Soc. Jap.*, **49**, 775 (1976).
237. D. Dodd and M. D. Johnson, *J. Chem. Soc., Chem. Commun.*, **1971**, 571.
238. S. N. Anderson, D. H. Ballard, J. Z. Chrzastowski, D. Dodd, and M. D. Johnson, *J. Chem. Soc., Chem. Commun.*, **1972**, 685.
239. J. Halpern, M. S. Chan, J. Hanson, T. S. Roche, and J. A. Topich, *J. Am. Chem. Soc.*, **97**, 1606 (1975).
240. R. H. Magnuson and J. Halpern, *J. Chem. Soc., Chem. Commun.*, **1978**, 44.
241. P. K. Wong and J. K. Stille, *J. Organomet. Chem.*, **70**, 121 (1974).
242. H. Shinozaki, M. Kubota, O. Yagi, and M. Tada, *Bull. Chem. Soc. Jap.*, **49**, 2280 (1976).
243. D. Dodd, M. D. Johnson, B. S. Meeks, D. M. Titchmarsh, K. N. V. Duong, and A. Gaudemer, *J. Chem. Soc., Perkin Trans.*, 2, **1976**, 1261.
244. (a) H. Westmijze, H. Kleijn, and P. Vermeer, *Tetrahedron Lett.*, **1977**, 2023; (b) A. B. Levy, P. Talley, and J. A. Dunford, *ibid.*, **1977**, 3545.
245. L. J. Dizikes and A. Wojcicki, *J. Am. Chem. Soc.*, **99**, 5295 (1977).
246. D. Dong, D. A. Slack, and M. C. Baird, *Inorg. Chem.*, **18**, 188 (1979).
247. H. L. Fritz, J. H. Espenson, D. A. Williams, and G. A. Molander, *J. Am. Chem. Soc.*, **96**, 2378 (1974).
248. T. C. Flood and D. L. Miles, *J. Organomet. Chem.*, **127**, 33 (1977).
249. H. Brunner and G. Wallner, *Chem. Ber.*, **109**, 1053 (1976).
250. D. L. Beach, M. Dattilo, and K. W. Barnett, *J. Organomet. Chem.*, **140**, 47 (1977).
251. J.-E. Bäckvall, *Tetrahedron Lett.*, **1975**, 2225.
252. P. M. Henry, M. Davies, G. Ferguson, S. Phillips, and R. Restivo, *J. Chem. Soc., Chem. Commun.*, **1974**, 112.
253. S.-K. Chung and A. I. Scott, *Tetrahedron Lett.*, **1975**, 49.
254. J.-E. Bäckvall, *Tetrahedron Lett.*, **1977**, 467.
255. R. A. Budnik and J. K. Kochi, *J. Organomet. Chem.*, **116**, C3 (1976).
256. I. J. Harvie and F. J. McQuillin, *J. Chem. Soc., Chem. Commun.*, **1977**, 241.
257. C. Fontaine, K. N. V. Duong, C. Merienne, A. Gaudemer, and C. Giannotti, *J. Organomet. Chem.*, **38**, 167 (1972).
258. C. Giannotti, C. Fontaine, and A. Gaudemer, *J. Organomet. Chem.*, **39**, 381 (1972).
259. F. R. Jensen and R. C. Kiskis, *J. Am. Chem. Soc.*, **97**, 5825 (1975).
260. H. Shinozaki and M. Tada, *Chem. Ind.*, **1975**, 178.
261. C. Bied-Charreton and A. Gaudemer, *J. Am. Chem. Soc.*, **98**, 3997 (1976).
262. E. Vedejs and M. F. Salomon, *J. Org. Chem.*, **37**, 2075 (1972).
263. D. Milstein and J. K. Stille, *J. Am. Chem. Soc.*, **101**, 4981, 4992 (1979).

Asymmetric Synthesis Mediated by Transition Metal Complexes

B. BOSNICH

Lash Miller Chemical Laboratories, University of Toronto,
Toronto, Ontario, Canada

MICHAEL D. FRYZUK

Chemistry Department, University of British Columbia,
Vancouver, British Columbia, Canada

I.	Introduction	119
II.	Preamble	120
III.	Asymmetric Hydrogenation	121
IV.	Asymmetric Cyclopropanation	126
V.	Asymmetric Codimerization	132
VI.	Asymmetric Allylic Alkylation	135
VII.	Asymmetric Epoxidation	138
VIII.	Asymmetric Hydroformylation	140
IX.	Asymmetric Hydroesterification	145
X.	Asymmetric Hydrosilylation	147
XI.	Asymmetric Synthesis Using Coordination Complexes	149
XII.	Conclusion	152
	References	152

I. INTRODUCTION

The state of sophistication of organic transformations mediated by transition metal complexes is revealed in the recent incorporation of this new methodology in the total synthesis of natural products (1,2). Organic transformations of this kind are particularly attractive because specific stoichiometric reagents can often be made catalytic.

A particular advantage of using metal complexes for organic reactions is that the reactions can be "tuned" by variation of the surrounding ligands. This tuning may take the form of altering the electronic distribution in the complex so that rate and specificity are maximized. A second form of tuning to which complexes are readily susceptible is the introduction of specific steric constraints on the reaction substrate. It is this second form of tuning that is the subject of this chapter. We shall show that the incorporation of chiral ligands (or the use of chiral complexes) can lead to asymmetric synthesis where the chiral discrimination rivals that of enzymes. Moreover, we hope to indicate that at present, chiral metal complexes are probably the most expeditious means of achieving high optical yields for a wide variety of substrates with the minimum of preliminary synthetic elaboration.

This chapter surveys what in our judgment have been the main achievements in the field. A notable lack of rationality and systematization pervades the approaches to asymmetric synthesis, and it seems useful to attempt to give a mechanistic basis for the reactions we describe. In so doing, we focus on the aspects that are crucial to successful asymmetric syntheses. The reader unfamiliar with the field, however, should regard some of the mechanistic proposals with circumspection because little is known about the intimate details of many of the reactions presented here.

II. PREAMBLE

It is useful to review a number of the terms that are used routinely throughout this chapter. An *asymmetric synthesis* has been defined as a "kinetically controlled asymmetric transformation" (4) in that normally it can be achieved only by the reaction proceeding through *diastereomeric* transition states or intermediates. There are a number of diastereomeric combinations of transition metal complex and substrate possible that can eventually lead to an asymmetric synthesis (5). This chapter is concerned only with the combination of a chirally modified transition metal complex and an achiral substrate molecule. This interaction between the chiral complex and the achiral substrate consists of two parts: the total interaction and that part of the total interaction which leads to the asymmetric synthesis, namely, the *diastereotopic interaction*. The latter interaction is defined as the free energy difference between the diastereomeric transition states of the asymmetric synthesis: $\Delta\Delta G^{\ddagger}$ (6). It follows that if the efficiency of the asymmetric synthesis is to be high, it is $\Delta\Delta G^{\ddagger}$, the diastereotopic interaction, that must be maximized, *not* necessarily the total interaction.

Although the terms "optical purity" (o.p.) and "enantiomeric excess" (e.e.) are often considered synonymous, the latter term is used here. It is simply defined as:

$$\% \text{ enantiomeric excess } (\% \text{ e.e.}) = |\%R - \%S|$$

III. ASYMMETRIC HYDROGENATION

The most spectacular asymmetric syntheses by transition metal complexes have been achieved in asymmetric homogeneous hydrogenation. Although much of this field has been recently reviewed (7–10), the continued input of new systems and refined mechanistic studies permits a more detailed discussion of the origins of the observed high asymmetric inductions with certain highly functionalized substrates.

The most efficient and most widely studied transition metal system for asymmetric hydrogenation is the rhodium-phosphine combination in which the phosphine ligands have been chirally modified in a variety of ingenious ways (8–10). The early work in asymmetric hydrogenation used chiral unidentate phosphines, which produced low and capricious enantiomeric excesses. In principle a unidentate phosphine is rotamerically labile (11) and can present an incoming prochiral substrate with a number of different diastereotopic interactions that might "wash out" any asymmetric induction. The importance of structural rigidity was substantiated by a number of reports (12–17) in which chiral chelating diphosphines coordinated to rhodium produced very high amounts of enantiomeric excess, in some cases greater than 95% e.e. (13,14). Though not having the same rotational freedom of unidentate systems, some of these bidentate ligands were potentially flexible (conformationally labile) and produced enantiomeric excess slightly lower than did the more rigid bidentate systems. This chapter deals only with a number of the more effective chiral chelating diphosphine rhodium systems from the points of view of mechanism of hydrogenation, type of substrate, and possible origins of the observed high asymmetric inductions.

Figure 1 illustrates some of the chiral chelating diphosphines that have been

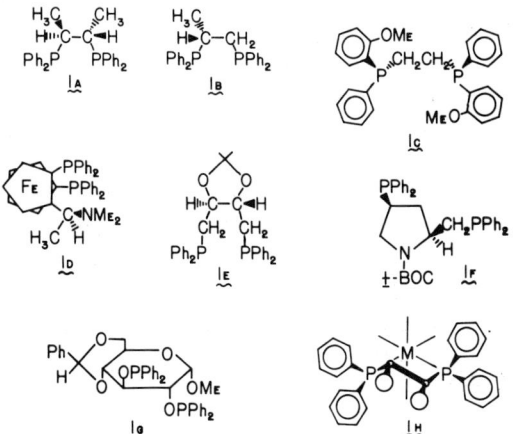

Fig. 1. Selected chiral diphosphines used in asymmetric synthesis. Also included is the rigid ring formation of **1a** (**1h**).

used in asymmetric synthesis. There are numerous other chelating diphosphines in the literature and all have been adequately reviewed (8–10); however the diphosphines in Figure 1 represent all the major innovations in the field of asymmetric hydrogenation to date with regard to ligand design.

Mechanistic studies by Halpern (18) on achiral cationic diphosphine rhodium systems have provided good evidence for the following catalytic cycle for the hydrogenation of olefinic substrates (Fig. 2). The catalyst precursors are usually

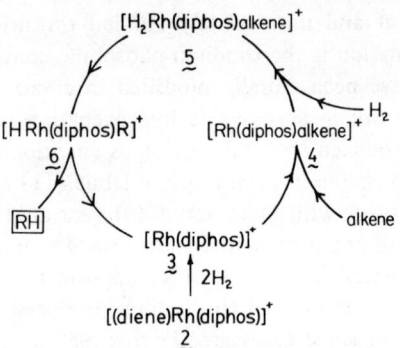

Fig. 2. Catalytic cycle for hydrogenation of olefins using *cationic* rhodium diphosphine complexes (coordinated solvent molecules not shown).

cationic diene complexes of the type $[(diene)Rh(diphos)]^+X^-$ [diene ≡ 1,5-cyclooctadiene (COD) or 2,5-norbornadiene (NBD); diphos ≡ chelating diphosphine; X ≡ noncoordinating anion]. In a suitable solvent S (S ≡ THF, alcohols, aqueous alcohols, benzene, etc.) the complexes **2** will absorb 2 equivalents of hydrogen at atmospheric pressure to give the new cationic species $[Rh(diphos)]^+$ (**3**) (coordinated solvent molecules not shown). Species **3** does *not* react further with hydrogen unless an excess of olefinic substrate is present, whereupon it acts as an efficient hydrogenation catalyst as shown in Figure 2.

The active catalyst **3** initially coordinates the substrate olefin to give **4** in a preequilibrium step; **4** then reacts with hydrogen, presumably to give the dihydride-olefin intermediate **5**. Hydride transfer to the coordinated olefin gives the alkyl hydride intermediate **6**, which undergoes reductive elimination to produce the hydrogenated product (RH) and regenerates the active catalyst **3**. The addition of hydrogen has been shown to be stereospecific, giving an overall cis-endo addition to the coordinated face of the olefin (14a). From Halpern's work (18) the following two equations summarize the catalytic cycle:

$$[Rh(diphos)]^+ + alkene \overset{K}{\rightleftharpoons} [Rh(diphos)alkene]^+ \qquad [1]$$
$$\mathbf{3}\mathbf{4}$$

$$[Rh(diphos)alkene]^+ + H_2 \overset{K}{\rightarrow} [Rh(diphos)]^+ + RH \qquad [2]$$
$$\mathbf{4}\mathbf{3}$$

Interestingly, the equilibrium between **3** and **4** (eq. [1]) was measured for a variety of olefins, and it was found that the substrate olefins that produced very high enantiomeric excesses upon asymmetric hydrogenation also gave very high binding constants ($K \geqslant 10^4$) (19). This must be considered coincidental, since the binding constant K from eq. [1] is a measure of the total interaction and need not necessarily give any indication of the diastereotopic interaction.

The general types of olefin that have been found to produce high amounts of enantiomeric excess upon asymmetric hydrogenation are shown in Figure 3. All

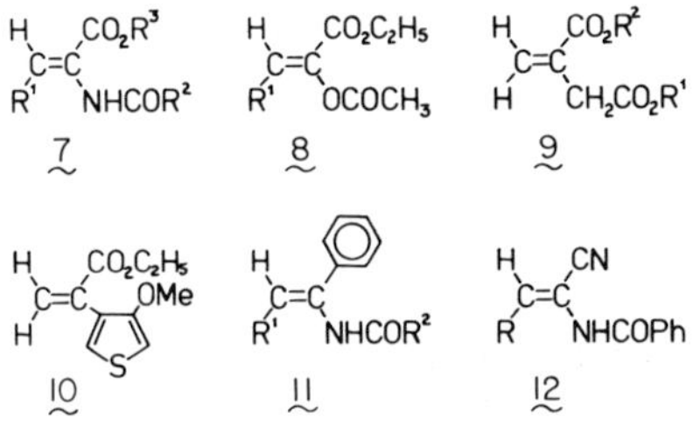

Fig. 3. General types of functionalized substrate olefin that produce high amounts of enantiomeric excess upon hydrogenation: R_1, R_2, and $R_3 \equiv$ alkyl, aryl, or H.

the substrates in Figure 3 are capable of generating a secondary interaction (olefin coordination being the primary interaction) with the "ligand-deficient" (18) [Rh(diphos)]$^+$ moiety. This is illustrated below (**13**) for the substrate type α-acylaminoacrylic acid (**7**).

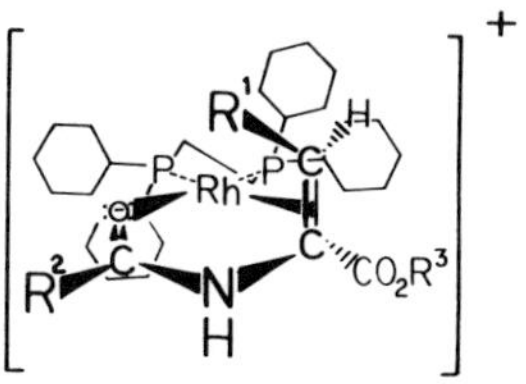

This bidentate interaction of the substrate olefin with the rhodium in the binding step would serve to reduce the number of rotamers that would otherwise exist for unidentate rhodium-olefin interactions. Olefins that cannot generate a secondary interaction with the rhodium invariably give low enantiomeric excess. Some representative results are shown in Table 1.

It is important that we clarify the possible origins of the observed high enantiomeric excess in these systems. The rigidity of the chelate ring formed by the

Table 1 Enantiomeric Excess[a] for a Variety of Substrates upon Asymmetric Hydrogenation Using the Indicated Chiral Chelating Diphosphine (Fig. 1) Coordinated to Rhodium

Substrate	Chiral Diphosphine						
	1a[b]	1b[c]	1c[d]	1d[e]	1e[f]	1f[g]	1g[h]
CH₂=C(CO₂H)(NHCOCH₃)	91 (R)	90 (S)	93.5 (S)	93 (S)	73 (R)	–	80 (S)
PhCH=C(CO₂H)(NHCOPh)	99 (R)	93 (S)	96 (S)	93 (S)	64 (R)	–	–
PhCH=C(CO₂H)(NHCOCH₃)	89 (R)	91 (S)	95 (S)	–	81 (R)	91 (R)	75 (S)
(AcO, OMe-C₆H₃)CH=C(CO₂H)(NHCOCH₃)	83 (R)	89 (S)	94 (S)	86 (S)	84 (R)	86 (R)	–
(CH₃)₂C=C(CO₂H)(NHCOCH₃)	100 (R)	87 (S)	–	–	–	–	–
CH₂=C(CO₂Et)(OCOCH₃)	84 (R)	81 (S)	–	–	–	–	–
CH₂=C(CO₂H)(CH₂CO₂H)	–	–	–	–	–	92 (S)[i]	–
(3-OMe-thien-2-yl)CH=C(CO₂Et)	–	–	–	–	88[j]	–	–
PhCH=C(CH₃)(NHCOCH₃)	–	–	–	–	92 (R)[k]	–	–
PhCH=CH(CO₂H)	–	–	–	–	64 (S)	–	–
PhCH=C(CO₂H)(CH₃)	–	–	1 (R)	–	25 (S)[l]	–	–

[a] Values given are highest published; absolute configurations are shown in parentheses.
[b] Reference 14a.

Table footnotes (*continued*)

[c] Reference 14b.
[d] Reference 13b.
[e] Reference 16a.
[f] Reference 12b.
[g] Reference 15.
[h] Reference 17.
[i] I. Ojima and T. Kogure, *Chem. Lett.*, **1978**, 567.
[j] A. P. Stoll and R. Süess, *Helv. Chim. Acta*, **57**, 2487 (1974).
[k] Reference 7b, p. 241, Table 4.
[l] Reference 9, p. 343, Table 3.

chiral diphosphines seems to be an important feature, since the chiral diphosphines that can form rigid five-membered chelate rings (**1a, 1b, 1c,** and **1d**) generally display higher enantiomeric excess than the chiral diphosphines that have more flexible seven- or higher membered chelate rings (**1e, 1f,** and **1g**). It is reasonable to trace this effect to the ubiquitous phenyl substituents of the phosphorus donors that are disposed in a rigid, chiral array by the rigid chelate ring (**1h**). This rigid, chiral array of phenyl groups interacts with the incoming prochiral substrate olefin to generate the necessary diastereotopic interaction such that one of the enantiotopic faces of the olefin becomes preferentially coordinated. Under hydrogenation conditions, the preferentially coordinated face of the olefin is then reduced to give the saturated chiral product in an enantiomeric excess that is a direct measurement of the degree of discrimination of the prochiral olefinic faces in the binding step. However it should be noted that the requirement for an asymmetric synthesis, namely, that the reaction proceed through diastereomeric transition states, can be fulfilled at the alkyl hydride stage (assuming that the rhodium-carbon bond forms at the more substituted carbon of the double bond) of the catalytic cycle (species **6** in Fig. 2). Thus the formation of the diastereomeric alkyl hydride intermediates and/or the reductive elimination could be the step or steps in which asymmetric induction occurs. This model cannot be easily ruled out. However ^{31}P NMR studies (20,21) on the interaction of some of these chelating substrate olefins with various chiral rhodium diphosphine complexes does support the former model in that diastereomers are detected in solution in a ratio very close to that observed for the enantiomeric excess of the resultant hydrogenated substrate. Another possibility is that the induction observed is a result of all the possible diastereomeric species present in the catalytic cycle (19).

A related area that is attracting more attention recently is in the asymmetric hydrogenation of ketones. Much of this work has been reviewed (10); however some recent research supports a number of aspects related to the hydrogenation of highly functionalized olefins.

The mechanism of hydrogenation of ketones is not understood, and many crucial questions on the nature of the intermediates remain to be answered. As in asymmetric hydrogenation, there seem to be two types of substrate that have been investigated: simple prochiral ketones such as acetophenone or propiophenone, where enantiomeric excess is generally low (22), and functionalized ketones or carbonyl containing substrates, where enantiomeric excess is high (75–95% e.e.) (23,29). One might speculate that the types of substrate that produce high enantiomeric excess (**14,15**) do so because they generate a secondary interaction with the rhodium (coordination through the carbonyl lone pair is

considered the primary interaction) to produce a chelating effect much like that previously described for some olefinic substrates (**16** and **17**).

It is clear that more mechanistic work must be accomplished in this area before further speculations on the origins of the asymmetric induction are undertaken.

IV. ASYMMETRIC CYCLOPROPANATION

That transition metal complexes can stabilize "carbenoid"-type ligands has been amply demonstrated (25). Relevant here is work reported by a number of groups on reactions that ultimately generate carbon-carbon bonds via asymmetric cyclopropanation, wherein there is postulated a transition metal stabilized carbene intermediate (26). There are two general methods for producing chiral cyclopropanes using transition metal complexes: a stoichiometric reaction utilizing chiral pseudotetrahedral iron precursors (27), and a catalytic sequence using chiral copper (28) or cobalt (29) complexes.

The stoichiometric reaction is not synthetically useful but is worth discussing to illustrate some important aspects of asymmetric synthesis mediated by transition metal complexes. The reaction of the pure diastereomeric complexes (+)- and (−)-[η^5-(C_5H_5)Fe(CO)PPh$_3$(CH$_2$OMen)] (OMen ≡ (−)-menthoxy) (**12**) with HBF$_4$ in the presence of *trans*-α-methylstyrene gives the chiral enatiomeric cyclopropane derivatives in 26 and 38.5% e.e. but in low chemical yield

(27a). The postulated intermediate (27b,30–32) is the cationic carbene complex **19**. Since this cationic carbene complex is an 18-electron species, prior coordination of a prochiral olefin is unlikely; therefore the mechanism of the asymmetric induction, that is, how the chiral iron complex **19** discriminates between the enantiotopic faces of the olefin, remains unclear. A related study (27b) concerns the thermal decomposition of the enantiomerically pure $(+)$-$[\eta^5=(C_5H_5)$-Fe(CO)PPh$_3$(CH$_2$Br)] (**20**) in the presence of *trans*-α-methylstyrene to give the analogous cyclopropane derivative in 9% e.e.

This result is anomalous in that the *identical* absolute configuration of **20** compared to $(+)$-**18** gives chiral cyclopropane derivatives of *opposite* absolute configuration. It is difficult to reconcile these results, since both routes should produce the same chiral iron intermediate **19**.

The more synthetically useful systems for asymmetric cyclopropanation reactions are those that utilize chiral copper or cobalt complexes as *catalysts*. These will be discussed in some detail.

Initial attempts (33,34) at achieving catalytic asymmetric cyclopropanation resulted in very low amounts of asymmetric induction, the highest being 8% e.e.

(34). More recently, two groups have achieved extremely high enantiomeric excess, up to 90% e.e., in the synthesis of chiral cyclopropane derivatives (28,29).

The general reaction entails the decomposition of a diazoalkane with a chiral transition metal complex (catalytic amount) in the presence of a suitable olefin to give the cyclopropane product, generally in excellent chemical yields (eq. [3]).

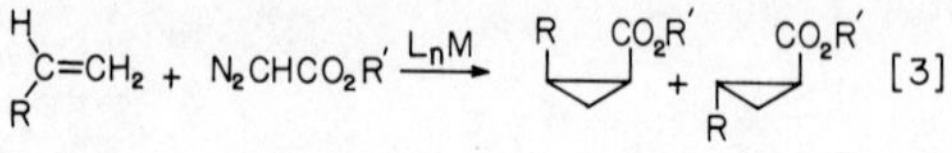

$$\overset{H}{\underset{R}{C}}=CH_2 + N_2CHCO_2R' \xrightarrow{L_nM} \underset{R}{\bigtriangleup}\overset{CO_2R'}{} + \bigtriangleup\overset{CO_2R'}{} \quad [3]$$

One study (28) systematically varied both the chiral ligands around a copper complex and the ester group of the diazoacetate in an attempt to prepare chrysanthemic acid via the reaction of 2,5-dimethyl-2,4-hexadiene as the prochiral substrate (eq. [4]).

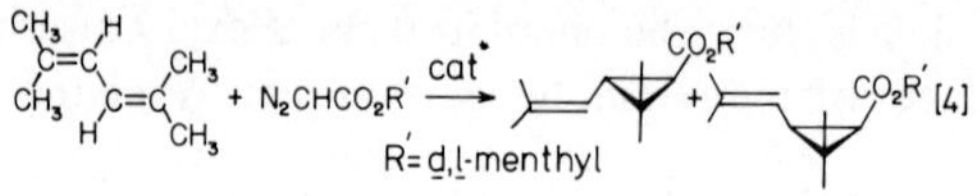

$$\underset{CH_3}{\overset{CH_3}{C}}=\underset{H}{\overset{H}{C}}\underset{CH_3}{\overset{CH_3}{C}}=\underset{CH_3}{\overset{CH_3}{C}} + N_2CHCO_2R' \xrightarrow[R=\underline{d,l}\text{-menthyl}]{cat^*} \quad [4]$$

The catalysts (cat*) were chiral copper complexes of the general formula:

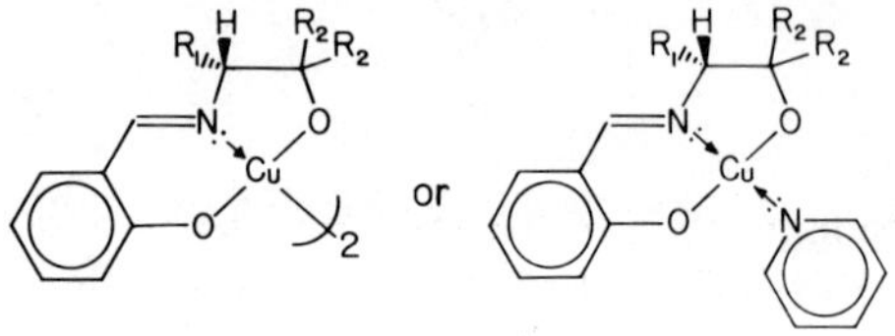

The best enantiomeric excess (90% e.e.) was achieved with $R_1 = CH_3$ and $R_2 =$ 2-octyloxy-5-*tert*-butylphenyl. Although this study is notable in that chrysanthemic acid was obtained in very high enantiomeric excess, the generality of this system has not been reported.

A more interesting catalytic system (29) utilizes bis[(−)-camphorquinone-α-dioximato] cobalt(II) hydrate (21) as the active catalyst for efficient cycloprop-

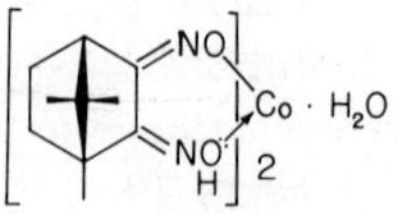

α-21

anation of a number of conjugated terminal olefins with alkyldiazoacetates. A few pertinent examples are shown in Table 2.

Table 2 Asymmetric Cyclopropanation Reactions [a]

Substrate	Diazoester	Chemical Yield (%) [b]	Product	Configuration	% e.e. ($[\alpha]_D$)
PhCH=CH$_2$	N$_2$CHCO$_2$Et	92	Ph-△-CO$_2$Et	(1S, 2R)	67
			Ph-△-CO$_2$Et	(1S, 2S)	75
PhCH=CH$_2$	N$_2$CHCO$_2$neopent	87	Ph-△-CO$_2$neopent	(1S, 2R)	81
			Ph-△-CO$_2$neopent	(1S, 2S)	88
Ph$_2$C=CH$_2$	N$_2$CHCO$_2$Et	95	Ph$_2$(Ph)-△-CO$_2$Et	(1S)	70
CH$_2$=CH–CH=CH$_2$	N$_2$CHCO$_2$Et	87	CH$_2$=CH-△-CO$_2$Et	—[c]	(+120°)
CH$_2$=CHCO$_2$Me	N$_2$CHCO$_2$Me	11	CO$_2$Me-△-CO$_2$Me	(1S, 2S)	33

[a] Reaction was performed in the neat substrate using α-**21** as catalyst (29).
[b] Based on diazoester.
[c] Absolute configuration unknown.

Thus styrene is converted to an approximately 1:1 mixture of neopentyl *cis*- and *trans*-2-phenylcyclopropanecarboxylate in up to 88% e.e. Diolefins such as butadiene or isoprene give products in excellent chemical yields and with very high optical rotations; however the absolute configuration of the products and their maximum rotations are unknown. Terminal olefins such as acrylic acid esters and acrylonitrile give low chemical yields and correspondingly low amounts of enantiomeric excess. One should note that the reaction works equally well for nonprochiral olefins such as 1,1-diphenylethylene, for which a 70% e.e. was observed.

The choice of the diazoalkane is important; diazomethane works only with terminal olefins and is therefore unsuitable for asymmetric synthesis. Bulky diazo compounds such as Ph_2CN_2 and 9-diazofluorene are inert. The reaction of $N_2C(CN)_2$ with styrene catalyzed by α-21 gives 2-phenyl-1,1-dicyanocyclopropane in low chemical yield and in only 4.6% e.e.

$$PhCH=CH_2 + N_2C(CN)_2 \xrightarrow[20\%]{\alpha\text{-}21} \underset{4.6\%\text{ e.e.}}{\overset{Ph\quad CN}{\triangle_{CN}}}$$

With diazoacetophenone and styrene, a mixture of *cis*- and *trans*-1-benzoyl-2-phenylcyclopropane is produced with 20% e.e. for the *trans* isomer, and an unknown enantiomeric excess for the cis isomer.

$$PhCH=CH_2 + N_2CHCOPh \xrightarrow[44\%]{\alpha\text{-}21} \underset{\underset{20\%\text{ e.e.}}{Ph}}{\overset{COPh}{\triangle}} + \underset{(-18.2°)}{\overset{Ph\quad COPh}{\triangle}}$$

A very curious result obtains with *cis*-[$^2H_2$]-styrene. It was found that both the *cis*- and *trans*-cyclopropane products had undergone deuterium scrambling, implying that the olefin undergoes geometrical isomerization during cyclopropanation.

$$\underset{D\quad D}{\overset{Ph\quad H}{C=C}} + N_2CHCO_2Et \xrightarrow[H_2O]{\alpha\text{-}21\quad NaOH} \underset{\text{trans-}d_2}{\begin{array}{c}\triangle + \triangle\\ \triangle + \triangle\end{array}} \quad \underset{\text{cis-}d_2}{}$$

The authors postulate (29b) a very reasonable mechanism involving initial formation of a cobalt carbene intermediate, which in the presence of a suitable terminal olefin, generates a cobaltacyclobutane complex that then collapses to the cyclopropane derivative and regenerates the active catalyst. This is outlined in the catalytic cycle shown in Figure 4. Although the original study (29b)

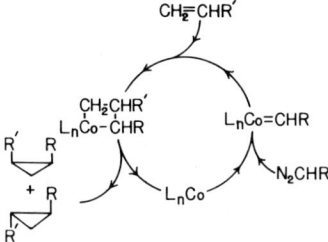

Fig. 4. Simplified catalytic cycle for the cyclopropanation reaction using cobalt complexes.

contains additional data on reaction rates, effects of added ligand, and oxidation state of the catalyst, this simplified catalytic cycle does outline the essential features of the cyclopropanation reaction.

The intriguing aspect of this system is the mechanism by which the chiral catalyst induces the observed high asymmetric induction to the cyclopropane products. That the chiral catalyst does *not* necessarily discriminate the enantiotopic faces of prochiral substrates is evident from the chiral product obtained from 1,1-diphenylethylene, which has equivalent faces but gives an enantiomeric excess comparable to styrene. In addition, olefins that do have enantiotopic faces, such as styrene, give approximately equal amounts of the cis and trans isomers, clearly indicating that both enantiotopic faces coordinate to the chiral catalyst during the cycle with little or no preference.

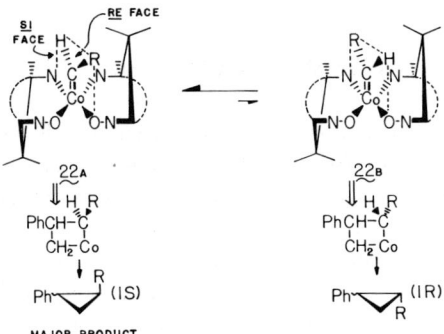

Fig. 5. Proposed chiral environment around the cobalt carbene intermediate. (*Note*: the geometrical isomerization of the ligands about the cobalt atom is assumed but has not been established.)

The postulated intermediate is shown in Figure 5. This cobalt carbene species containing an "sp^2-like" carbon atom directly bonded to the cobalt has diastereotopic faces; thus attack at the *si*-face of **22a** by both faces of the olefin results in the observed excess of the *S*-enantiomer at the carbenoid carbon. Attack at the *re*-face of **22a** is blocked by the steric bulk of the chiral camphorquinone-α-dioxime ligand. Why **22a** is formed preferentially to **22b** is not clear; however the authors (29b) invoke a hydrogen bonding argument between the carboxylic ester of the carbene and the oxygen donor of the dioxime ligand that would serve to stabilize **22a** with respect to **22b**.

The cobaltacyclobutane intermediate can be used to explain the deuterium scrambling observed with *cis*-[$^2H_2$]-styrene if one assumes that the lifetime of the cobaltacyclobutane is sufficiently long to allow the realization of the equilibrium shown in Figure 6. Simple olefin rotation of **24** to give **25** generates **26**,

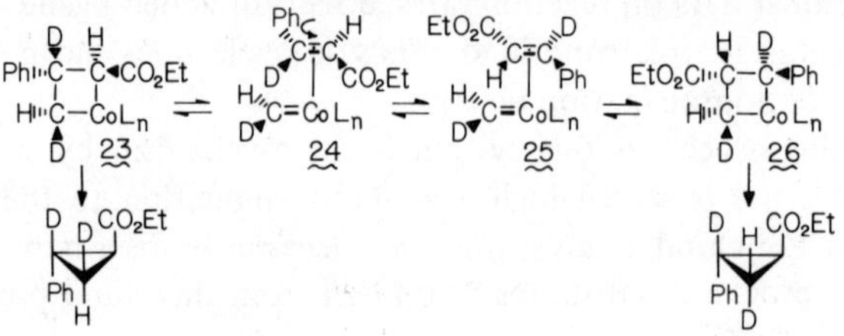

Fig. 6. Possible mechanism for deuterium scrambling.

which collapses to give the scrambled *trans*-[$^2H_2$] product *without* loss of chirality at the carbenoid carbon.

The details of the reactive intermediates are by no means conclusive in this system; however the evidence does suggest that a cobalt carbene moiety having diastereotopic faces is the major source of the asymmetric induction.

V. ASYMMETRIC CODIMERIZATION

The organometallic chemistry of nickel is rich, as evidenced by its numerous catalytic reactions and also its unusual stoichiometric transformations (35); however, relatively little work has been done on chiral modifications of nickel systems for the purpose of asymmetric synthesis. The only extensive study in this area has been performed by the Mülheim group, and this work is discussed here. The material has been reviewed elsewhere in detail (36,37).

The general reaction types of interest in asymmetric codimerization are shown in Figure 7. The catalyst is prepared by the reaction of [Ni(η^3-allyl)Cl]$_2$ with an alkylchloroaluminum compound [$(C_2H_5)_3Al_2Cl_3$] in the presence of an optically active phosphine. The Mülheim group has experimented with a

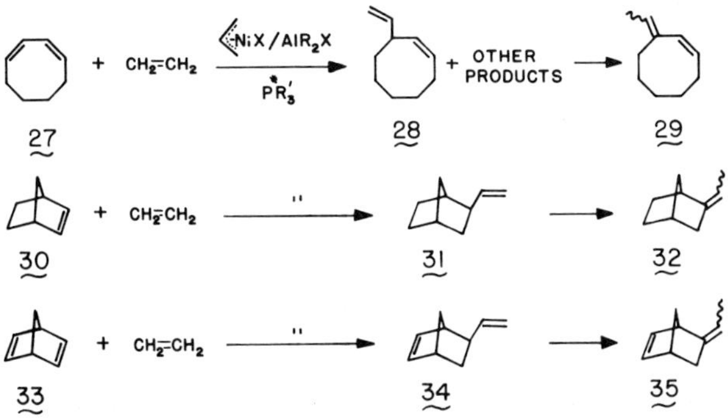

Fig. 7. Asymmetric codimerization reactions of ethylene and various olefinic substrates.

variety of chiral monophosphine ligands, and it appears that the best results are obtained with (−)-isopropyldimenthylphosphine (36) or (−)-methyldimenthylphosphine (37).

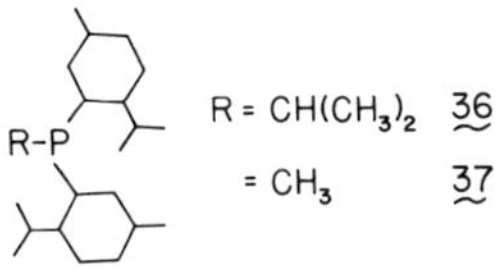

Thus (−)-(3S)-vinyl-1-cyclooctene (28) is produced in 70% e.e. using excess 36 (Ni : P = 3.8) at 0°C and in 53% e.e. using 37 at −75°C. In the codimerization of norbornene and ethylene (30 → 31) the highest enantiomeric excess achieved was 80.6% at −97°C using 36. A temperature variation study clearly showed that the enantiomeric excess of 31 increased linearly with decreased reaction temperature. In the analogous codimerization of norbornadiene and ethylene (33 → 34), an enantiomeric excess of 77.5% at −75°C was achieved, again using 36. The last two codimerization reactions, 30 → 31 and 33 → 34, are notable in that high chemical yields are obtained, whereas in the codimerization of 1,3-cyclooctadiene and ethylene other oligomers can predominate, depending on the nature of the phosphine being used. Therefore we focus our mechanistic arguments on the codimerization of norbornene and norbornadiene with ethylene; the codimerization of 1,3-cyclooctadiene and ethylene is not considered further.

A mechanism has been proposed for this reaction having as the active catalyst a nickel(II) hydride species as shown in Figure 8. The nickel hydride, 38, inserts into the cyclic olefin and is trapped by ethylene to give 39; this species undergoes olefin insertion to give 40 after being trapped by the cyclic olefin; a simple β-elimination regenerates 38 and produces the codimerization product 41. The transformation involving the olefin insertion into the nickel-alkyl bond, 39 → 40,

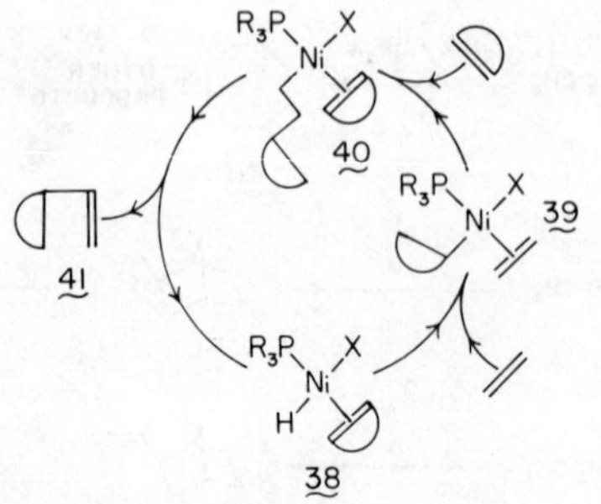

Fig. 8. Proposed mechanism for codimerization involving a nickel hydride species: $PR_3 \equiv$ chiral monophosphine; $X \equiv$ alkylchloroaluminum derivative; ⬡ ≡ cyclic olefin, norbornene, or norbornadiene.

did not have precedence (38,39a) but has recently been demonstrated for cobalt alkyls (39b). Based on some recent results by Grubbs and co-workers (40,41), one might speculate that an alternate mechanism is occurring, involving a Ni(0) ⇌ Ni(II) equilibrium (Fig. 9). The essential feature of this alternate mechanism is

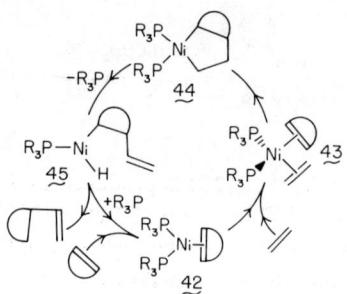

Fig. 9. Alternative mechanism for codimerization reaction.

that the product forms via transformation of a nickel(0)bis(olefin) species, **43**, into a nickelocyclopentane complex, **44**, without incurring olefin insertion into a metal-carbon bond. When coordinatively unsaturated, species such as **44** can undergo β-elimination to give **45** followed by facile reductive elimination (40) to give the product and regenerate **42**. The data collected on these systems (36,37, 42) do not clearly support or exclude either mechanism.

An analysis of the stereoisomeric relationships of the substrates (norbornene and norbornadiene) and the resulting codimerized products leads to the following conclusions:

1. The olefinic faces of norbornene (and for a particular double bond of norbornadiene) are diastereotopic.
2. The olefinic carbon atoms of norbornene (and for a particular double bond of norbornadiene) are enantiotopic.
3. The exo product is formed exclusively.

Therefore the origin of the asymmetry in the product is *not* due to the discrimination of the diastereotopic olefinic faces of norbornene (or norbornadiene) because the exo isomer is the exclusive product; rather, asymmetry arises according to which enantiotopic carbon atom forms the nickel-alkyl bond. Both mechanisms are consistent with this observation. In the former mechanism (Fig. 8) the nickel(olefin)alkyl species (**39**) are diastereomeric, whereas in the alternate mechanism (Fig. 9) the nickelocyclopentane derivatives (**44**) are diastereomeric.

A very interesting feature about the mechanism as illustrated in Figure 9 is the accessibility of a preequilibrium step between the nickelbis(olefin) species (**43**) and the nickelocyclopentane complex (**44**) (41). With norbornene and ethylene, the equilibrium in Figure 10 can be visualized. This equilibrium allows

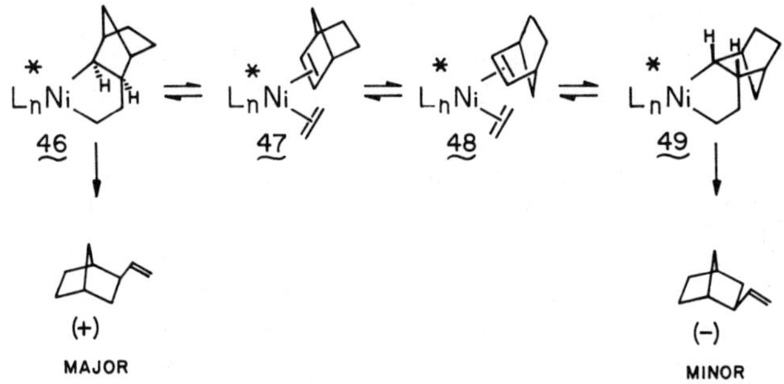

Fig. 10. Possible interconversion of diastereomers in codimerization reactions: $L_n^* \equiv$ ancillary ligands; includes at least one chiral phosphine.

the interconversion of the diastereomers **46** and **49** and implies that at lower temperatures, where high enantiomeric excess is observed, the equilibrium shifts to the left so that **46** is the major intermediate.

VI. ASYMMETRIC ALLYLIC ALKYLATION

A new and interesting methodology for the formation of carbon-carbon bonds using allylpalladium complexes has been recently reported by Trost (43). He has investigated two modifications: a stoichiometric reaction between preformed cationic allylpalladium complexes and the appropriate nucleophiles (M^+Nu^-), and a catalytic reaction in which an allylic acetate is alkylated by an appropriate nucleophile in the presence of a catalytic amount of a palladium(0) phosphine complex (eqs. [5] and [6]).

Trost has examined the regiochemistry, chemoselectivity, and stereochemistry of both reactions and has shown that once the allylpalladium complex is formed, it is activated to attack by certain soft nucleophiles ($pK_a \sim 14-16$) such that attack occurs exclusively at the exo face, that is, the face distal to the palladium. In addition, the attack usually occurs at the least substituted end of the allyl

$$R\underset{2}{\overset{}{\diagdown}}Pd\underset{PR_3}{\overset{Cl}{\diagup}} \xrightarrow{M^+Nu^-} R\diagdown\diagup^{Nu} + Pd(PR_3)_n \quad [5]$$

$$R\underset{Pd(PR_3)_n}{\overset{OAc}{\diagdown}} \xrightarrow{M^+Nu^-} R\diagdown\diagup^{Nu} + MOAc \quad [6]$$

ligand. In the catalytic reaction, the stereochemistry of the oxidative addition of the allylic acetate to give the allylpalladium species is purported to proceed with complete inversion at the carbon bearing the acetate group (44); however this may be not be the general rule (see below).

An extension of this work is in the use of chiral phosphines to induce asymmetry in these transformations. We examine the catalytic modification in some detail.

Two different types of substrate have been examined in the catalytic asymmetric alkylation reaction. The first was cis-3-acetoxy-5-carbomethoxycyclohexene (50), which was treated with the sodium salt of methylphenysulfonylacetate in the presence of a catalytic amount of $Pd(PPh_3)_4$ and (+)-diop (see Sect. III) to give the alkylation product 51 after routine desulfonylation (eq. [7]). The (R,R) diastereomer of 51 was produced with 24% e.e. The oxidative

[Reaction scheme showing compound 50 + NaCH(CO$_2$CH$_3$)(SO$_2$Ph) with Pd(PPh$_3$)$_4$, (+)-DIOP giving intermediate and then compound 51]   [7]

addition of racemic 50 to palladium(0) gives 52, the intermediate allylpalladium complex. When the phosphines are achiral, the allylic carbons C_3 and C_5 are enantiotopic, whereas if the phosphines are chiral, C_3 and C_5 become diastereotopic. The asymmetric synthesis that results is therefore due to preferential attack by the nucleophile at either C_3 or C_5. It is not clear how the coordinated phosphine can direct exo attack preferentially to C_3 or C_5; however one might speculate that the diastereomeric olefin complexes that are formed after exo attack are substantially different in energy, so that formation of one diastereomer is preferred. That the asymmetric induction is due to preferential formation of either diastereomic olefin complexes 53 or 54 does have precedent in a study on the regioselectivity of some related achiral systems (46).

The second type of substrate studied was 1-methyl-1-acetoxymethylcyclo-

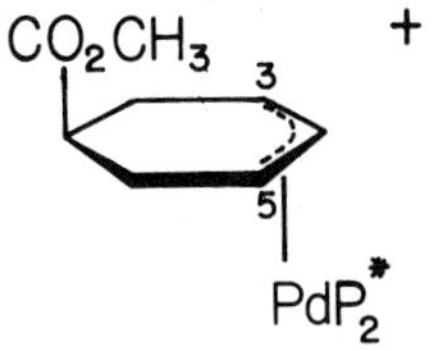

structures 52, 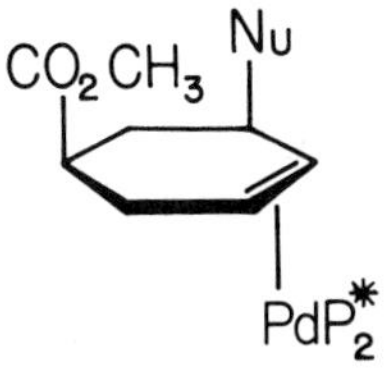 53, 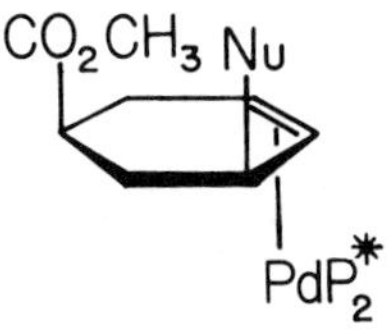 54

pentene (**55**). Similarly as described above, this racemic allylic acetate was allowed to react with either the sodium salt of dimethylmalonate or the sodium salt of methylphenysulfonylacetate in the presence of catalytic amounts of various chiral palladium(0) phosphine complexes. Thus **55** was catalytically alkylated to give **56**, which was then transformed to **57** using standard procedures.

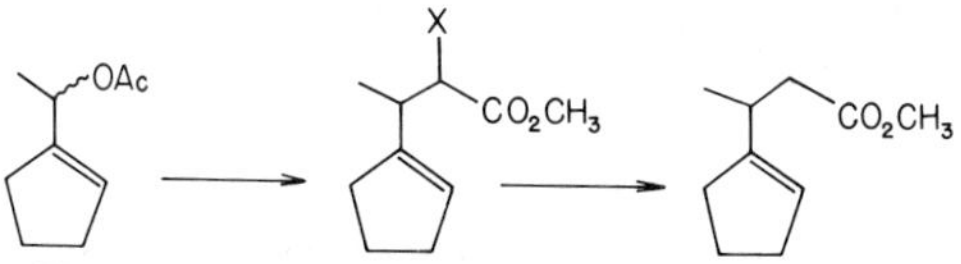

structures 55, 56, 57

The enantiomeric excess of **57** varied from 16 to 46% as a function of the chiral phosphine used and the bulkiness of the incoming nucleophile. The best result was obtained with $X = SO_2Ph$ and the chiral diphosphine (+)-diop.

The oxidative addition of **55** to palladium(0) presents a situation different from that for **50**. If we assume that the oxidative addition of racemic **55** is completely stereospecific, the complexes **58** and **59** would be formed in equal amounts regardless of whether the phosphines are chiral. Exo attack by the

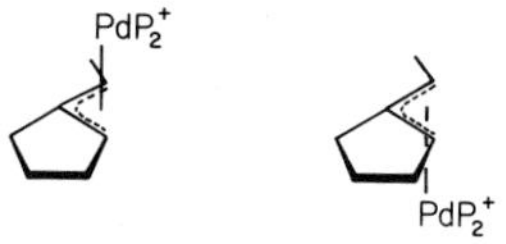

structures 58, 59

nucleophile at the exocyclic carbon atoms of **58** and **59** will then produce racemic **56**, therefore 0% enantiomeric excess. Since the enantiomeric excess is, in fact, nonzero, there must be a mechanism whereby an excess of **58** or **59** is generated, since it has been previously shown that the exo attack by the nucleophile is completely stereospecific. The well known syn-anti isomerization does not lead to racemization at the exocyclic carbon and can be disregarded as far as asymmetric synthesis is concerned (Fig. 11). One could invoke a nonstereospecific

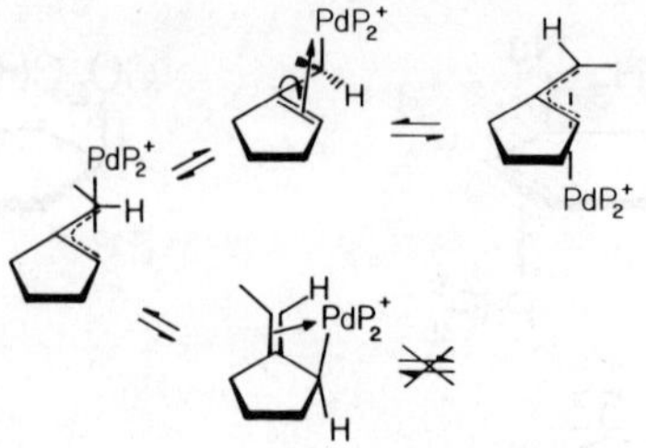

Fig. 11. Syn-anti isomerization sequence. The chirality at the exocyclic carbon atom remains constant throughout.

oxidative addition (47) that might presumably produce one of the diastereomers, **58** or **59**, in excess as a function of the chiral phosphine being used. The possibility of an acetate-induced reversible oxidative addition sequence has some support: Trost (48) has observed cases of rapid preliminary scrambling of the allylic acetates occurring in the presence of palladium(0) species. This scrambling mechanism is not yet understood.

VII. ASYMMETRIC EPOXIDATION

A potentially useful synthetic transformation is the catalytic epoxidation of allylic alcohols by high oxidation state molybdenum- and vanadium-oxo complexes in the presence of alkylhydroperoxides (49). Modification of this reaction for asymmetric synthesis has been recently reported by two groups (50,51). The first report (50) used the chiral molybdenum complex acetylacetonato[(−)-N-alkylephedrinatodioxomolybdenum(VI)] , (**60,61**).

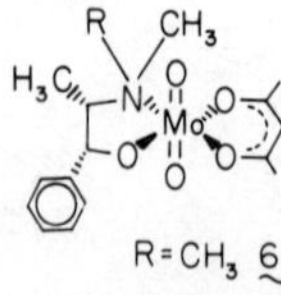

R = CH_3 **60**
R = C_2H_3 **61**

The general reaction illustrated in Table 3 gives the chiral oxiranes in 30 to 60% chemical yields with enantiomeric excess of up to 33% (i.e., **62a** → **63a** using **61** and **62c** → **63c** using **61**). The authors also report that enantiomeric excess in-

Table 3 Asymmetric Epoxidations by Chiral Molybdenum Complexes (50).

$$R_1R_2C=CR(CH_2OH)H \xrightarrow[\text{CH}_3\text{CH}_3\text{C}_6\text{H}_4\text{OOH, 40-45°C}]{27A \text{ OR } B} R_1R_2C(O)C(CH_2OH)H$$

62 → **63**

Compound	R_1	R_2	Catalyst	% e.e.	Absolute configuration
63a		CH_3	60	29	2R
			61	33	2R
63b	$(CH_3)_2C=CH-(CH_2)_2-$	CH_3	60	18	2R,3R
			61	24	2R,3R
63c	CH_3	$(CH_3)_2C=CH-(CH_2)_2-$	60	18	2R,3S
			61	33	2R,3S

creased with decreased reaction temperatures; however the reaction times at lower temperatures become inconveniently long.

The second report (51) used a chiral vanadium complex as the catalyst prepared *in situ* from VO(acac)$_2$ and a chiral hydroxamic acid (**64**).

$$\text{R-N(OH)-C(=O)-[pyrrolidine]} + VO(ACAC)_2$$

R = CH$_3$ **64A**
R = Ph   **64B**
R = 2,6-(CH$_3$)$_2$Ph **64C**

The best enantiomeric excess reported with this system was 50% for the epoxidation of **65** by *tert*-butylhydroperoxide using **64b** as the catalyst (Table 4). A

Table 4 Asymmetric Epoxidations by a Chiral Vanadium Complex (51)

$$R_1R_2C=CR_3(CH_2OH) \xrightarrow[\text{(CH}_3)_3\text{COOH}]{64B} R_1R_2C(O)CR_3(CH_2OH)$$

Compound	R_1	R_2	R_3	% e.e
65	H	Ph	Ph	50
66	CH_3	$-CH_2CH_2CH_2-$		44
67	CH_3	$-(CH_2)_2CH=C(CH_3)_2$	H	30

44% enantiomeric excess for **66** and a 30% enantiomeric excess for **67** were observed using the catalyst **64b**.

The mechanism of this reaction is still speculative (51,52); however there is little doubt that the transition metal plays a crucial role in the catalytic cycle. It is believed that the transition metal activates both the hydroperoxide and the allylic alcohol via coordination of their respective hydroxyl functions in a transition state resembling **68**. Although both groups investigated allylic alcohols having enantiotopic olefinic faces, it is unlikely that discrimination of these

68

prochiral faces through direct coordination to the chiral metal complex is occurring, since olefin coordination to d^0 transition metal complexes is rare. Since the asymmetric synthesis is occurring outside the inner coordination sphere, it is not surprising that the enantiomeric excess observed is relatively low. Further success in this area will undoubtedly come with the advent of chiral bidentate (or tridentate) ligands that are stable to the reaction conditions and are able to transmit their chiral influence to the reactive site more efficiently.

VIII. ASYMMETRIC HYDROFORMYLATION

The hydroformylation or "oxo" reaction converts an olefin to a saturated aldehyde using CO and H_2 (synthesis gas) in the presence of a homogeneous transition metal catalyst according to the following equations:

$$R^1R^2C=CH_2 \xrightarrow{CO/H_2, CAT} R^1CH_2CHR^2 + R^1CHCH_2R^2$$
$$\qquad\qquad\qquad\qquad\qquad CHO \qquad\quad CHO$$

$$R^1R^2C=CH_2 \xrightarrow{CO/H_2, CAT} R^1R^2CHCH_2CHO + R^1R^2CCH_3$$
$$\qquad\qquad\qquad\qquad\qquad\qquad\qquad\qquad\qquad CHO$$

The reaction is of interest in synthetic chemistry because a carbon-carbon bond is formed and results in an overall one-carbon homologation of the olefin to a saturated aldehyde.

Attempts to modify this catalytic reaction for the purpose of asymmetric synthesis have been only moderately successful. Three transition metal based systems have been investigated: cobalt complexes in the presence of chiral Schiff base ligands (i.e., N-α-methylbenzylsalicylaldimine), which led to extremely low enantiomeric excess, and rhodium complexes containing chiral phosphine ligands and platinum complexes also containing chiral phosphine ligands, both of which produced higher enantiomeric excess than did the cobalt system. Only the rhodium and platinum systems are discussed here, since the majority of research has been performed with these two systems. Some of this work has been reviewed (53,54).

In the rhodium system a number of different chiral monophosphines have been used for asymmetric hydroformylation with moderate success; however the majority of work has utilized the chiral diphosphine (−)-diop (see Sect. III) and related derivatives (53,55) in the presence of the known hydroformylation catalyst $HRh(CO)(PPh_3)_3$ (56). Some of the reported extents of enantiomeric excess are shown in Table 5.

Table 5 Asymmetric Hydroformylation in the Presence of $HRh(CO)(PPh_3)_3$ and (−)-Diop (1 : 4)[a]

Olefin	Total Pressure (atm), $pH_2/pCO = 1$	Product	% e.e (absolute configuration)
$CH_2=CH(CH_2CH_3)$	1	$CH_3CHCH_2CH_3$ | CHO	18.8(R)
$CH_2=CH(CH_2CH_3)$	80	$CH_3CHCH_2CH_3$ | CHO	3.8(R)
cis-$CH_3CH=CHCH_3$ (CH₃, CH₃ on same side)	1	$CH_3CHCH_2CH_3$ | CHO	27.0(S)
cis-$CH_3CH=CHCH_3$	84	$CH_3CHCH_2CH_3$ | CHO	8.1(S)
trans-$CH_3CH=CHCH_3$	84	$CH_3CHCH_2CH_3$ | CHO	3.2(S)
C_6H_5-$CH=CHCH_3$ (styrene deriv.)	1	C_6H_5-$CHCH_3$ | CHO	22.7(R)

[a]Condensed from Ref. 53.

In this rhodium system the enantiomeric excess is consistently rather low and depends on parameters like pressure of CO and H_2, and solvent; however it is worthwhile to review some of the theories on the mechanism of the asymmetric induction that have been reported (53,54).

The mechanism for the rhodium-catalyzed hydroformylation reaction has been well studied (56); a simplified catalytic cycle is shown in Figure 12. The

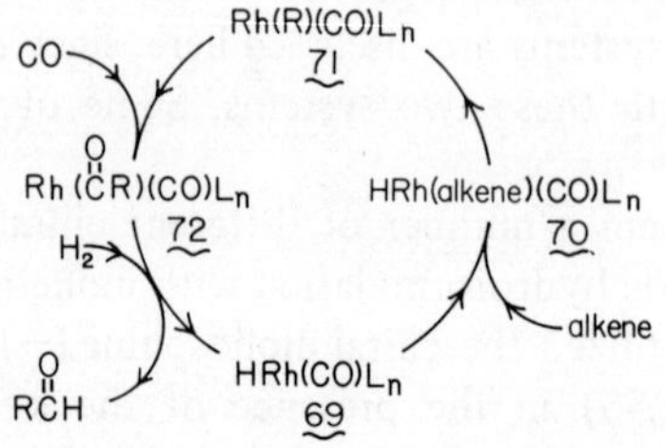

```
           CO        Rh(R)(CO)Lₙ
              ↘       71       ↘
         O
     Rh(CR)(CO)Lₙ              HRh(alkene)(CO)Lₙ
   ↗      72                        70      ↘
 H₂                                         ← alkene
      ↘                             ↗
        O        HRh(CO)Lₙ
       RCH          69
```

Fig. 12. Simplified catalytic cycle for rhodium-catalysed hydroformylation reaction: $L_n \equiv$ chiral phosphines, solvent molecules, and CO ligands.

olefin coordinates to **69** to give **70**, which undergoes olefin insertion into the rhodium-hydride bond to give the alkylrhodium species **71**. Migratory insertion of CO into this rhodium alkyl gives the acyl species, which is trapped by CO to give **72**. Oxidative addition of H_2 and subsequent reductive elimination of the aldehyde regenerates **69**.

The stereochemistry of the hydroformylation reaction has been determined to be cis; that is, the carbon-hydrogen bond and the carbon-carbon bond are formed on the same olefinic face. This important fact, coupled with the knowledge that the insertion of CO into a metal-alkyl bond proceeds with complete retention (47), facilitates discussion of the origin of the asymmetric induction.

There are three reasonable steps in the catalytic cycle that in principle could control the asymmetric induction: (1) olefin coordination to **69** to give **70**; this is simply discrimination of the enantiotopic faces of an olefin by the chiral metal fragment; (2) the olefin insertion step to generate the diastereomeric rhodium alkyl species, and (3) the insertion of CO into the diastereomeric rhodium alkyl species. The oxidative addition of hydrogen to **72** and reductive elimination of the chiral aldehyde are assumed to be unimportant in determining the asymmetric induction. Pino (53) believes that the asymmetric induction occurs in the formation of the diastereomeric alkyl rhodium species, that is, step (2), the olefin insertion step. He based this conclusion on his study using 1-butene, *cis*-butene, and *trans*-butene as substrates. The results are shown in Table 5. Since the identical product, 2-methylbutanal, is produced from all three olefinic substrates but in varying degrees of enantiomeric excess and differing absolute configurations, he reasoned that the asymmetric induction could not be occurring in the olefin binding, step (1). The result with *cis*-butene is especially interesting in this regard because this olefin does not have enantiotopic faces, and therefore asymmetric induction cannot possibly occur in step (1). As Pino suggested, the asymmetric induction must occur in step (2), the olefin insertion step, whereby the enantiotopic olefinic carbon atoms of *cis*- and *trans*-olefins are discriminated. However a very recent report by Stefani and Tatone (57) conflicts with the generality of Pino's conclusions. When α-[$^2H_1$]-styrene was asymmetrically hydroformylated using RhH(CO)(PPh$_3$)$_3$ and (−)-diop, two regioisomeric aldehyde products were produced as shown in Figure 13. The absolute configuration and enantiomeric excess of **74** was determined by conversion to **75** in the manner indicated.

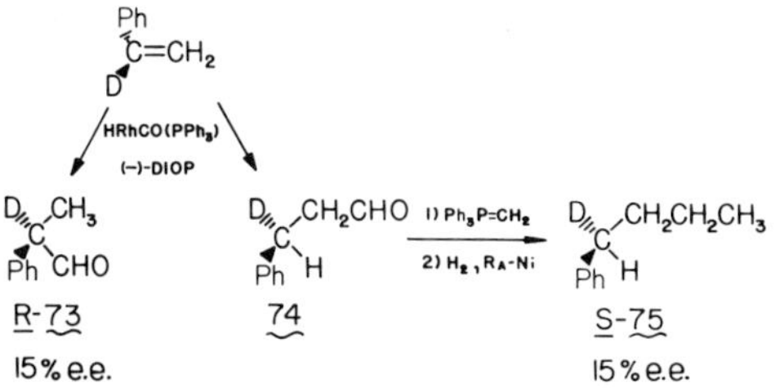

Fig. 13. Asymmetric hydroformylation of α-[$^2H_1$]-styrene.

This experiment suggests that step 1, the discrimination of the enantiotopic faces of α-[$^2H_1$]-styrene, is the step whereby the asymmetric induction occurs, since both products have the same enantiomeric excess and the absolute configuration for each product is consistent with the identical enantiotopic face of the substrate being hydroformylated.

If, as it seems, the origin of the asymmetric induction for these simple substrate olefins is dependent on the nature of the substrate, then it is unlikely that a single step in the catalytic cycle can account for the observed enantiomeric excess. It may well be that each step in the cycle makes a finite contribution, either positive or negative, to the asymmetric induction resulting in the observed enantiomeric excess. Such a system would defy unambiguous analysis.

A recent patent (58) suggests that work with more highly functionalized substrates capable of a bidentate interaction with the metal may increase the observed enantiomeric excess.

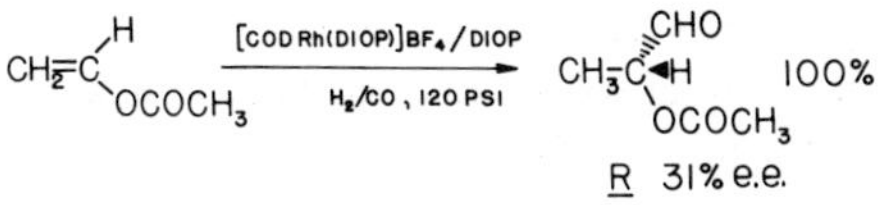

Another system that is reported to be an active asymmetric hydroformylation catalyst is [(−)-diop-PtCl$_2$]-SnCl$_2$ (59–63). This platinum-based system is not as synthetically useful as the rhodium system, since hydrogenation is competitive with hydroformylation even at 50 atm CO; in the analogous rhodium system, hydrogenation products are not detected under "oxo" conditions (60). Some typical enantiomeric excesses reported by Pino for this platinum system are shown in Table 6.

Table 6 Asymmetric Hydroformylation^a Using [(−)-Diop-PtCl$_2$]SnCl$_2$ (1 : 5) as Reported by Pino

Olefin	Product	% e.e. (absolute configuration)	Reference
CH$_2$=C(H)(CH$_2$CH$_3$)	CH$_3$CH(CHO)–CH$_2$CH$_3$	1.2 (S)	59
(CH$_3$)(H)C=C(CH$_3$)(H) cis	CH$_3$CH(CHO)–CH$_2$CH$_3$	1.2 (S)	59
(H)(CH$_3$)C=C(CH$_3$)(H) trans	CH$_3$CH(CHO)–CH$_2$CH$_3$	1.2 (S)	59
(Ph)(H)C=CH$_2$	PhCH(CH$_3$)–CHO	2.2 (S)	61
(Ph)(CH$_3$)C=CH$_2$	PhCH(CH$_2$CHO)–CH$_3$	13.2 (S)	61
(Ph)(CH$_3$CH$_2$)C=CH$_2$	PhCH(CH$_2$CHO)–CH$_2$CH$_3$	20.7 (S)	60

^{a} Highest enantiomeric excess reported.

Based on the results shown in Table 6, Pino concluded that the asymmetric induction occurs *after* the formation of the alkylplatinum intermediates, since all three hydroformylation products from 1-butene, *cis*-butene, and *trans*-butene have identical enantiomeric excess and absolute configuration (59,60). A Japanese group (63), however, has repeated Pino's work using the same [(−)-diop-PtCl$_2$]SnCl$_2$ system and substrates; they measured different extents of enantiomeric excess and different absolute configurations as shown in Table 7. The origins of the discrepancy are not clear.

It would appear that the origin of the asymmetric induction in the platinum system is complicated, as it is in the rhodium system. The low and capricious nature of the enantiomeric excess observed suggests that the step or steps that control the asymmetric induction are extremely sensitive to numerous parameters. For example, a modified derivative of (−)-diop (62,63) (Fig. 14), which has the same absolute configuration as (−)-diop on the backbone but has the 5*H*-dibenzophospholyl group instead of diphenylphosphine, gives dramatically different enantiomeric excess and opposite absolute configuration of the 2-methylbutanal products (Table 7) than (−)-diop.

Table 7 Asymmetric Hydroformylation Using
[(Chiral Diphosphine)PtCl$_2$]SnCl$_2$ (1 : 5) from Ref. 63

Olefin	Product	% e.e. (absolute configuration)	
		(−)-Diop	(−)-DBP-Diop
CH$_2$=C(H)(CH$_2$CH$_3$)	CH$_3$CH(CHO)−CH$_2$CH$_3$	2.8 (R)	12.1 (S)
(CH$_3$)(H)C=C(CH$_3$)(H) cis	CH$_3$CH(CHO)−CH$_2$CH$_3$	9.9 (S)	0.6 (R)
(H)(CH$_3$)C=C(CH$_3$)(H) trans	CH$_3$CH(CHO)−CH$_2$CH$_3$	12.8 (S)	1.8 (R)
(Ph)(H)C=CH$_2$	PhCH(CHO)CH$_3$	18.0 (S)	22.1 (S)

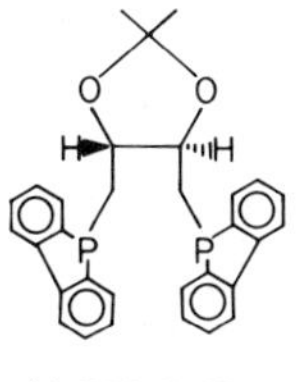

(−)-DBP-DIOP

Figure 14.

IX. ASYMMETRIC HYDROESTERIFICATION

A relatively recent but promising area of asymmetric synthesis mediated by transition metal complexes is *hydroesterification* (64). This reaction is similar to *hydroformylation* in that an olefin undergoes a one-carbon homologation to give a saturated ester in the presence of CO, an alcohol, and a palladium(II) phosphine catalyst.

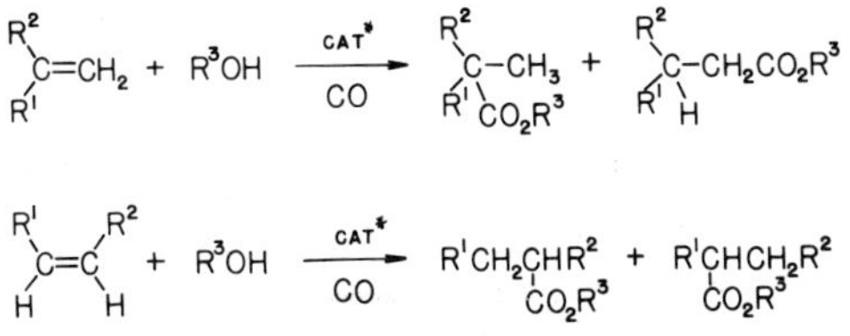

$$R^1R^2C=CH_2 + R^3OH \xrightarrow[CO]{CAT^*} R^1R^2C(CH_3)(CO_2R^3) + R^1R^2CH-CH_2CO_2R^3$$

$$R^1(H)C=C(R^2)(H) + R^3OH \xrightarrow[CO]{CAT^*} R^1CH_2CHR^2(CO_2R^3) + R^1CH(CO_2R^3)CH_2R^2$$

Little is known with certainty about the mechanism of the reaction, although different theories have been proposed (65,66). The stereochemistry of this reaction, like that of hydroformylation, involves a cis addition (67) across the olefin. Studies have been performed on varying the pressure of CO, solvent composition, and phosphine concentration on the regioselectivity (68) (straight-chain vs. branched isomers) and the extent of the asymmetric induction (69–72).

The chiral phosphines investigated to date to modify the palladium complexes are illustrated in Figure 15. Table 8 lists the enantiomeric excess measured in

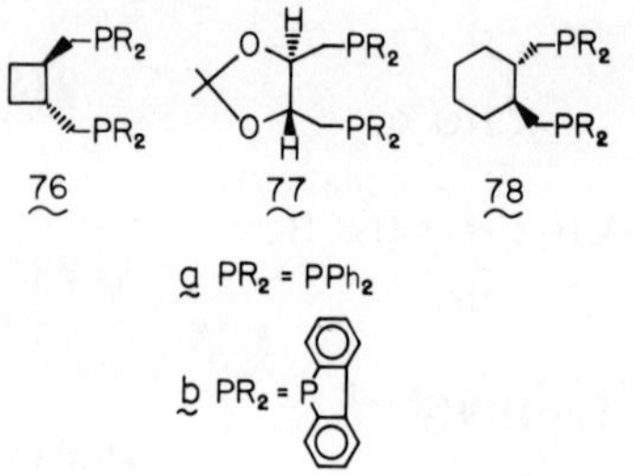

Fig. 15. Chiral phosphines used to date in catalytic asymmetric hydroesterification.

some asymmetric hydroesterifications using α-methylstyrene as the prochiral substrate. Other substrates have been investigated, but with little success (70).

Table 8 Asymmetric Hydroesterification of α-Methylstyrene using Chiral Palladium Phosphine Catalysts

Entry	Chiral phosphine[a]	Alcohol	P/Pd (atom)	Pressure (atm)	% e.e. (absolute configuration)	Reference
1	77a	$(CH_3)_3COH$	4	400	37.0 (S)	69
2	77a	$(CH_3)_3COH$	4	50	3.5 (S)	69
3	77a	$(CH_3)_3COH$	4	700	50.4 (S)	69
4	77a	$(CH_3)_3COH$	0.8	700	58.6 (S)	69,71
5	77b	$(CH_3)_3COH$	2	238	69 (S)	72
6	76a	$(CH_3)_2CHOH$	2	240–220	9.3 (S)	72
7	76b	$(CH_3)_2CHOH$	0.8	240–220	40 (S)	72
8	76b	$(CH_3)_2CHOH$	2	240–220	40 (S)	72
9	77b	$(CH_3)_2CHOH$	2	240–220	44 (S)	72
10	78a	$(CH_3)_2CHOH$	2	240–220	22 (R)	72
11	78b	$(CH_3)_2CHOH$	2	240–220	7.3 (S)	72

[a] See Figure 15.

As can be seen from Table 8, the enantiomeric excess for asymmetric hydroesterification of α-methylstyrene is reasonably high (up to 69% e.e); this suggests that this reaction will be of some synthetic interest. Since the mechanism is still speculative, a discussion of the origin of the asymmetric induction is not attempted. However there is one interesting mechanistic study in the literature.

Consiglio (71) varied the ratio of (−)-diop (**77a**) to palladium (in the presence of PPh$_3$ to maintain a constant ratio of phosphorus to palladium of 2) and found that when 0.5 equivalent of (−)-diop was present (P/Pd = 1), the enantiomeric excess reached a maximum of 58.6% (entry 4, Table 8). The implication here is that the chiral ligand must coordinate in a monodentate fashion to palladium during the catalytic cycle to maximize the enantiomeric excess. However other work (72) indicates that the concentration of the chiral ligand is not important (entries 7 and 8 in Table 8). Further work on the mechanism, especially with respect to the nature of the intermediates in the catalytic cycle, will undoubtedly shed light on the origins of the asymmetric induction and perhaps will indicate alternative ways to further increase the amounts of enantiomeric excess observed to date.

X. ASYMMETRIC HYDROSILYLATION

The field of asymmetric hydrosilylation has enjoyed a parallel existence with that of asymmetric hydrogenation. These fields share a number of common catalysts and catalyst precursors, and both areas have used similar types of substrate for reduction.

$$R^1CH=CHR^2 + R_3SiH \xrightarrow{CAT} \underset{SiR_3}{R^1CH_2\overset{|}{C}HR^2} + \underset{SiR_3}{R^1\overset{|}{C}HCH_2R^2} \quad [8]$$

$$R^1R^2C=O + R_3SiH \xrightarrow{\;\;''\;\;} \underset{H}{R^1R^2\overset{|}{C}OSiR_3} \xrightarrow{H^+} \underset{H}{R^1R^2\overset{|}{C}OH} \quad [9]$$

$$R^1R^2C=NR^3 + R_3SiH \xrightarrow{\;\;''\;\;} R^1R^2CHNR^3SiR_3 \quad [10]$$

Moreover, new chiral phosphine ligands that have proved to be efficient in asymmetric hydrogenation have been applied to asymmetric hydrosilylation with interesting results, though not as spectacular as in asymmetric hydrogenation. The asymmetric hydrosilylation of carbonyl-containing compounds (eq. [9]) catalyzed by chiral rhodium phosphine complexes is by far the most studied of the known systems capable of asymmetric hydrosilylation and is discussed briefly here. Other more comprehensive reviews are available (73,74).

Although a mechanism for the homogeneous hydrogenation of carbonyl compounds catalyzed by rhodium phosphine complexes is still speculative (75), there is less doubt about the mechanism for the homogeneous hydrosilylation of carbonyl compounds (76). A reasonable catalytic cycle is shown in Figure 16. The active catalyst **79** undergoes an oxidative addition of the silane to give the silyl hydride **80**, which then coordinates the carbonyl compound to give **81**; the carbonyl substrate inserts into the rhodium-silyl bond to give the silyloxyalkyl species **82**, which reductively eliminates the hydrosilylated product and regener-

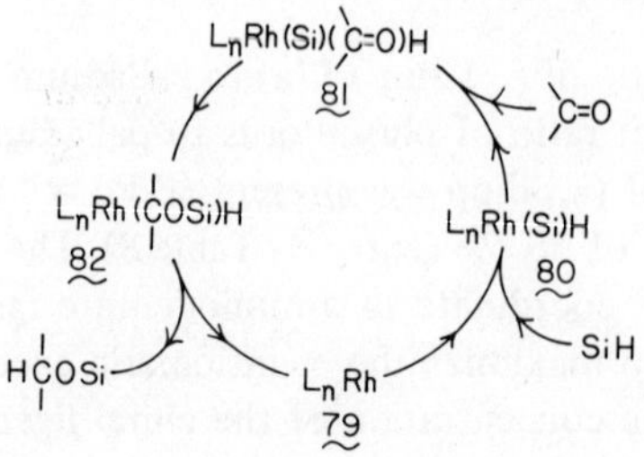

Fig. 16. Mechanism of hydrosilylation of carbonyl compounds; $L_n \equiv$ ancillary ligands and solvent molecules.

ates **79**. It should be noted that oxidative addition *followed* by coordination of the carbonyl substrates may be the sequence when L_n includes monophosphines; when L_n includes diphosphines, the order of these two transformations may be reversed as in the hydrogenation of olefins (18).

A variety of carbonyl compounds have been asymmetrically hydrosilylated with moderate to good enantiomeric excess. For example, simple prochiral ketones have been asymmetrically hydrosilylated in up to 61.8% e.e. Unlike the asymmetric hydrogenation of certain olefinic substrates, where chiral monophosphines were found to be less effective than chiral diphosphines, in asymmetric hydrosilylation chiral monophosphines such as benzylmethylphenylphosphine (BMPP) and (R)-α-[(S)-2-dimethylphosphinoferrocenyl]ethyldimethylamine ((R)-(S)-MPFA) were as effective as the chiral diphosphines diop and (S)-(R)-BPPFA (cf. Sect. III, Fig. 1, **1d**) (73). In addition, the enantiomeric excess

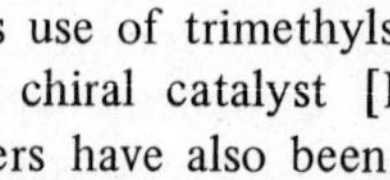

(R)-(S)-MPFA

was very sensitive to the nature of the silane used; in the case of phenyl *tert*-butyl ketone, asymmetric hydrosilylation using dimethylphenylsilane gave 61.8% e.e. of the *S* enantiomer, whereas use of trimethylsilane gave 28.1% e.e. of the *R* enantiomer using the same chiral catalyst $[Rh(H)_2\{(R)\text{-BMPP}\}_2(S)_2]^+ClO_4^-$ (73). A variety of α-ketoesters have also been asymmetrically hydrosilylated (77). The enantiomeric excess is in general higher than for simple ketones (e.g., as high as 85.4% e.e. for *n*-propyl pyruvate using the (−)-diop-based catalyst and α-naphthylphenylsilane). Enantiomeric excess in asymmetric reduction of a number of acetoacetates via hydrosilylation was rather low, typically in the range of 10–30% e.e. In the case of the asymmetric reduction of levulinates (γ-ketoesters), the enantiomeric excess is usually much higher, in the range of 75–85% e.e.

There are two schools of thought as to which step in the catalytic cycle

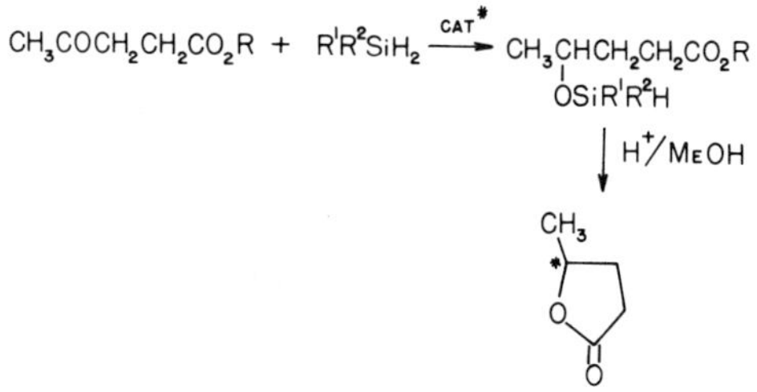

$$CH_3COCH_2CH_2CO_2R + R^1R^2SiH_2 \xrightarrow{CAT^*} CH_3\underset{\underset{OSiR^1R^2H}{|}}{C}HCH_2CH_2CO_2R$$

$$\downarrow H^+/MeOH$$

controls the asymmetric induction. Glaser (78) believes that the induction occurs in the carbonyl substrate binding step, thus invoking steric approach control. His proposal is based on a purely arbitrary model of a conformation for coordinated diop. Ojima (73,77), on the other hand, believes that the asymmetric induction occurs during the insertion step of the carbonyl unit into the rhodium-silyl bond to produce the diastereomeric α-silyloxyalkyl-rhodium intermediates. Both Glaser and Ojima have formulated models to explain and predict absolute configurations of the hydrosilylated products; however these models are purely empirical and should be used with extreme caution. It is interesting to point out that in the case of the levulinates (γ-ketoesters) Ojima postulates (77) that the observed high degree of enantiomeric excess (up to 85% e.e.) may be due to an interaction of the ester carbonyl group during the catalytic cycle as shown below.

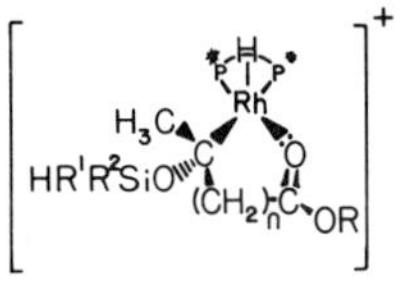

This is consistent with the secondary interaction postulated to explain the very high enantiomeric excess observed in the asymmetric hydrogenation of certain highly functionalized olefins and carbonyl compounds (Sect. III).

XI. ASYMMETRIC SYNTHESIS USING COORDINATION COMPLEXES

Although the majority of the material covered in this chapter has concerned organometallic aspects of asymmetric synthesis, there is an active interest in performing asymmetric synthesis with the more classical coordination compounds of cobalt(III) and copper(II). Since this does fall into the realm of asymmetric synthesis mediated by transition metal complexes, we shall briefly discuss some of the highlights of this field involving cobalt(III). A more comprehensive review has recently been published (79).

In the case of cobalt(III), the basis methodology has been to use the chirality inherent in certain chelate complexes to induce asymmetric synthesis on a coordinated ligand. For the most part amino acids have been used. In general, coordination of an otherwise unreactive and optically stable α-amino acid to a metal ion activates it toward racemization, presumably via a carbanionic reaction at the α center. In the case of achiral glycine, coordination to a cobalt(III) fragment, resolution of the octahedral complex followed by synthetic elaboration at the α center should, in principle, permit the stereospecific synthesis of other more complex α-amino acids. Unfortunately this has proved to be feasible in only a few of the systems studied. The problem seems to be one of choosing the appropriate reaction conditions, since side reactions tend to dominate.

The reaction of the resolved complex Λ-$[(en)_2 Co(gly)]^{2+}$ (en ≡ $NH_2 CH_2 CH_2$-NH_2; gly ≡ $NH_2 CH_2 CO_2^-$) with acetaldehyde under base catalysis is one of the exceptions. If appropriate reaction conditions are met, the coordinated glycine is asymmetrically converted to (S)-threonine and (S)-allothreonine as shown below (80).

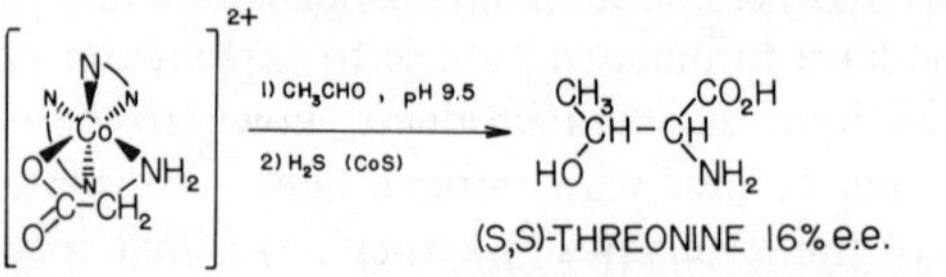

(S,S)-THREONINE 16% e.e.

(R,S)-ALLOTHREONINE 35% e.e.

This does seem to be a special case, since the reaction with other aldehydes and alkyl halides gives rise to numerous side reactions and only very low yields of the desired new amino acids (79).

The stereochemical course of the reaction between coordinated glycine and acetaldehyde has been very elegantly unraveled by Phipps and Sargeson. Although it is tempting to ascribe the origin of the asymmetric induction to the intrinsically different diastereotopic methylene protons of the glycine ligand, work by Sargeson (81) has shown that in weak aqueous alkali both these protons readily exchange at the same rate. However, upon substitution of the amino group of glycine (e.g., in coordinated N-benzylglycine) thereby creating a new chiral center at the nitrogen atom, there was induced a measurable difference in the rate of exchange of the two diastereotopic methylene protons. Thus it was observed by NMR in the complex Λ-$[(en)_2 Co(R$-N-benzylgly$)]^{2+}$ in $D_2 O$ (pH 10.5), that the *pro-S* hydrogen exchanged at a faster rate than the *pro-R* hydrogen. It is believed (81) that the Λ configuration of the cobalt complex stabilizes the R absolute configuration at the nitrogen atom by minimizing nonbonding interactions, which in turn sterically shields the *pro-R* hydrogen to attack by base.

Using these results and others of Sargeson, Phipps has formulated the mechanism illustrated in Figure 17 to explain the asymmetric synthesis of threonine and allothreonine from coordinated glycine. He postulates that the first step is sub-

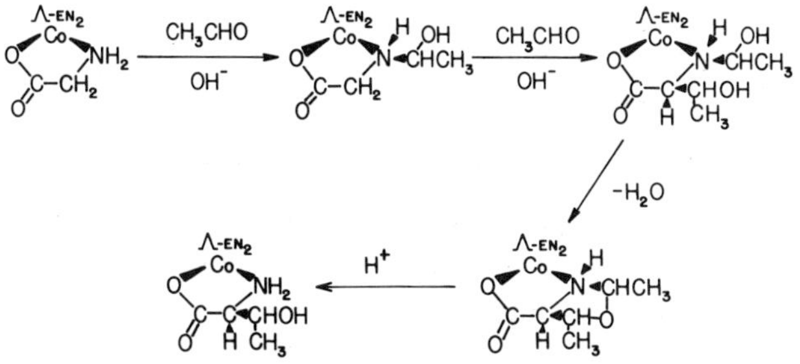

Fig. 17. Suggested mechanism for asymmetric synthesis of threonine and allothreonine from coordinated glycine.

stitution at the amino group of the glycine to generate a chiral center at this nitrogen atom; this, as in the case of coordinated N-benzylglycine, induces a stereochemical preference for the removal of the *pro-S* methylene proton to produce the coordinated S-threonine and S-allothreonine derivatives. Phipps also suggests that there is a cyclization to give the oxazolidine ring, which locks the chiral α center to the chiral nitrogen atom. This is necessary because under these basic reaction conditions, complexes of the type Λ-[(en)$_2$Co(S-AA)]$^{2+}$ (S-AA $\equiv$ S-amino acid) equilibrate to give a mixture of diastereomers containing an excess of the thermodynamically more stable Λ-[(en)$_2$Co(R-AA)]$^{2+}$ (79). Only when S-AA is S-proline does this not occur. In this case the complex Λ-[(en)$_2$Co(S-

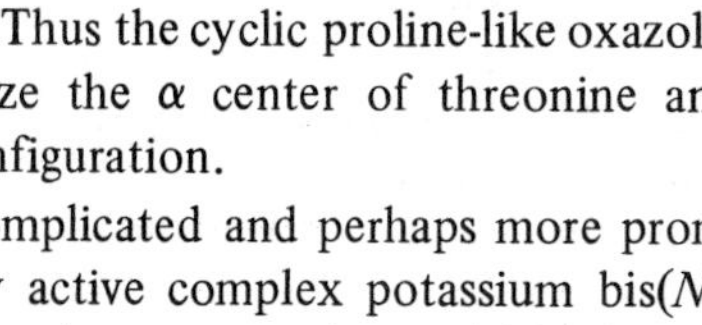

(S)-PROLINE   OXAZOLIDINE RING

proline)]$^{2+}$ undergoes a stereospecific exchange at the α center to retain its S-absolute configuration. Thus the cyclic proline-like oxazolidine ring is a necessary intermediate to stabilize the α center of threonine and allothreonine in the observed S absolute configuration.

A somewhat less complicated and perhaps more promising system has been reported. The optically active complex potassium bis(N-salicylideneglycinato)-cobaltate has been induced to react with acetaldehyde at pH 11.2 (82) (Fig. 18). The amino acid ratio of threo to allo is 1.25, with an overall chemical yield of 86%. The two L-amino acids, (3R,2S)-threonine and (3S,2S)-allothreonine, are produced in 45 and 65% e.e., respectively. It is clear that in this case the nitrogen atoms are not directly involved in the reaction, which presumably proceeds

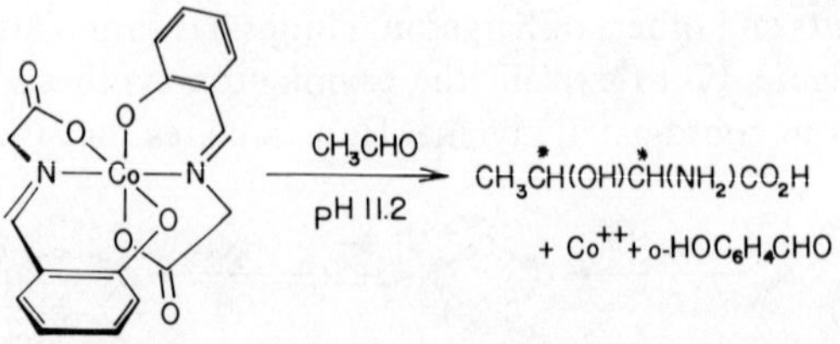

Fig. 18. Asymmetric synthesis using glycine–Schiff base ligands on cobalt(III).

by the conventional Knoevenagel mechanism. It seems to us that the use of amino acids imines is a promising approach to asymmetric amino acid synthesis using coordination compounds.

XII. CONCLUSION

This chapter surveys what we believe to be the more important aspects of asymmetric synthesis mediated by transition metal ions. Judgments of this kind depend on both the authors and the time of writing. Inevitably our perceptions of the direction of the field will change and some aspects will supersede others, but we hope we have given a balanced sketch of the present status of the field.

Whatever the limitations of this chapter, it will be clear that the use of metal complexes provides a diverse and potentially potent asymmetric synthetic method. The key to asymmetric synthesis is to arrange that the prochiral substrate and the chiral mediating agent associate in a fixed, rigid array. To achieve this with purely organic systems requires considerable elaboration, whereas rather simple systems suffice for some metal systems. In addition to the facile stereochemical adjustments, metal systems undergo numerous exotic reactions that may be catalytic. It would appear, therefore, that the field is limited largely by the ingenuity of the participants.

REFERENCES

1. A. P. Kozikowski and H. F. Wetter, *Synthesis*, **1976**, 561.
2. D. W. Slocum, Ed., *Ann. N.Y. Acad. Sci.*, **295** (1977).
3. K. Mislow and M. Raban, *Top. Stereochem.*, **1**, 1 (1967).
4. J. D. Morrison and H. S. Mosher, *Asymmetric Organic Reactions*, American Chemical Society, Washington, D.C., 1976, p. 280.
5. M. D. Fryzuk, Ph.D. dissertation, University of Toronto, 1978, p. 23.
6. Ref. 4, p. 40.
7. (a) H. B. Kagan, *Pure Appl. Chem.*, **43**, 401 (1975); (b) H. B. Kagan and J. C. Fiaud, *Top. Stereochem.*, **10**, 175 (1978).
8. J. D. Morrison, W. F. Masler, and M. K. Neuberg, *Adv. Catal.*, **25**, 81 (1976).

9. D. Valentine, Jr., and J. W. Scott, *Synthesis*, **1978**, 329.
10. B. R. James, *Adv. Organomet. Chem.*, **17**, 319 (1979).
11. H. A. Boucher and B. Bosnich, *J. Am. Chem. Soc.*, **99**, 6253 (1977).
12. (a) H. B. Kagan and T. P. Dang, *J. Chem. Soc., Chem. Commun.*, **1971**, 481; (b) H. B. Kagan and T. P. Dang, *J. Am. Chem. Soc.*, **94**, 6429 (1972).
13. (a) W. S. Knowles, M. J. Sabacky, B. D. Vineyard, and D. J. Weinkauff, *J. Am. Chem. Soc.*, **97**, 2567 (1975); (b) B. D. Vineyard, W. S. Knowles, M. J. Sabacky, G. L. Bachman, and D. J. Weinkauff, *ibid.*, **99**, 5946 (1977).
14. (a) M. D. Fryzuk and B. Bosnich, *J. Am. Chem. Soc.*, **99**, 6262 (1977); (b) M. D. Fryzuk and B. Bosnich, *ibid.*, **100**, 5491 (1978).
15. K. Achiwa, *J. Am. Chem. Soc.*, **98**, 8265 (1976).
16. (a) T. Hayashi, T. Mise, S. Mitachi, K. Yamatnoto, and M. Kumada, *Tetrahedron Lett.*, **1976**, 1133; (b) T. Hayashi, T. Mise, and M. Kumada, *ibid.*, **1976**, 4351.
17. W. R. Cullen and Y. Sugi, *Tetrahedron Lett.*, **1978**, 1635.
18. J. Halpern, D. P. Riley, A. C. S. Chan, and J. P. Pluth, *J. Am. Chem. Soc.*, **99**, 8055 (1977).
19. For a discussion of the origin of the asymmetric induction in asymmetric hydrogenation, see A. S. C. Chan, J. J. Pluth, and J. Halpern, *J. Am. Chem. Soc.*, **102**, 5952 (1980).
20. J. M. Brown and P. A. Chalconer, *Tetrahedron Lett.*, **1978**, 1877.
21. K. Achiwa, Y. Ohga, and Y. Iitaka, *Tetrahedron Lett.*, **1978**, 4683.
22. T. Hayashi, T. Mise, and M. Kumada, *Tetrahedron Lett.*, **1976**, 4357.
23. I. Ojima, T. Kogure, and K. Achiwa, *Chem. Soc., J. Chem. Commun.*, **1977**, 428.
24. T. Hayashi, A. Katsumura, M. Knoishi, and M. Kumada, *Tetrahedron Lett.*, **1979**, 425.
25. R. R. Schrock, *Acc. Chem. Res.*, **12**, 98 (1979).
26. P. Yates, *J. Am. Chem. Soc.*, **74**, 5376 (1952).
27. (a) A. Davison, W. C. Krussell, and R. C. Michaelson, *J. Organomet. Chem.*, **72**, C7 (1974); (b) T. C. Flood, F. J. Disanti, and D. L. Miles, *Inorg. Chem.*, **15**, 1910 (1976).
28. T. Aratani, Y. Yoneyoshi, and T. Nagase, *Tetrahedron Lett.*, **1975**, 1707; **1977**, 2599.
29. (a) A. Nakamura, A. Konishi, Y. Tatsuno, and S. Otsuka, *J. Am. Chem. Soc.*, **100**, 3443, (1978); (b) A. Nakamura, A. Konishi, R. Tsujitani, M. Kudo, and S. Otsuka, *ibid.*, **100**, 3449 (1978).
30. P. W. Jolly and R. Pettit, *J. Am. Chem. Soc.*, **88**, 5044 (1966).
31. M. L. H. Green, M. Ishzq, and R. N. Whitely, *J. Chem. Soc. A*, **1967**, 1508.
32. M. Brookhart and G. O. Nelson, *J. Am. Chem. Soc.*, **99**, 6099 (1977).
33. W. R. Moser, *J. Am. Chem. Soc.*, **91**, 1135, 1141 (1969).
34. H. Nozaki, H. Takaya, S. Moriuti, and R. Noyori, *Tetrahedron*, **24**, 3655 (1968).
35. P. W. Jolly and G. Wilke, *The Organic Chemistry of Nickel*, Vol. 2, Academic Press, New York, 1975.
36. B. Bogdanović, *Angew. Chem.*, **85**, 1013 (1973); *Angew. Chem., Int. Ed. Engl.*, **12**, 954 (1973).
37. B. Bogdanović, *Adv. Organomet. Chem.*, **17**, 105 (1979).
38. Ref. 35, p. 25.
39. (a) K. J. Ivin, F. J. Rooney, C. D. Stewart, M. L. H. Green, and R. Mahtab, *J. Chem. Soc. Chem. Commun.*, **1978**, 604; (b) E. R. Evitt and R. G. Bergman, *J. Am. Chem. Soc.*, **101**, 3973 (1979).
40. R. H. Grubbs, A. Miyashita, M. M. Liu, and P. L. Burk, *J. Am. Chem. Soc.*, **99**, 3863 (1977).
41. R. H. Grubbs and A. Miyashita, *J. Am. Chem. Soc.*, **100**, 1300 (1978).
42. Ref. 35, p. 1–49.

43. B. M. Trost, P. E. Strege, L. Weber, T. J. Fullerton, and T. J. Dietsche, *J. Am. Chem. Soc.*, **100**, 3407, 3416, 3426 (1978); (b) B. M. Trost and T. R. Verhoeven, *ibid.*, **100**, 3435 (1978).
44. B. M. Trost and L. Weber, *J. Am. Chem. Soc.*, **97**, 1611 (1975).
45. B. M. Trost and P. E. Strege, *J. Am. Chem. Soc.*, **99**, 1649 (1977).
46. B. M. Trost and P. E. Strege, *J. Am. Chem. Soc.*, **97**, 2534 (1975).
47. J. K. Stille and K. S. Y. Lau, *Acct. Chem. Res.*, **10**, 434 (1977).
48. B. M. Trost, T. R. Verhoeven, and J. M. Fortunak, *Tetrahedron Lett.*, **1979**, 2301.
49. K. B. Sharpless and R. C. Michaelson, *J. Am. Chem. Soc.*, **95**, 6136 (1973).
50. S. Yamada, T. Mashiko, and S. Terashima, *J. Am. Chem. Soc.*, **99**, 1990 (1977).
51. R. C. Michaelson, R. E. Palermo, and K. B. Sharpless, *J. Am. Chem. Soc.*, **99**, 1990 (1977); spectacular asymmetric epoxidations have been reported by T. Katsuki and K. B. Sharpless, *J. Am. Chem. Soc.*, **102**, 5976 (1980).
52. T. Itoh, K. Jitsukawa, K. Kaneda, and S. Teranishi, *J. Am. Chem. Soc.*, **101**, 159 (1979).
53. P. Pino, G. Consiglio, C. Botteghi, and C. Saloman, *Adv. Chem. Ser.*, **132**, 295 (1974).
54. P. Pino and G. Consiglio, in *Fundamental Research in Homogeneous Catalysis*, M. Tsutsui and R. Ugo, Eds. Plenum Press, New York, 1977.
55. M. Tanaka, Y. Ikeda, and I. Ogata, *Chem. Lett.*, **1975**, 1115.
56. D. Evans, J. A. Osborn, and G. Wilkinson, *J. Chem. Soc. A*, **1968**, 3133.
57. A. Stefani and D. Tatone, *Helv. Chim. Acta*, **60**, 518 (1977).
58. H. B. Tinker and A. J. Solodar, Canadian patent 1,027,41; *Chem. Abstr.*, **89**, 42440m (1978).
59. G. Consiglio and P. Pino, *Helv. Chim. Acta*, **59**, 642 (1976).
60. G. Consiglio and P. Pino, *Isr. J. Chem.*, **15**, 221 (1976–1977).
61. G. Consiglio, W. Arber, and P. Pino, *Chem. Ind. (Milan)*, **60**, 396 (1978).
62. M. Tanaka, Y. Ikeda, and I. Ogata, *Chem. Lett.*, **1975**, 1115.
63. Y. Kawabata, T. M. Suzuki, and I. Ogata, *Chem. Lett.*, **1978**, 361.
64. J. Tsuji, *Acc. Chem. Res.*, **2**, 144 (1969).
65. K. Bittler, N. V. Kutepow, D. Neubauer, and H. Reis, *Angew Chem.*, **80**, 352 (1968); *Angew Chem., Int. Ed. Engl.*, **7**, 329 (1968).
66. D. M. Fenton, *J. Org. Chem.*, **38**, 3192 (1973).
67. G. Consiglio and P. Pino, *Gazz. Chim. Ital.*, **105**, 1133 (1975).
68. G. Consiglio and P. Pino, *Chimica*, **30**, 26 (1976).
69. G. Consiglio and P. Pino, *Chimica*, **30**, 193 (1976).
70. C. Botteghi, G. Consiglio, and P. Pino, *Chimica*, **27**, 477 (1973).
71. G. Consiglio, *J. Organomet. Chem.*, **132** C26 (1977).
72. T. Hayashi, M. Tanaka, and I. Ogata, *Tetrehedron Lett.*, **1978**, 3925.
73. I. Ojima, K. Yamamoto, and M. Kumada, in *Aspects of Homogeneous Catalysis*, Vol. 3, R. Ugo, Ed., Reidel., Dordrecht, Netherlands, 1977, p. 186.
74. J. L. Speier, *Adv. Organomet. Chem.*, **17**, 407 (1979).
75. R. R. Schrock and J. A. Osborn, *J. Chem. Soc., Chem. Commun.*, **1970**, 567.
76. J. F. Peyronel and H. B. Kagan, *Nouv. J. Chimie*, **2**, 211 (1978).
77. I. Ojima, T. Kogure, and M. Kumagai, *J. Org. Chem.*, **42**, 1671 (1977).
78. R. Glaser, *Tetrahedron Lett.*, **1975**, 2127.
79. D. A. Phipps, *J. Mol. Catal.*, **5**, 81 (1979), and references therein.
80. J. D. Dabrowiak and D. W. Cooke, *Inorg. Chem.*, **14**, 1305 (1975).
81. B. T. Golding, G. J. Gainesford, A. J. Herlt, and A. M. Sargeson, *Tetrahedron*, **32**, 389 (1976), and references therein.
82. Tu. Belokon, M. M. Dolgoya, N. I. Kuznetsova, S. V. Vitt, and V. M. Belikov, *Izv. Akad. Nauk SSSR, Ser. Khim.*, **1973**, 156.

Structures of Metal Nitrosyls

ROBERT D. FELTHAM and JOHN H. ENEMARK

Department of Chemistry, University of Arizona, Tucson, Arizona

I.	Introduction	156
	A. Historical Perspective	156
	B. The $\{M(NO)_m\}^n$ Formalism	158
	C. Organization	158
II.	Mononitrosyl Complexes, MNO	158
	A. Six-Coordination	169
	B. Five-Coordination	174
	1. Tetragonal Pyramidal $\{MNO\}^{4,5,\text{and }6}$	174
	2. Tetragonal Pyramidal $\{MNO\}^7$	175
	3. Tetragonal Pyramidal $\{MNO\}^8$	177
	4. Trigonal Bipyramidal $\{MNO\}^8$	178
	5. Distorted Five-Coordinate $\{MNO\}^8$	179
	C. Four-Coordination	180
	1. Tetrahedral $\{MNO\}^{10}$	181
	2. Distorted Four-Coordinate $\{MNO\}^{10}$	181
	D. Three-Coordination	184
	E. Higher Coordination Numbers	184
III.	Dinitrosyl Complexes	185
	A. Six-Coordination	185
	B. Five-Coordination	187
	C. Four-Coordination	190
	D. Polynitrosyl Complexes	190
IV.	Bridging and Polynuclear Nitrosyls	195
	A. $M_2(\mu_2\text{-NO})$ Complexes	195
	B. $M_2(\mu_2\text{-NO})_2$ Complexes	198
	C. Triply Bridging NO Complexes $M_3(\mu_3\text{-NO})$	198
	D. Polynuclear Metal Nitrosyls	200
V.	Nitrosyl Complexes Containing SO_2 and C_3H_5 Ligands . .	200
	A. $M(NO)(SO_2)$	203
	B. $M(NO)(C_3H_5)$	203
VI.	Summary	206
	Acknowledgments	206
	Abbreviations	207
	References	209

I. INTRODUCTION

A. Historical Perspective

The nitric oxide molecule (NO) is a radical with one unpaired electron. As early as 1934 Sidgwick (222) recognized that nitric oxide could lose or gain one electron in its bonding interactions with transition metals to give species that have been variously classified as complexes of NO^+ and NO^- (183), or as complexes in which NO serves as a three-electron and a one-electron donor (115). The valence bond structures for $|N{\equiv}O|^+$ and $\overset{\frown}{N{=}O}\,^-$ have sp and sp^2 hybridization, respectively, at the nitrogen atom, implying the possibility of both linear and strongly bent MNO geometries. Linear MNO complexes were established in 1937 (33), but it was not until 1968 that the first definitive structural evidence for strongly bent MNO complexes was provided by Ibers and his associates (129). This discovery, coupled with the demonstration by Eisenberg and co-workers (198) that strongly bent and linear MNO groups can be simultaneously attached to the same metal, helped to stimulate the extensive structural and theoretical investigations of nitrosyl complexes that have been published since that time.

From the valence bond structures $|N{\equiv}O|^+$ and $\overset{\frown}{N{=}O}\,^-$ it would seem that there should be a direct correlation between the NO stretching frequency (ν_{NO}) and the M–N–O angle. However, such is not the case. Metal nitrosyl complexes do exhibit a wide range of NO stretching frequencies ($\sim$1500–2000 cm^{-1}), but the M–N–O angle cannot be generally correlated with ν_{NO} unless several empirical corrections are assumed (125). Therefore structure determination has been of major importance in understanding the chemistry of metal nitrosyl complexes. In recent years the bonding in metal nitrosyl complexes has generally been described within the framework of molecular orbital theory (see Sect. I–B).

The burgeoning information in the field of nitrosyl complexes has led to the publication of several recent reviews (Table 1). However the two extant reviews of nitrosyl structures (91,105) were written in 1972 and 1974, and are now outdated. A general theoretical base has been developed since 1971 that provides a framework for correlating the chemical properties and structural features of transition metal nitrosyls. Consequently, a new summary and review of their structural chemistry is apropos.

This review is restricted to a discussion of metal nitrosyl complexes whose structures have been determined by X-ray crystallography, neutron diffraction, electron diffraction, or microwave spectroscopy. Structural data available to us as of December, 1979, are included in this review. The chemical reactions and other physical properties of metal nitrosyl complexes are not discussed here, but reviews of those aspects of metal nitrosyls are available elsewhere (Table 1).

Table 1 Other Reviews of Transition Metal Nitrosyls

Title	Authors	Reference
Inorganic Nitrosyl Compounds	T. Moeller (1946)	183
Chemistry of the Nitrosyl Group (NO)	C. C. Addison and J. Lewis (1955)	2
Nitric Oxide Compounds of Transition Metals	B. F. G. Johnson and J. A. McCleverty (1966)	142
The Nitroprusside Ion	J. H. Swinehart (1967)	233
Organometallic Nitrosyls	W. P. Griffith (1968)	116
Electrophilic Reactivity and Structure of Co-ordinated Nitrosyl Group	J. A. Masek (1969)	170
Recent Developments in Transition Metal Nitrosyl Chemistry	N. G. Connelly (1972)	68
Structural Chemistry of Transition Metal Complexes: (1) 5-Coordination, (2) Nitrosyl Complexes.	B. A. Frenz and J. A. Ibers (1972)	105
Complexes of Nitrogen and Oxygen	J. A. McGinnety (1972)	173
Principles of Structure, Bonding and Reactivity for Metal Nitrosyl Complexes	J. H. Enemark and R. D. Feltham (1974)	91
Synthetic Methods in Transition Metal Nitrosyl Chemistry	K. G. Caulton (1975)	46
Nitrosyl, Dinitrogen and Dioxygen Complexes	J. E. Fergusson and G. A. Rodley (1975)	252
The Coordination Chemistry of Nitric Oxide	R. Eisenberg and C. D. Meyer (1975)	88
Electrophilic Behavior of Coordinated Nitric Oxide	F. Bottomley (1978)	21
Nitrosyl Complexes of Ruthenium	F. Bottomley (1978)	22
Reactions of Nitric Oxide Coordinated to Transition Metals	J. A. McCleverty (1979)	171

B. The $\{M(NO)_m\}^n$ Formalism

Several authors (91,101,126,132,177,193) have pointed out that the variables of importance for determining the structures adopted by nitrosyl complexes include the coordination number of the metal, and the total number of electrons associated with the metal d orbitals and the $\pi^*(NO)$ orbitals. Since the bonding in most nitrosyl complexes is dominated by covalent interaction with the metal atom, assignment of oxidation states to the metal atom and the NO is undesirable.

Metal nitrosyl complexes are conveniently classified as $\{M(NO)_m\}^n$ species, where n is the total number of electrons associated with the metal d and/or $\pi^*(NO)$ orbitals (91). The value of n corresponds to the familiar number of d electrons on the metal when the nitrosyl ligand is formally considered to be bound as NO^+. However the $\{M(NO)_m\}^n$ bonding notation makes no assumption about the actual distribution of electrons between the metal and the NO group, and no assumption about the M—N—O angle. A more detailed discussion of this bonding notation, which is used throughout this chapter, can be found elsewhere (91).

C. Organization

Mononitrosyl complexes are tabulated according to the total number of electrons, n, of the $\{MNO\}^n$ group. Each compound has been assigned a number to uniquely place it in the tables. The discussion of compounds is organized according to the coordination number of the metal. The coordination number of the central metal atom is unambiguous for most complexes. All η^2 ligands are regarded as occupying two coordination sites. The cyclopentadienide anion $C_5H_5^-$ and its derivatives have three pairs of π- electrons and therefore can replace up to three monodentate electron pair donor ligands from a metal atom. In subdividing the compounds by coordination number, the η^5-Cp ligand is taken to occupy three coordination sites of the metal. Dinitrosyl complexes, bridging nitrosyl complexes, and complexes containing other ambidentate ligands such as C_3H_5 and SO_2 are discussed separately.

Numerous abbreviations have been used to conserve space. The key to the abbreviations is given at the end of the chapter. Other abbreviations, symbols, and so on, follow IUPAC rules or established journal conventions.

II. MONONITROSYL COMPLEXES, MNO

Some important metrical details for mononitrosyl complexes including the M—N distance, the N—O distance, and the M—N—O angle and coordination number are summarized in Tables 2 to 10. Other pertinent structural data will be

Table 2 {MNO}^{n} Geometry in Six-Coordinate Complexes

n	M–N–O Angles (deg)	Average[a]
4	171–178 ⎫	
5	169–178 ⎬	175(7)
6	170–180 ⎭	
7	138–159	146(11)
8	119–141	126(9)
9	Unknown	
10	Unknown	

[a] The average is the unbiased estimate of the mean. The number in parentheses is the estimated standard deviation of the mean.

Table 3 Structural Data for Metal Nitrosyls: {MNO}^{3} Complexes

Compound	M–N (Å)	N–O (Å)	M–N–O (deg)	Coordination number	Reference
No Known Examples.					

159

Table 4 Structural Data for Metal Nitrosyls: {MNO}^{4} Complexes

	Compound	M–N (Å)	N–O (Å)	M–N–O (deg)	Coordination Number and Approximate Geometry	Reference
1.	$K_3Na[V(NO)(CN)_6]$	1.806(6)	1.235(8)	175.2(5)	PBP-7	137
2.	$K_3[V(NO)(CN)_5]$[a]	1.662(38)	1.294(46)	171.4(31)	TP-6	139,140
3.	$K_4[V(NO)(CN)_6]$[a]	1.680(16)	1.165(38)	164.2(15)	PBP-7	84
4.	$Cr(NO)[N(SiMe_3)_2]_3$	1.738(20)	1.191(28)	180[b]	T-4	27
5.	$Cr(NO)(Cp)(NPh_2)I$	1.676(3)	1.183(4)	172.7(3)	6	223
6.	$Mo(NO)(S_2CNBu^n_2)_3$	1.731(8)	1.154(9)	173.2(7)	7	28,29
7.	$Mo(NO)(Cp)_2(\eta^1$-$Cp)$	1.751(3)	1.207(4)	179.2(2)	See Sect. II-E	38
8.	$Mo(NO)(Cp)_2(Me)$	1.75(10	1.23(1)	178(2)	See Sect. II-E	70
9.	$[Mo(NO)(Cp)(NH_2NHPh)I][BF_4]$	1.780(4)	1.188(5)	170.6(3)	7	11
10.	$[PPh_4]_2[Mo(NO)(OCNMe_2)(NCS)_4]$	1.767(6)	1.179(8)	n.r.[c]	7	188
11.	$[Mo(NO)(Cp)I]_2(\mu_2$-$NNMe_2)$	1.765(12)	n.r.[c]	n.r.[c]	6 and 7	155
12.	$[Mo(NO)(OPr^i)_2]_2(\mu_2$-$OPr^i)_2$	1.747(9)	1.205(11)	178(1)	5	49
		1.761(10)	1.184(11)	177(1)		
13.	$Mo(NO)Cl(OPr^i)\{HB(PzMe_2Cl)_3\}$	1.764(8)	1.176(11)	179.4(8)	6	172
14.	$W(NO)(OBu^t)_3(Py)$	1.732(8)	1.250(10)	179.2(8)	5	50
15.	$Cs[V(NO)(\eta^2$-$ONH_2)(DIPIC)(H_2O)] \cdot 2H_2O$	1.693(17)	1.196(23)	175.6(16)	7	246

[a] Disordered.
[b] Estimated standard deviation not reported.
[c] Data not reported

Table 5 Structural Data for Metal Nitrosyls: {MNO}^{5} Complexes

	Compound	M–N (Å)	N–O (Å)	M–N–O (deg)	Coordination Number and Approximate Geometry	Reference
1.	cis- and trans-[Cr(NO)(Cp)]$_2$(μ_2-NMe$_2$)$_2$	1.63(2)	1.23(2)	169(1)	6	34,36
2.	[Cr(NO)(Cp)]$_2$(μ_2-OMe)$_2$	1.689(8)	1.199(10)	166.3(7)	6	121
3.	Cr(NO)(NO$_2$)$_2$(Py)$_3$	1.68(1)	1.15(1)	180^a	6	166
4.	[Cr(NO)(NO$_2$)(Me$_6$[14]4,11-diene N$_4$)][PF$_6$]	1.679(5)	1.193(6)	180^a	TP-6	245
5.	[Cr(NO)(Cp)]$_2$(μ_2-SPh)$_2$	1.662(7)	1.19(1)	169.9(7)	6	174
6.	Mo(NO)(MeOH)(TTP)	1.746(6)	n.r.^{b}	179.8(4)	TP-6	79
7.	[NEt$_4$][Re(NO)Br$_4$(EtOH)]	1.723(1)	1.19(2)	169(3)	TP-6	56
8.	[NEt$_4$][Re(NO)Br$_4$(MeCN)]	1.771(11)	0.99(2)	178(6)	TP-6	56
9.	[NEt$_4$][Re(NO)Cl$_4$Py]	1.749(6)	1.171(9)	178.9(7)	TP-6	57
10.	[Co(en)$_3$][Cr(NO)(CN)$_5$]	1.71(1)	1.21(1)	176(1)	TP-6	96

^{a} Estimated standard deviation not reported.
^{b} Data not reported.

Table 6 Structural Data for Metal Nitrosyls: {MNO}⁶ Complexes

	Compound	M–N (Å)	N–O (Å)	M–N–O (deg)	Coordination Number and Approximate Geometry	Reference
1.	$Cr(NO)(CO)_2(PPh_2Me)_2I$	1.705(14)	1.117(20)	n.r.[a]	6	69
2.	$Cr(NO)(CO)_2(Fl)$	1.687(7)	1.169(9)	178.9(7)	6	10
3.	$Cr(NO)(CO)_2(Cp)$[b]	1.801(5)	1.161(5)	178.7(4)	6	10
4.	$[Cr(NO)(Cp)]_2(\mu_2\text{-NO})(\mu_2\text{-NH}_2)$	1.637(23)	1.185(27)	171.7(24)	6	47
		1.672(23)	1.212(25)	172.6(25)		
5.	$Mo(NO)(CO)(Cp)(MPADP)$	1.809(10)	1.193(13)	176.8(11)	6	206
6.	$Mo(NO)(CO)_2(HBPz_3)$[b]	1.930(4)	1.150(6)	178(4)	6	133
7.	$K_4[Mo(NO)(CN)_5]$	1.95(3)	1.23(4)	175(3)	6	231
8.	$Mo(NO)(CO)(BPz_4)(PPh_3)$	1.849(5)	1.182(8)	176.0(5)	6	76
9.	$W_2(NO)(CO)_9H$[b]	1.917(2)	1.144(2)	177.0(2)	6	8,191
10.	$W_2(NO)(CO)_8H[P(OMe)_3]$	1.818(4)	1.172(4)	178.4(3)	6	164
11.	$Mn(NO)(CO)(MeCp)(PPh_3)$	1.674(5)	1.200(7)	178.4(5)	5	99
12.	$Mn(NO)(4\text{-MePi})(TPP)$	1.644(5)	1.176(7)	176.5(5)	TP-6	192,214
13.	$K_3[Mn(NO)(CN)_5]$	1.66(1)	1.21(2)	174(1)	TP-6	235,236
14.	$[Mn(NO)(Cp)](\mu_2\text{-NO})_2[Mn(Cp)(\eta^1\text{-Cp})]$	1.656(5)	1.194(5)	175.4[c]	6	41
15.	$[Mn(NO)(Cp)](\mu_2\text{-NO})_2[Mn(Cp)(NO_2)]$	1.652[c]	1.180[c]	175.4[c]	6	37
16.	$Mn(NO)(TTP)$	1.641(2)	1.160(3)	177.8(3)	TP-5	214
17.	$Re(NO)H_2(PPh_3)_3$	1.77(2)	1.25(3)	175(2)	6	58
		1.73(2)	1.24(3)	177(2)		
18.	$Re(NO)(Cp)(CHO)(PPh_3)$	1.777(8)	1.190(1)	178.0(8)	6	251
19.	$Na_2[Fe(NO)(CN)_5]$	1.653(4)	1.124(7)	175.7(5)	TP-6	25,168
20.	$Ba[Fe(NO)(CN)_5]$	1.71(4)	1.11(5)	166(4)	TP-6	162
21.	$Sr[Fe(NO)(CN)_5]$	n.r.[a]	n.r.[a]	n.r.[a]	—	45
22.	$[Fe(NO)(OH)(TMC)][ClO_4]_2$	1.621(6)	1.143(6)	178.3(6)	TP-6	128

#	Compound				Ref.
23.	cis-Fe(NO)(NO$_2$)(S$_2$CNEt$_2$)$_2$	1.695(5)	1.136(6)	174.9(5)	6	136
24.	[Et$_4$N][Fe(NO)(MNT)$_2$]	1.611(8)	n.r.^{a}	180(0)	5	237
25.	K$_2$[Ru(NO)Cl$_5$]	1.747(6)	1.112(7)	176.8(9)	TP-6	239
26.	[NH$_4$]$_2$[Ru(NO)Cl$_5$]	1.738(2)	1.131(3)	176.7(5)	TP-6	238
27.	trans-[Ru(NO)(NH$_3$)$_4$(OH)]Cl$_2$	1.735(3)	1.159(5)	173.8(3)	TP-6	19
28.	[Ru(NO)(NH$_3$)$_5$]Cl$_3$	1.770(9)	1.172(14)	172.8(9)	TP-6	19
29.	Na$_2$[Ru(NO)(NO$_2$)$_4$(OH)]	1.748(4)	1.127(7)	180.0(4)	TP-6	224
30.	[NH$_4$]$_2$[Ru(NO)Cl$_4$(H$_2$O)]Cl	1.656(16)	1.165(26)	177.5(14)	TP-6	152
31.	Ru(NO)(NO$_2$)(salen)	1.713^c	n.r.^{a}	n.r.^{a}	TP-6	42
32.	Ru(NO)(S$_2$CNEt$_2$)$_3$	1.72^c	1.17^c	170^c	TP-6	82
33.	Ru(NO)Cl$_3$(PPh$_2$Me)$_2$	1.744(6)	1.132(6)	176.4(6)	6	221
34.	Ru(NO)Cl(PPh$_3$)$_2$(SO$_4$)	1.80(3)	1.07(4)	175(3)	6	204
35.	[Ru(NO)Br$_2$(Et$_2$SO)]$_2$(μ_2-Br)$_2$	1.71(1)	1.16(1)	177.8(11)	6	102
36.	[Ru(NO)Cl{(EtO)$_2$PO}$_2$H]$_2$(μ_2-Cl)$_2$	n.r.^{a}	n.r.^{a}	177.6(2)	6	227
37.	Ru(NO)Cl$_3$(PPh$_3$)$_2$	1.737(7)	1.142(8)	180^c	TP-6	125
38.	Os(NO)(PPh$_3$)$_2$(CF$_3$CO$_2$)(CF$_3$CONO)	1.739(7)	1.207(10)	176.3(7)	6	123
39.	K[Ir(NO)Cl$_5$]	1.760(11)	1.124(17)	174.3(11)	TP-6	24
40.	K[Ir(NO)Br$_5$]	1.710(25)	1.166(42)	170.3(24)	TP-6	24

^{a} Data not reported.
^{b} Disordered.
^{c} Estimated standard deviation not reported.

Table 7 Structural Data for Metal Nitrosyls: {MNO}^{7} Complexes

	Compound	M–N (Å)	N–O (Å)	M–N–O (deg)	Coordination Number and Approximate Geometry	Reference
1.	Fe(NO)(S$_2$CNEt$_2$)$_2$	1.69(4)	1.16(5)	174(4)	TP-5	66
2.	Fe(NO)(S$_2$CNMe$_2$)$_2$	1.71(2)	1.02(2)	173(2)	TP-5	74,75
3.	[Fe(NO)(DAS)$_2$][ClO$_4$]$_2$	1.655(18)	1.141(27)	172.8(17)	TP-5	92
4.	[Fe(NO)(DAS)$_2$(NCS)][BPh$_4$]	1.729(9)	1.246(13)	158.6(9)	TP-6	92
5.	[Et$_4$N]$_2$[Fe(NO)(CN)$_4$]	1.669(7)	1.157(7)	174.7(7)	TP-5	157,217
6.	Fe(NO)(TPP)	1.717(7)	1.122(12)	149.2(6)	TP-5	213
7.	Fe(NO)(1-MeIm)(TPP)	1.743(4)	1.133(11)	140.1(8)	TP-6	192,216
8.	Fe(NO)(4-MePi)(TPP)·CHCl$_3$	1.721(10)	1.141(13)	138.5(11)	TP-6	212
9.	Fe(NO)(4-MePi)(TPP)	1.740(7)	1.112(9)	143.7(6)	TP-6	212
10.	[Fe(NO)(TMC)][BF$_4$]$_2$	1.737(6)	1.137(6)	177.5(5)	TP-5	128
11.	Fe(NO)(MEEN)	1.693(5)	1.170(6)	155.2(5)	TP-5	149
12.	Fe(NO)(salen)					120
	23°C	1.783(16)	1.10(3)	147(4)	TP-5	
	−175°C	1.80(13)	1.15^a	127(6)	TP-5	
13.	[Et$_4$N]$_2$[Fe(NO)(MNT)$_2$]	1.56(3)	1.06(6)	168(6)	TP-5	202

^{a} Estimated standard deviation not reported.

Table 8 Structural Data for Metal Nitrosyls: {MNO}^{8} Complexes

	Compound	M–N (Å)	N–O (Å)	M–N–O (deg)	Coordination Number and Approximate Geometry	Reference
1.	Mn(NO)(CO)$_4$	1.80(1)	1.15(2)	180(0)	TBP-5	104
2.	Mn(NO)(CO)$_3$(PPh$_3$)[a]	1.78(2)	1.15(1)	178(1)	TBP-5	95
3.	Mn(NO)(CO)$_2$(PPh$_3$)$_2$	1.73(1)	1.18(1)	178(1)	TBP-5	94
4.	[Fe(NO)(NP$_3$)][BPh$_4$]	1.60(7)	1.19(9)	164(7)	TBP-5	80
5.	Hg[Fe(NO)(CO)$_2$(PEt$_3$)]$_2$[a]	1.729(15)	1.156(16)	176.6(12)	TBP-5	228
6.	[Fe(NO)(PP$_3$)][BPh$_4$]	1.67(1)	1.16(1)	177.4(7)	TBP-5	81
7.	[Ru(NO)(DPPE)$_2$][BPh$_4$]	1.74(1)	1.20(1)	174(1)	TBP-5	196,197
8.	Ru(NO)H(PPh$_3$)$_3$	1.792(11)	1.183(11)	176(1)	TBP-5	195,197
9.	[Ru(NO)(DPPP)$_2$][BPh$_4$]	1.72(2)	1.23(3)	169.5(25)	TBP-5	18
10.	Ru(NO)(CO)I(PPh$_3$)$_2$[a]	1.80(4)	1.15(5)	159(2)	TBP-5	117
11.	[Ru(NO)(PPh$_2$Me)]$_2$(μ_2-PPh$_2$)$_2$	1.697(12)	1.23(1)	174.1(10)	T-4	203
12.	[Os(NO)(CO)$_2$(PPh$_3$)$_2$][ClO$_4$][a]	1.84(1)	1.16(1)	178(1)	TBP-5	59,62
13.	[Os(NO)(CO)(MePhNC)(PPh$_3$)$_2$][ClO$_4$][a]	1.76(3)	1.19(4)	174(3)	TBP-5	63
14.	Co(NO)(S$_2$CNMe$_2$)$_2$	1.746(7)	1.167(23)	135.1(15)	TP-5	6,7,90
15.	[Co(NO)(NH$_3$)$_5$]Cl$_2$	1.871(6)	1.154(7)	119(1)	TP-6	199
16.	[Co(NO)Cl(en)$_2$][ClO$_4$]	1.820(11)	1.043(17)	124.4(11)	TP-6	225,226
17.	Co(NO)(salen)	1.809(4)	1.071(4)	128.0(3)	TP-5	118
18.	[Co(NO)(en)$_2$(ClO$_4$)][ClO$_4$]	1.806(6)	1.015(11)	138.0(14)	TP-6	144
19.	[Co(NO)(DAS)$_2$][ClO$_4$]$_2$	1.68(3)	1.16(2)	178(2)	TBP-5	93
20.	[Co(NO)(DAS)$_2$(NCS)][NCS]	1.87(1)	1.18(2)	132.3(14)	TP-6	93
21.	Co(NO)(TPP)	1.833(52)	1.01(2)	135.2(8)	TP-5	215
22.	Co(NO)Cl$_2$(PPh$_2$Me)$_2$	1.705(5)	1.076(6)	164.5(6)	TBP-5	31
23.	Co(NO)(benacen)	1.831(11)	1.136(26)	122.9(8)	TP-5	247,248

Table 8 (continued)

	Compound	M–N (Å)	N–O (Å)	M–N–O (deg)	Coordination Number and Approximate Geometry	Reference
24.	Co(NO)(acacen)	1.821(9)	1.093(16)	122.4(9)	TP-5	247,248
25.	Co$_4$(NO)$_3$Na$_4$Si$_{12}$Al$_{12}$O$_{48}$	2.23(6)	1.47(11)	141(3)	TP-5	72
26.	Rh(NO)(SO$_4$)(PPh$_3$)$_2$	1.91(1)	1.11(1)	122(1)	TP-5	165
27.	Rh(NO)Br$_2${P(OPh)$_3$}$_2$	2.04(4)	1.26(9)	109(5)	TP-5	97
28.	[Rh(NO)(NCMe)$_3$(PPh$_3$)$_2$][PF$_6$]$_2$	2.026(8)	1.159(10)	118.4(6)	TP-6	150
29.	Rh(NO)Cl$_2$(PPh$_3$)$_2$	1.912(10)	1.15^b	124.8(16)	TP-5	113
30.	[Rh(NO)Cl(P$_3$)][PF$_6$]	1.909(15)	1.081(16)	131.0(1.4)	TP-5	189
31.	[Ir(NO)Cl(PPh$_3$)]$_2$O	1.72(1)	1.16(2)	175.7(10)	SP-4	44,48
32.	[Ir(NO)H(PPh$_3$)$_3$][ClO$_4$]					
	Equatorial NO	1.80(2)	1.14(3)	174.5(21)	TBP-5	60
	Axial NO	1.68(3)	1.21(3)	175.0(3)	TBP-5	181
33.	[Ir(NO)(CO)I(PPh$_3$)$_2$][BF$_4$]	1.89(3)	1.17(4)	125(3)	TP-5	130
34.	Ir(NO)I(Me)(PPh$_3$)$_2$	1.91(2)	1.23(2)	120(2)	TP-5	182
35.	Ir(NO)Cl$_2$(PPh$_3$)$_2$	1.94(2)	1.03(2)	123(2)	TP-5	180
36.	[Ir(NO)(CO)Cl(PPh$_3$)$_2$][BF$_4$]	1.972(11)	1.16(1)	124.1(9)	TP-5	129,131
37.	[Ir(NO)(PPh$_3$)]$_2${μ-(CF$_3$C-CCF$_3$)$_2$}	1.61(1)	1.21(1)	157(2) 171(2)	SP-4	65
38.	[{Ir(NO)(PPh$_3$)}$_2$(μ$_2$-N$_2$PhNO$_2$)(μ$_2$-O)][PF$_6$]	n.r.^{c}	n.r.^{c}	n.r.^{c}	SP-4	86
39.	[QH]$_2$[Pt(NO)Cl$_3$]$_2$(μ$_2$-Cl)$_2$	2.15^b	1.18^b	112^b	TP-6	153

^{a} Disordered.
^{b} Estimated standard deviation not reported.
^{c} Data not reported.

Table 9 Structural Data for Metal Nitrosyls: {MNO}⁹ Complexes

Compound	M–N (Å)	N–O (Å)	M–N–O (deg)	Coordination Number and Approximate Geometry	Reference
[Co(NO)(NP$_3$)][BPh$_4$]	1.83(6)	1.14(10)	165(7)	TBP-5	80

Table 10 Structural Data for Metal Nitrosyls: {MNO}¹⁰ Complexes

	Compound	M–N (Å)	N–O (Å)	M–N–O (deg)	Coordination Number and Approximate Geometry	Reference
1.	Co(NO)(CO)$_3$ [a]	1.76(3)	1.10(4)	180[b]	T-4	33
2.	Co(NO)(CO)$_2$(SbPh$_3$)	1.698(5)	1.150(7)	176.1(5)	T-4	110
3.	Co(NO)(CO)$_2$(PPh$_3$) [c]	1.740(7)	1.134(8)	178.2(7)	T-4	4,243
4.	Co(NO)(CO)(PPh$_3$)$_2$ [c]	1.718(8)	1.153(11)	177.4(7)	T-4	4
5.	Co(NO)(CO)$_2$(AsPh$_3$) [c]	1.740(20)	1.142(30)	177.0(2)	T-4	111
6.	Rh(NO)(PF$_3$)$_3$ [a]	1.848(21)	1.147(37)	180[b]	T-4	30

Table 10 (continued)

	Compound	M–N (Å)	N–O (Å)	M–N–O (deg)	Coordination Number and Approximate Geometry	Reference
7.	Rh(NO)(PPh$_3$)$_3$	1.759(13)	1.27(2)	156.7(26)	T-4	147
8.	Ir(NO)(CO)(PPh$_3$)$_2$	1.787(8)	1.180(9)	174.1(7)	T-4	32
9.	Ir(NO)(PPh$_3$)$_3$	1.67(2)	1.24(3)	180[b]	T-4	5
10.	Ni(NO)(N$_3$)(PPh$_3$)$_2$	1.686(7)	1.164(8)	152.7(7)	T-4	89
11.	Ni(NO)(NCS)(PPh$_3$)$_2$	1.648(5)	1.159(6)	161.5(5)	T-4	119
12.	[Ni(NO)(P$_3$C)][BF$_4$]	1.579(10)	1.199[b]	180[b]	T-4	15
13.	[Ni(NO)(POC)$_3$][BF$_4$]	1.58(1)	1.12(1)	176.8(18)	T-4	175
14.	[Ni(NO)(NP$_3$)][BPh$_4$]	1.59(2)	1.14(3)	167.7(21)	T-4	80
15.	Ni(NO)(Cp)	1.58(1)[a]	1.17(2)[a]	180[b]	—	209
		1.626(5)[d]	1.165(5)[d]	180[b]		71
16.	[Ni(NO)(Me$_2$Pz)]$_2$	1.616(4)	1.158(4)	178.9(4)	PT-3	51
17.	Ni[Ni(NO)(Me$_2$Pz)]$_2$]$_2$	1.625(3)	1.153(4)	168.9(3)	PT-3	51
18.	[Et$_4$N][{Ni(NO)}$_2$(μ_2-I)(μ_2-Me$_2$Pz)$_2$]	1.650(6)	1.118(8)	174.3(6)	T-4	52
		1.641(5)	1.139(7)	170.7(7)		
19.	[(C$_4$H$_8$O)$_2$Na][{Ni(NO)}$_2$(μ_2-Me$_2$Pz)$_3$]	1.580(7)	1.175(8)	175.8(9)	T-4	52
		1.597(6)	1.191(7)	178.0(7)		
20.	Ni(NO)GAP	1.627(4)	1.147(4)	162.3(4)	T-4	53

[a] Electron diffraction data.
[b] Estimated standard deviation not given.
[c] Disordered.
[d] Microwave study.

listed separately as required or can be found in the references. The general structural features of the mononitrosyl complexes show that the metal-nitrogen bond distances are rather short, indicating multiple bonding between the metal and nitrosyl ligand. The N—O distances range from 1.02 to 1.27 Å. The average value for N—O from the entire set is 1.159 Å. The M—N—O angles range from 180° to 109° with clustering near the values of 120° and 180° (Fig. 1).

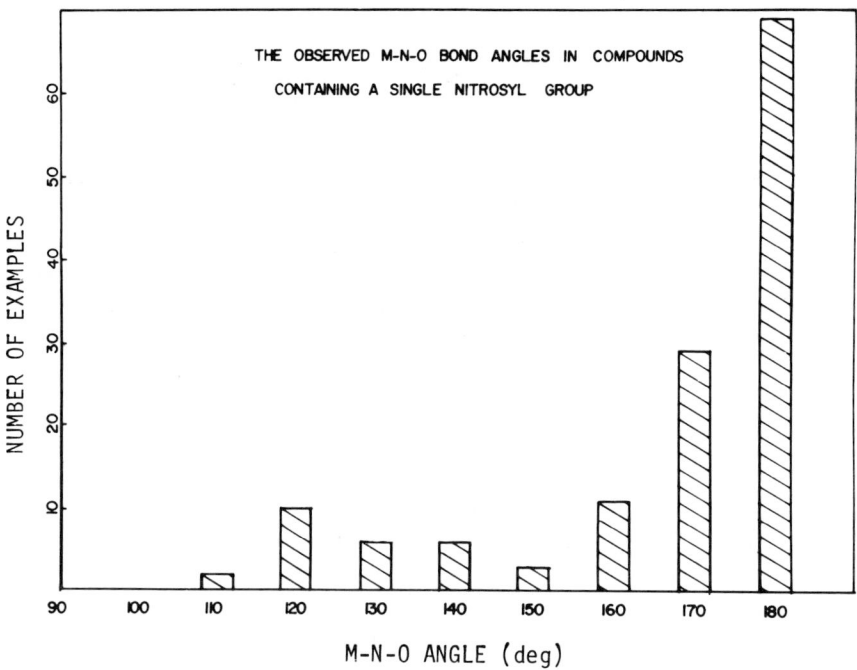

Fig. 1. Observed bond angles for MNO complexes. The bond angles are grouped in 10° ranges: 110 ± 5°, 120 ± 5°, 130 ± 5°, etc.

A. Six-Coordination

By far the most common coordination number found for the mononitrosyl complexes thus far structurally characterized is 6 (53 examples). Six-coordination occurs for $\{MNO\}^n$ complexes with $n = 4, 5, 6, 7$, and 8. The range of M—N—O bond angles encountered for each of these six-coordinate $\{MNO\}^n$ complexes is shown in Table 2. It is clear that the MNO group is approximately linear (±10°), provided that $n \leqslant 6$. The range of bond angles observed for $\{MNO\}^7$ and $\{MNO\}^8$ is somewhat larger, but the average bond angles of 146(11)° and 126(9)°, respectively, show that the M—N—O angle of six-coordinate complexes decreases as n increases to 7 and to 8.

The variation in M–N–O bond angle with total number of electrons, n, documented in Table 2 is reminiscent of the variation of the O–N–O angles observed for NO_2^+, NO_2, and NO_2^- (Fig. 2), which was analyzed in molecular

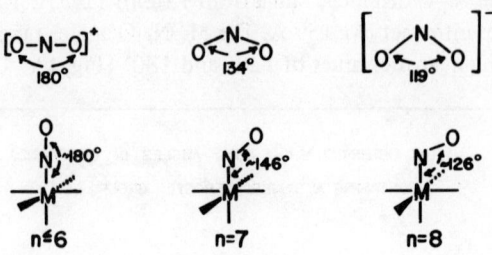

Fig. 2. Comparison of the geometries of $[ONO]^{+1,0,-1}$ and six-coordinate $\{MNO\}^n$ complexes.

orbital terms by Walsh (242). Several molecular orbital descriptions of six-coordinate mononitrosyl complexes with linear MNO groups and C_{4v} symmetry have been proposed (91,101,167,177). These molecular orbital descriptions differ in some important details, but there is general agreement that in complexes with electron configurations with $n = 0$–6 the electrons reside in the $d_{xy}(2B_2)$ and $d_{xz},d_{yz}(8E)$ orbitals (Fig. 3) (101). Since the character of the $2B_2$

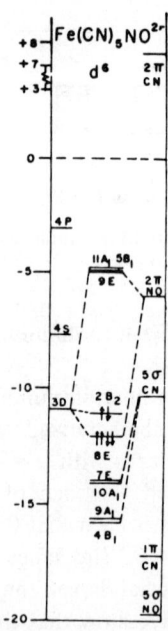

Fig. 3. Molecular orbital diagram for $[Fe(NO)(CN)_5]^{2-}$. (Reproduced from ref. 101.)

and $8E$ orbitals ranges from bonding to nonbonding, no major distortions of the MNO groups results from electrons occupying these orbitals. However, in $\{MNO\}^n$ complexes having electron configurations with $n \geq 7$, the excess electrons occupy the π-type orbitals such as $9E$ of a six-coordinate complex, which are totally antibonding with respect to M, N, and O (Fig. 4) (91, 177). The destabilizing effects of electrons in such antibonding orbitals can be reduced by bending the MNO group (91,177), and/or by reduction of the coordination number of the metal.

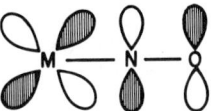

Fig. 4. The totally antibonding π-type orbital of a linear $\{MNO\}^n$ group (e.g., $9E$ of Fig. 3).

The precise stereochemistry of distinctly bent $\{MNO\}^7$ and $\{MNO\}^8$ groups is sometimes difficult to determine by X-ray crystallography because of rotational disorder of the bent MNO group in the crystal. Both twofold and fourfold crystallographic disorder of bent MNO groups have been observed for six-coordinate complexes (144,199,216). The various positions for the oxygen atoms of the disordered bent nitrosyl ligand can usually be determined, but the coordinates of the nitrogen atom are often impossible to define uniquely. For example, a twofold disorder model with a single N atom position and two O atom positions may lead to an unrealistically short N–O distance (≤ 1.0 Å) and abnormally large thermal parameters for the N atom (135). Such a result suggests an alternative model in which the N atoms are also disordered. However the close proximity of the two N atom positions (~ 0.4 Å apart) requires that some constraints be imposed on the disordered NO groups if the least-squares refinement is to be well behaved. It is not uncommon for two different disorder models to give essentially identical crystallographic R factors (180). Consequently, the apparent molecular parameters for the bent nitrosyl will reflect the model chosen. Figure 5 illustrates the difficulties presented by disorder for $[Co(NO)(en)_2(ClO_4)]^+$ (8: 18).* The choice of model has no effect on the Co–N(O) distance, but the Co–N–O angles differ by 16° for the two models that were explored.

The relatively large thermal motion of the oxygen atoms of nitrosyl groups requires that some caution be exercised whenever M–N–O angles are discussed in detail – even if the estimated standard deviations of the M–N–O angles are a fraction of a degree. Root-mean-square amplitudes of vibration of 0.4 to 0.5 Å

---

*Number in parentheses preceding colon designates a table number; number following colon indicates entry in that table corresponding to the referenced compound; thus (8: 18) refers to compound 18 in Table 8.

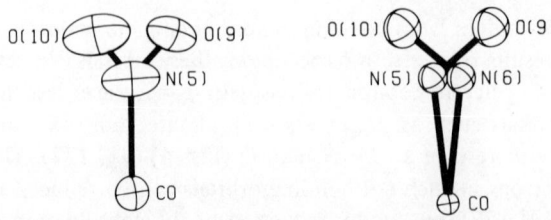

Fig. 5. Two different disorder models for the nitrosyl group in *trans*-[Co(NO)(en)$_2$(ClO$_4$)]-[ClO$_4$]. *Left*: the CoNO fragment after refinement of a single anisotropic N atom and two disordered anisotropic O atoms: Co–N(5) = 1.806(6), N(5)–O(9) = 1.005(9), N(5)–O(10) = 0.911(10) Å; Co–N(5)–O(9) = 135.2(11), Co–N(5)–O(10) = 140.8(14)°. *Right*: the CoNO fragment after refinement as two disordered NO groups with N–O = 1.15 Å: Co–N(5) = 1.805(13), Co–N(6) = 1.803(17) Å; Co–N(5)–O(9) = 122.7(8), Co–N(6)–O(10) = 122.0(11)°. (Reproduced from ref. 144.)

are not uncommon. Recent increased use of low-temperature X-ray crystallography may help to define the geometry of coordinated nitrosyl ligands more precisely, but this technique cannot eliminate the problems presented by disordered crystals.

Effects of the NO ligand on the other ligands in the coordination sphere of the metal can be discerned in the present structural data for six-coordinate complexes (Table 11) that have identical ligands both cis and trans to the nitrosyl group. Comparison of the differences in metal-ligand bond lengths for the cis and trans ligands shows that the complexes with linear MNO groups with $\nu_{NO} >$ 1800 cm^{-1} have *trans*-M–L bonds that are slightly shorter (0.013–0.075 Å) than those of the cis ligands (19,20,25,238,239). On the other hand, [Cr(NO)(CN)$_5$]$^{3-}$ (5: 10) is a {CrNO}^{5} complex with a nearly linear nitrosyl group and a low value of ν_{NO}(1643 cm^{-1}), and it has *trans*-M–L distances that are 0.042 Å *longer* than the *cis*-M–L distances (96). No definitive explanation of these differences has been offered. However it has been pointed out that complexes with high NO frequencies have physical and chemical characteristics similar to NO$^+$, whereas those with low NO stretching frequencies have features related to those of NO$^-$ (23). The data in Table 11 indicate trans bond lengthening by a linear "NO$^-$" ligand and the trans bond shortening by a linear "NO$^+$" ligand (19,20,25,238,239).

The six-coordinate {CoNO}^{8} complex, [Co(NO)(NH$_3$)$_5$]$^{2+}$ (8: 15) has a bent nitrosyl group, and the *trans*-M–L distance is 0.24 Å longer than the *cis*-M–L distances. The large structural trans influence of a bent NO group in six-coordinate {MNO}^{8} complexes is well recognized (226). In fact, many {MNO}^{8} complexes with a strongly bent nitrosyl group have no ligand trans to the NO group. Such five-coordinate complexes are discussed in Section II-B.

In each of the complexes for which the information is reported and a meaningful equatorial plane can be defined, the metal is displaced from the plane toward the nitrosyl group by as much as 0.28 Å. This effect has been attributed to the strong multiple bonding between the metal and the nitrosyl group.

Table 11 Effects of the NO Group on trans Metal-Ligand Distances

	Compound	ν_{NO} (cm^{-1})	trans M–X Distance (Å)	cis M–X Distance (Å)	cis – trans (Å)	Reference
1.	Na$_2$[Fe(NO)(CN)$_5$]	1944	1.918(6)	1.929(4) 1.936(4)	0.015	25
2.	[NH$_4$]$_2$[Ru(NO)Cl$_5$]	1916	2.357(1)	2.378(2) 2.378(1) 2.373(1) 2.374(1)	0.019	238
3.	K$_2$[Ru(NO)Cl$_5$]	1916	2.359(2)	2.377(2) 2.363(7) 2.371(2) 2.375(8)	0.013	239
4.	[Ru(NO)(NH$_3$)$_5$]Cl$_3$	1845	2.017(1)	2.133(1) 2.042(11) 2.100(8) 2.093(9)	0.075	19
5.	K[Ir(NO)Br$_5$]	1953	2.419(4)	2.475(3) 2.485(3)	0.061	20, 24
6.	K[Ir(NO)Cl$_5$]	2008	2.286(3)	2.335(2) 2.342(2)	0.053	20, 24
7.	[Co(en)$_3$][Cr(NO)(CN)$_5$]	1643	2.075(14)	2.047(13) 2.021(13) 2.022(14) 2.043(13)	−0.042	96
8.	[Co(NO)(NH$_3$)$_5$]Cl$_2$	1614	2.220(4)	1.984(5) 1.978(6)	−0.239	176, 199

In spite of some of the ambiguities and experimental difficulties outlined above, the general structural features of six-coordinate $\{MNO\}^n$ complexes have now been delineated. From the experimental results and the theoretical description of the complexes, it is concluded that for *six-coordinate mononitrosyl complexes* the M—N—O angle is $180 \pm 10°$ for $n \leq 6$; $145 \pm 10°$ for $n = 7$; and $125 \pm 10°$ for $n = 8$.

B. Five-Coordination

The next most numerous group of mononitrosyl complexes that have been structurally characterized are the five-coordinate species. The interest in this class of mononitrosyls stems from the two possible idealized geometries, tetragonal pyramidal (TP) and trigonal bipyramidal (TBP), which the central metal can adopt in addition to the possible variation in M—N—O angles discussed in Sect. II-A. The interplay between the M—N—O bond angle and the coordination geometry of the central metal atom has stimulated much research in both structural and theoretical aspects of this problem. In fact, a five-coordinate complex provided the first definitive example of a strongly bent nitrosyl group (131), and the first molecular orbital explanation proposed for linear vs. bent MNO geometry dealt with the five-coordinate complexes (193).

Regular trigonal bipyramidal geometry cannot be realized for five-coordinate MNO complexes. However we have adopted the view extant in the literature that when an equatorial plane with three bond angles of approximately 120° and an axis consisting of two ligands perpendicular to that plane are present, the complex is designated as having TBP geometry. When the NO group lies in the equatorial plane the maximum symmetry is C_{2v}, but deviations of the bond angles within that plane toward 90° from 120° will transform the complex into tetragonal pyramidal geometry, leaving the distinction between TBP and TP ambiguous (132). This approximation of a TBP should be borne in mind by the reader when examining the tables and perusing the text. When the NO group occupies an axial position of a TBP, the designation of coordination geometry is less subjective, since threefold symmetry (C_{3v}) can be realized. Finally, there are a few examples of five-coordinate complexes in which the M—N—O angle is intermediate between 120° and 180° and/or the coordination geometry about the metal is distorted. Complexes in this category are discussed separately.

1. Tetragonal Pyramidal $\{MNO\}^{4, 5, and\ 6}$

The structures of TP complexes follow the same basic pattern as the six-coordinate complexes. This class of complexes can be viewed as six-coordinate complexes with a vacant coordination site. The TP complexes with $n \leq 6$ are "electron deficient"; that is, the metal atom has fewer than the 18 electrons required by the Effective Atomic Number (EAN) rule. The highest occupied molecular orbitals (HOMO) (126,132,193) consist of $d_{xy}(b_2)$ and $d_{xz}d_{yz}(e)$ as is the case in the six-coordinate complexes. Thus the two known examples of such complexes, Mn(NO)(TPP) and [Fe(NO)(MNT)$_2$]$^-$ (Table 6, compounds 16 and 24) have essentially linear MNO linkages.

2. Tetragonal Pyramidal {MNO}^{7}

The absence of a sixth ligand for complexes with $n > 6$ provides a larger variety in stereochemistry than is possible in six-coordination. However the examples of {MNO}^{7} moieties that have thus far been structurally characterized are all TP complexes of iron (Table 7). The Fe–N–O bond angles range from 177.5(5)° to 127(6)° and appear to be markedly dependent on environmental factors such as temperature, crystal packing, and the nature of the other ligands attached to the iron atom (Fig. 6).

The variable bond angle in {FeNO}^{7} complexes with TP geometry is due to the presence of one electron in antibonding orbitals of the FeNO group. Compared with the six-coordinate {FeNO}^{7} complexes, the d_{z^2} orbital is less antibonding in these five-coordinate TP complexes. Consequently, d_{z^2} can be sufficiently low in energy that $^2A_1(d_{z^2})$ can become the ground state and consequently linear geometry can be maintained (83,91,126). However, the presence of a σ antibonding electron makes the geometry of these complexes more susceptible to small perturbations than are other metal nitrosyl complexes. Moreover, the energy of d_{z^2} and the totally antibonding π-type orbital (Fig. 4) are similar, with the consequence that this π-type orbital can also sometimes be populated, resulting in distinctly bent {FeNO}^{7} groups. This simple molecular orbital view is in accord with the experimental results. A characteristic feature of {FeNO}^{7} complexes is the small energy required to bend the MNO group. Infrared studies using ^{15}NO labeling (200) show very low bending frequencies for the {FeNO}^{7} group in Fe(NO)(S$_2$CNMe$_2$)$_2$ (7: 2). A theoretical calculation of the potential energy for bending the {FeNO}^{7} group in [Fe(NO)(CN)$_4$]$^{2-}$ shows a very flat curve with no clearly defined minimum (126). The "floppy" nature of the NO ligand in {FeNO}^{7} complexes is also manifested by the relatively large thermal parameters for the NO group (91), which in turn decrease the reliability of the determination of the Fe–N–O angle, as noted in Sect. II-A.

In principle {MNO}^{7} complexes can have either TP or TBP coordination geometry, but to date only approximate TP geometry has been observed in the solid state. However, Fe(NO)(MEEN) (7: 11) is somewhat distorted toward TBP geometry. To further explore the relationships between molecular and electronic structure of {FeNO}^{7} complexes, it would be useful to prepare complexes that are constrained to TBP geometry.

The interpretation of the structural features of complexes of the {FeNO}^{7}

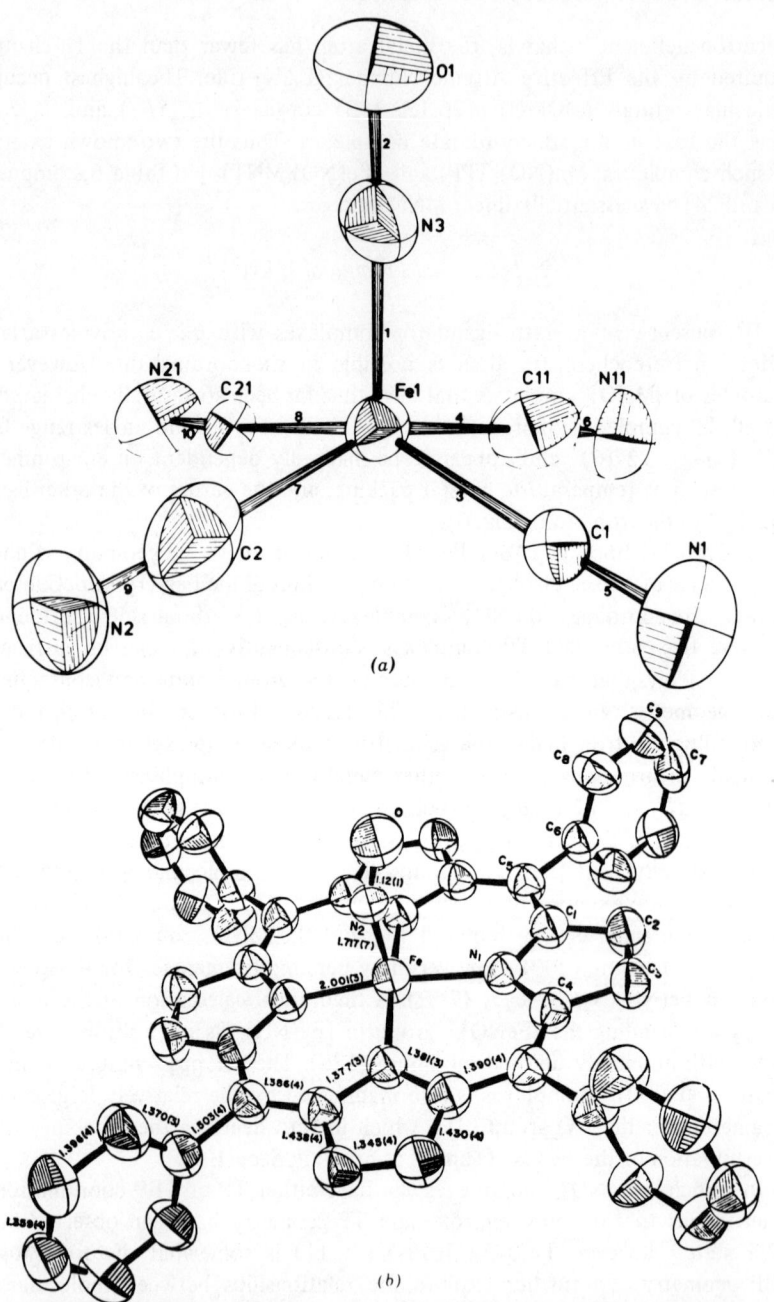

Fig. 6. Molecular structures of five-coordinate $\{FeNO\}^7$ complexes: (a) $[Fe(NO)(CN)_4]^{2-}$ (linear Fe–N–O); (b) Fe(NO)(TPP) (strongly bent Fe–N–O). (Reproduced from refs. 157 and 213.)

group is further complicated by the fact that not all complexes are low-spin ($S = 1/2$) at room temperature. In particular, complexes Fe(NO)(TMC)$^{2+}$ (7: 10) and Fe(NO)(salen) (7: 12) exhibit temperature-dependent magnetic properties. For Fe(NO)(salen) (7: 12) the $S = 1/2$ state can be obtained at liquid nitrogen temperature ($-196°$C), but Fe(NO)(TMC)$^{2+}$ (7: 10) requires liquid helium temperatures ($-268°$C) to achieve the $S = 1/2$ state. The X-ray structure of Fe(NO)(salen) (7: 12) has been determined at room temperature (23°C) and at $-175°$C (120). The differences between the room-temperature and low-temperature structures are of marginal significance, but they are consistent with a change in spin state from $S = 3/2$ to $S = 1/2$ upon cooling. The Fe–salen distances are shorter at low temperature, and the Fe–N–O angle is smaller (127° vs. ~148°) Additional structural studies of {FeNO}^{7} complexes at various temperatures are required to establish definite relationships between the molecular geometry and magnetic ground state.

The range of Fe–N distances for TP complexes of the {FeNO}^{7} group is rather large [1.56–1.80(13) Å]. Although the experimental errors are somewhat larger than for other classes of compounds, the total spread in Fe–N distances is nonetheless outside the errors reported for the experiments. Complexes with "softer" ligands such as As and S have significantly shorter Fe–N distances (1.56–1.70 Å) and larger Fe–N–O angles (168°–174°), whereas complexes with harder ligands such as N and O have longer Fe–N distances (1.72–1.80 Å) and smaller Fe–N–O angles (122°–150°). One possible explanation of these observations is that the harder ligands produce more antibonding character in d_{z^2}, which leads to a longer Fe–N bond and a more strongly bent FeNO group.

3. Tetragonal Pyramidal {MNO}^{8}

The structural chemistry of the {MNO}^{8} nitrosyl complexes has been studied extensively. Most five-coordinate {MNO}^{8} complexes fall into two categories: TP with a strongly bent MNO group and TBP with a linear MNO group. The structural data for {MNO}^{8} complexes are set out in Table 8 along with the approximate stereochemistry assigned to them by the original authors. Each of the {MNO}^{8} complexes with TP geometry (13 examples) has a strongly bent MNO group with the reported M–N–O bond angles ranging from 141° to 109° and NO distances from 1.03 to 1.47 Å. The average M–N–O angle is 125(7)° and the average NO distance is 1.18 Å. These general structural features are similar to those of six-coordinate {MNO}^{8} complexes discussed in Sect. II-A. In contrast with {MNO}^{7} complexes with TP geometry, the presence of *two* electrons in the antibonding orbitals of a TP complex of a linear MNO group is sufficiently destabilizing that the group will be distorted by bending. The structural relationship between coordination geometry and M–N–O angle in {MNO}^{8} complexes has been extensively discussed by several authors (91,126, 132,193).

As was noted for the {FeNO}[7] complexes (Sect. II-B-2), the M–N distance and the M–N–O angle of these {MNO}[8] complexes appear to depend on the properties of the ligands in the equatorial plane, with the "soft" ligands favoring a somewhat shorter M–N bond and larger M–N–O angle, and the "hard" ligands favoring longer M–N distances and smaller M–N–O angles: compare Co(NO)(S_2CNMe$_2$)$_2$ (8: 14) with Co(NO)(salen) (8: 17) and Co(NO)(TPP) (8: 21). Some authors have noted (132,180) that the strongly bent {MNO}[8] group has preferred orientations with respect to the in-plane ligands for certain M(NO)(PPh$_3$)$_2$X$_2$ complexes. However the barrier to rotation appears to be low, and the energy for site preference small.

4. Trigonal Bipyramidal {MNO}[8]

Table 8 lists 16 examples of complexes of the {MNO}[8] group that have TBP-type geometry. The M–N–O angles range from 159° to 180°, with an average of 173(6)°. Twelve of the complexes in Table 8 have the NO group in an equatorial position (Fig. 7), consistent with the hypothesis that strong π-accepting

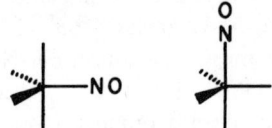

Fig. 7. Equatorial (*left*) and axial (*right*) isomers of nitrosyl complexes with TBP geometry.

ligands should occupy the equatorial site in five-coordinate complexes of a formally d^8 metal atom (114,132). Two complexes, [Fe(NO)(NP$_3$)][BPh$_4$] (8: 4) and [Fe(NO)(PP$_3$)][BPh$_4$] (8: 6) have tetradentate ligands that force the NO group to occupy the axial position. The other two examples of TBP complexes with axial NO ligands are the closely related complexes Ru(NO)H(PPh$_3$)$_3$ (8: 8) and one form of [Ir(NO)H(PPh$_3$)$_3$][ClO$_4$] (8: 32), which contain three bulky equatorial phosphine ligands.

The iridium compound [Ir(NO)H(PPh$_3$)$_3$][ClO$_4$] (8: 32) provides a graphic example of the small energy differences between the various possible five-coordinate geometries. Crystals of [Ir(NO)H(PPh$_3$)$_3$][ClO$_4$] with two different morphologies were isolated. One form of the complex adopts TBP geometry with the NO group in an axial position (181). The other form is also assigned TBP geometry but with the NO group and an H atom in the equatorial plane (60). Neither structure determination directly revealed the position of the hydride ligand.

For many TBP complexes containing an equatorial NO group, the details of the coordination geometry in the equatorial plane are obscured by disorder between the NO group and other ligands of similar size. However, 7 of the 12 complexes with equatorial NO groups exhibit ordered structures. For Mn(NO)(CO)$_4$ (8: 1) and Mn(NO)(CO)$_2$(PPh$_3$)$_2$ (8: 3) the equatorial L—M—L angles are all within 5° of the L—M—L angle in a complex with regular TBP geometry (120°).

5. Distorted Five-Coordinate {MNO}^{8}

Four TBP complexes with ordered equatorial NO groups show some distortions in the equatorial plane: [Ru(NO)(DPPE)$_2$][BPh$_4$] (8: 7), [Ru(NO)(DPPP)$_2$][BPh$_4$] (8: 9), [Co(NO)(DAS)$_2$][ClO$_4$]$_2$ (8:19), and Co(NO)Cl$_2$(PPh$_2$Me)$_2$ (8: 22). The equatorial distortions are very similar to one another but are *not* in the direction of a TP complex with an axial NO group (Fig. 8).

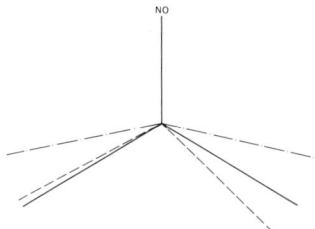

Fig. 8. Distortions from TBP geometry in the equatorial plane of {MNO}^{8} complexes. The solid lines (———) show the positions of the equatorial ligands for idealized TBP geometry. The dashed lines (- - - -) show the position observed for the equatorial ligands in Co(NO)Cl$_2$-(PPh$_2$Me)$_2$ (8: 22); similar positions are adopted by the equatorial ligands in [Co(NO)(DAS)$_2$]$^{2+}$ (8: 19), [Ru(NO)(DPPE)$_2$]$^+$ (8: 7), and Ru(NO)(DPPP)$_2$]$^+$ (8: 9). For comparison, the positions of the Cl ligands in the TP complex Ir(NO)Cl$_2$(PPh$_3$)$_2$ (8: 35) are shown by broken lines (——·——·).

The irregular stereochemistry of compound 22 in Table 8 has been discussed (31,91), but the close relationship among the equatorial angles of compounds 7, 9, 19, and 22 in Table 8 was not noted in the earlier reviews (91,105).

In a TBP molecule with an equatorial NO group, the true symmetry can be no higher than C_{2v}. This low symmetry lifts the degeneracy of the d orbitals (132). If the other ligands in the equatorial plane are grossly different from NO in their σ- and π-bonding abilities, the molecules can distort to accommodate such differences. The coordinate system used in previous discussions of equatorial bonding by NO is shown in Figure 9 (91,132). In a regular TBP, the d_{yz} orbital is π-bonding with respect to NO and primarily σ-antibonding with respect to the other two ligands in the equatorial plane. In complexes 7, 9, 19, and 22 in Table 8 the other ligands in the equatorial plane are good σ donors and poor π acceptors.

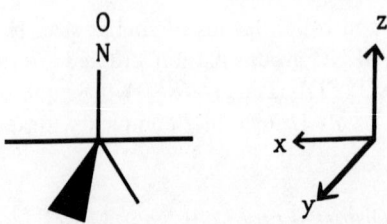

Fig. 9. The coordinate system used to describe the bonding in TBP complexes with the NO group in the equatorial plane. The d_{yz} orbital on the metal atom is π-bonding with respect to the NO ligand and primarily σ-antibonding with respect to the other two ligands in the equatorial plane.

Decreasing the L—M—L angle enhances direct transfer of σ electron density from the other ligands to the NO group. The d_{z^2} orbital is σ-antibonding with respect to the NO group and essentially nonbonding with respect to the other ligands. TBP complexes in which all the ligands in the equatorial plane are strong π-acceptors are not distorted because such a distortion would destabilize the bonding interactions of the other two π-bonding ligands.

In summary, we conclude that the five-coordinate $\{MNO\}^n$ complexes will have linear MNO groups when $n \leq 6$ as was found for the six-coordinate complexes. When $n = 7$, the MNO group in TP geometry will be very sensitive to the other ligands and to environmental effects, with a minimum bond angle in the range of 135°–140° being anticipated. Although there are no examples of $\{MNO\}^7$ complexes with TBP geometry yet, they would be expected to have nearly linear MNO groups. The $\{MNO\}^8$ complexes will have strongly bent MNO groups in TP geometry, but more nearly linear MNO groups in TBP geometry. When no steric restraints are imposed, the NO ligand prefers the equatorial site of the TBP, but its presence in the equatorial plane is likely to cause some distortion from regular TBP geometry when the other ligands have dissimilar σ- and π-bonding ability.

C. Four-Coordination

Structurally characterized examples of four-coordinate mononitrosyl complexes have now been obtained for $n = 10, 8$, and 4. Complexes with $n = 10$ are by far the most numerous (18 examples). There is only one example of four-coordinate $\{MNO\}^4$: $Cr(NO)[N(SiMe_3)_2]_3$ (4: 4). This unique complex has space group imposed C_{3v} symmetry and a linear CrNO group. The complex is diamagnetic and has no electrons in antibonding orbitals (Fig. 10). The unusually low coordination number for this complex is dictated by the steric requirements of the bistrimethylsilylamide ligand (26,27). There is also just a single structural example, $[Ir(NO)Cl(PPh_3)]_2O$, (8: 31) of an $\{MNO\}^8$ complex, although this category of complexes should be rather numerous. This dimeric complex has square planar geometry about each iridum atom and a linear IrNO group [175.7(10)°].

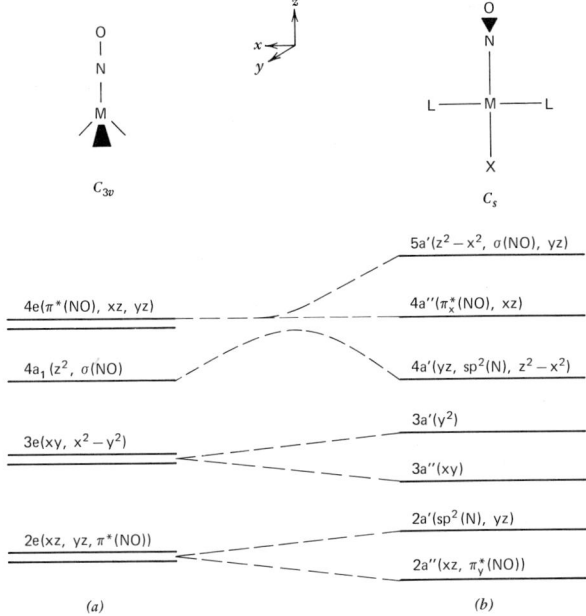

Fig. 10. Correlation diagram for four-coordinate {MNO}^{n} complexes in C_{3v} and C_s symmetry. (Reproduced from ref. 91.)

1. Tetrahedral {MNO}^{10}

With but one exception, to be discussed later, the {MNO}^{10} complexes with three *equivalent* ligands have pseudotetrahedral coordination geometry, and linear MNO groups (Fig. 11). The observed range in M—N—O bond angles for this class of complexes is 174° to 180°. The L—M—N(O) angles are ~5° greater than those of a regular tetrahedron unless larger angles are imposed by a chelating ligand as in [Ni(NO)(P$_3$C)]$^+$ (10: 12) or by ligands of grossly different steric requirements, as in Ir(NO)(CO)(PPh$_3$)$_2$ (10: 8), where the L—M—N(O) angle opens to 129°. These structural features have been discussed in terms of the molecular orbital schemes for {MNO}^{10} complexes in C_{3v} symmetry (91,177).

2. Distorted Four-Coordinate {MNO}^{10}

In the case of the {MNO}^{10} complexes in symmetry lower than C_{3v}, the MNO angle may deviate markedly from 180°. The three examples that have thus far been reported are the closely related Ni(NO)(N$_3$)(PPh$_3$)$_2$ (10: 10), Ni(NO)-(NCS)(PPh$_3$)$_2$ [(10: 11), Fig. 12] and Ni(NO)(GAP) (10: 20) complexes. The Ni—N—O angles in these three complexes are 152.7(7)°, 161.5(5)°, and 162.3(4)°, respectively. The coordination sphere consists of a slightly distorted tetrahedron that is flattened toward planar geometry.

One of the most intriguing structures to be reported for a four-coordinate

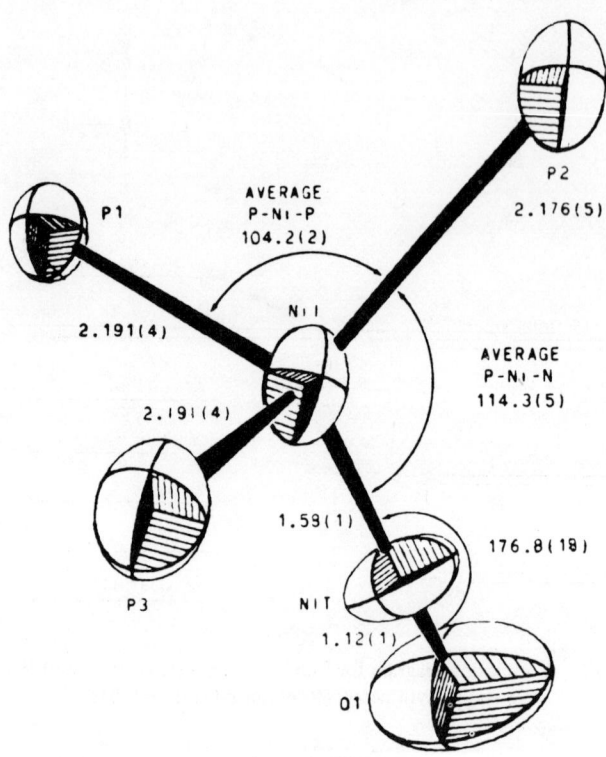

Fig. 11. The coordination geometry of the {NiNO}[10] complex with a threefold axis: [Ni(NO)(POC)$_3$] [PF$_6$]. (Reproduced from ref. 175.)

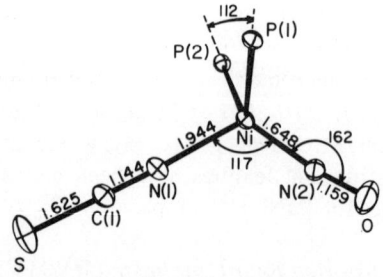

Fig. 12. The coordination geometry of Ni(NO)(NCS)(PPh$_3$)$_2$. (Reproduced from ref. 119.)

{MNO}[10] complex is that of Rh(NO)(PPh$_3$)$_3$ (10:7). This complex crystallizes in space group *P3* with three independent molecules per unit cell, each of which has crystallographically imposed C_3 symmetry in the solid state. However, in spite of the imposed threefold axis, the data indicate that the RhNO groups are distinctly nonlinear. The Rh—N—O angles found for the three independent molecules are 157.3(7)°, 159.0(7)° and 153.8(9)°. Of course, the space group requires a threefold disorder for each bent RhNO group in the crystal, as shown in Fig. 13. The oxygen atoms of the bent disordered nitrosyl ligand and the

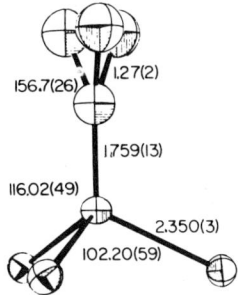

Fig. 13. The average coordination geometry of Rh(NO)(PPh$_3$)$_2$ showing the three disordered oxygen positions of the NO group. (Reproduced from ref. 147.)

phosphorus atoms of each RhP$_3$ unit adopt a staggered configuration relative to the Rh—N bond. The torsional angles along the Rh—N bond are 32° to 49° for the three independent molecules in the unit cell. Kaduk and Ibers (147) also note that the Rh—N distance [1.759(13) Å] is the shortest yet observed for a rhodium nitrosyl complex, but it is longer than the Ir—N distance of 1.67(2) Å in the isoelectronic complex Ir(NO)(PPh$_3$)$_3$ (10:9), which has a linear IrNO group.

Analysis of the intermolecular contacts in Rh(NO)(PPh$_3$)$_3$ suggests that the bending of the nitrosyl group is *not* due to packing effects. No electronic rationalization for the bending of the {RhNO}^{10} group in Rh(NO)(PPh$_3$)$_3$ has previously been offered. In fact, earlier theoretical discussions emphasize that {MNO}^{10} groups should be linear in rigorous trigonal symmetry (89,91,177). In principle the molecular orbital scheme of Fig. 10(a) can explain a nonlinear {MNO}^{10} group, provided the $4a_1(z^2, \sigma(NO))$ and $4e(\pi^*(NO), xz, yz)$ orbitals are nearly degenerate. The states arising from a pair of electrons in two such orbitals have been analyzed in detail previously (91). Suffice it to say that the degeneracy of the $^1E(4a)^1(4e)^1$ state is lifted by coupling with the M—N—O bending vibration. Our previous analyses of vibronic coupling in transition metal nitrosyl complexes assumed that changes in the coordination geometry would accompany changes in the M—N—O angle. The structure observed for Rh(NO)(PPh$_3$)$_3$ indicates that this need not be the case. Moreover, the bending of the {RhNO}^{10} unit is completely consistent with Renner's analysis (207) of the behavior of linear molecules with degenerate electronic states. The novel structure of Rh(NO)(PPh$_3$)$_3$ further underscores the utility of the simple {MNO}^{n} bonding framework for discussing the properties of mononitrosyl complexes.

D. Three-Coordination

Very recently three-coordinate $\{MNO\}^{10}$ complexes have been prepared (51). The structure of one example, $[Ni(NO)(Me_2Pz)]_2$ (10: 16) is shown in Fig. 14.

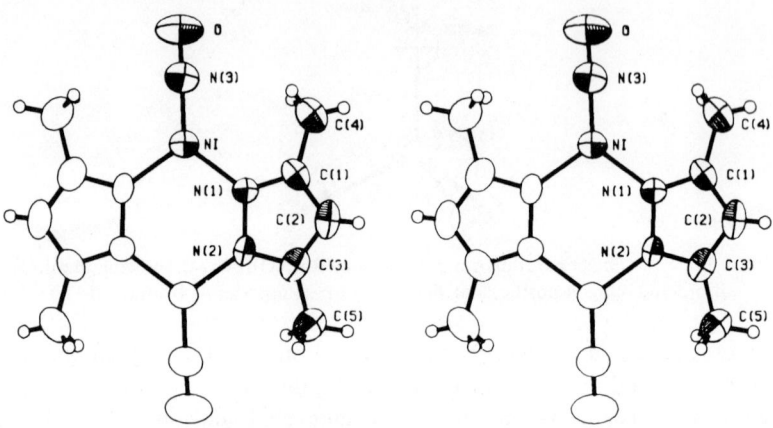

Fig. 14. Stereoview of the geometry of the three-coordinate $\{MNO\}^{10}$ complex, $[Ni(NO)(Me_2Pz)]_2$. (Reproduced from ref. 51.)

The stereochemistry about each Ni atom is approximately trigonal planar, and the two Ni atoms of the dimer are 3.673(1) Å apart. It seems likely that the pyrazoyl ligand plays an important role in the stability of such complexes. The complexes react with neutral and anionic donors to form binuclear four-coordinate $\{NiNO\}^{10}$ complexes with pseudotetrahedral coordination geometry about each Ni atom (52) (10: 18).

E. Higher Coordination Numbers

Examples of nitrosyl complexes with coordination numbers greater than six are rare and thus far are limited to $\{MNO\}^4$ species (Table 4). The complexes $[V(NO)(CN)_6]^{4-}$ (4: 1), $Mo(NO)(S_2CNBu_2^n)_3$ (4: 6), and $[Mo(NO)(OCNMe_2)$-$(NCS)_4]^{2-}$ (4: 10) are each seven-coordinate with pentagonal bipyramidal (PBP) geometry. In each complex the NO ligand is linear and occupies the axial position. When not distorted because of the bite angle of chelating ligands, the seven-coordinate complex has regular PBP geometry, with the L–M–L angles in the equatorial plane being within 1° of the idealized angle of 72°. The atoms in the equatorial plane exhibit only slight deviations from coplanarity (0.02–0.07 Å). However the presence of bidentate ligands in $Mo(NO)(S_2CNBu_2^n)_3$ (4: 6) results in puckering of the equatorial plane by as much as 0.4 Å. The sulfur atom trans to the NO group is also displaced by 12° from the axis of the PBP because of the small bite angle of the dithiocarbamate ligand. Although a complete report of the structural details for the carbamido complex, [Mo(NO)-(OCNMe$_2$)(NCS)$_4$]$^{2-}$ (4: 10) is not yet available, PBP geometry with an axial NO ligand also occurs in this complex.

Each of the other examples of complexes with higher coordination numbers contains the ambidentate cyclopentadienyl ligand, and consequently, the coordination number is more difficult to define. In its η^5 mode, the cyclopentadienyl ligand normally replaces three monodentate electron pair donor ligands, and consequently could be considered as a tridentate ligand. However, if the five Cp atoms are not equidistant from the metal atom, it is difficult, if not impossible, to assess the number of electrons donated by the Cp ligand and hence the effective coordination number of the metal.

Each of the {MNO}^{4} complexes in Table 4 that has cyclopentadienyl ligands attached to the central metal atom has a linear MNO group. Symmetrical coordination of the cyclopentadienyl ligands would lead to a maximum coordination number of 7 in [Mo(NO)(Cp)NH$_2$NHPh)I]$^+$ (4: 9) and [Mo(NO)(Cp)I]$_2$-(μ_2-NNMe$_2$) (4: 11) and to a linear MNO group. However similar symmetrical coordination of two cyclopentadienyl rings in Mo(NO)(Cp)$_2$(η^1-Cp) (4: 7) and Mo(NO)(Cp)$_2$Me (4: 8) would lead to an effective coordination number of 8, for which a strongly bent {MNO}^{4} group is expected. However, the MNO group in the latter two complexes is linear, suggesting that the effective coordination number is 7 or less, consistent with the observed inequivalence of their Mo–C distances. Other examples of nitrosyl complexes of ambidentate ligands are given in Sect. V.

III. DINITROSYL COMPLEXES

Compared with the large number of structural data available for mononitrosyl complexes, few dinitrosyl complexes have been structurely characterized. The structures of dinitrosyl complexes are tabulated using the {M(NO)$_2$}^{n} notation, and the discussion in the text is subdivided according to the coordination number of the central metal atom.

A. Six-Coordination

All six-coordinate complexes structurally characterized (Table 12) have $n = 6$ with the nitrosyl groups cis to one another, including Mo(NO)$_2$TTP (12: 4). The cis geometry of the nitrosyl ligands in the latter compound is probably due in part to the large size of the Mo atom relative to the size of the hole in the dianion of TTP and in part to the well-known propensity of π-accepting ligands such as NO to adopt cis geometry. The structures of Cr(NO)$_2$(Cp)(NCO) (12: 1) and

Table 12 Structural Data for Metal Dinitrosyl Complexes: $\{M(NO)_2\}^6$

Compound	M–N (Å)	N–O (Å)	M–N–O (deg.)	N–M–N (deg.)	Reference
1. $Cr(NO)_2(Cp)(NCO)$	1.716(3)	1.157(3)	171.0(2)	94.9(2)	34,35
2. $Cr(NO)_2Cl(Cp)$	1.717(12) 1.704(13)	1.128(19) 1.152(19)	170.8(13) 166.4(13)	94.3(6)	43
3. $Mo(NO)_2Cl_2(PPh_3)_2$	1.818(9) 1.905(40)	1.223(12) 1.158(50)	163.1(10) 160.4(10)	n.r.[a]	240
4. cis-$Mo(NO)_2(TTP)$	1.70(1)	n.r.[a]	158.0(8)	78.4(5)	79

[a] Data not reported.

Cr(NO)₂Cl(Cp) (12: 2) are nearly identical and have Cr—N—O angles of 166°–171°. As was first noted in 1958 (141), slightly bent nitrosyl groups are a common feature of polynitrosyl complexes; an electronic rationalization for this was given by Kettle (151) several years ago. In the compounds Cr(NO)₂(Cp)NCO (4: 1) and Cr(NO)₂Cl(Cp) (4: 2) the bending of the nitrosyl oxygen atoms is normal to the CrN₂ plane and away from the anionic ligand (Fig. 15).

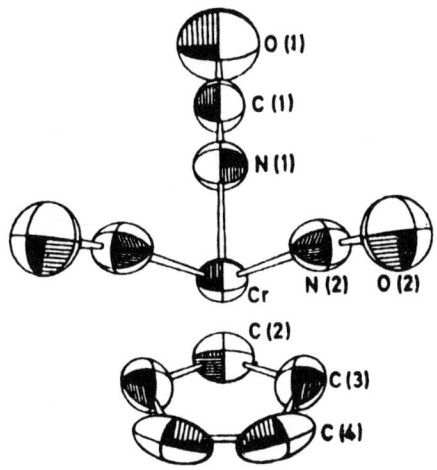

Fig. 15. The structure of the $\{M(NO)_2\}^6$ complex Cr(NO)₂(Cp)(NCO), showing the slight bending of the nitrosyl ligands. (Reproduced from ref. 35.)

B. Five-Coordination

Most five-coordinate dinitrosyl complexes characterized to date have an $\{M(NO)_2\}^8$ electron configuration. This class of compounds (Table 13) is particularly interesting because compounds with equivalent nitrosyl groups Mn(NO)₂Cl(P(OMe)₂Ph)₂ and [Mn(NO)₂(P(OMe)₂Ph)₃]⁺ (Table 13, compounds 1–3) and with nonequivalent nitrosyl groups [Ru(NO)₂Cl(PPh₃)₂]⁺ and [Os(NO)₂(OH)(PPh₃)₂]⁺ (compounds 4 and 5) are known. One of the two examples of complexes with strikingly different nitrosyl ligands attached to the same metal is [Ru(NO)₂Cl(PPh₃)₂]⁺ (13: 4), first reported by Eisenberg and co-workers (194,198) (Fig. 16). One of the MNO groups is linear and the other is strongly bent (138°). Similar results have been found for the related osmium complex (13: 5).

On the other hand, the $\{M(NO)_2\}^8$ complexes Mn(NO)₂Cl(P(OMe)₂Ph)₂ (13: 1) and [Mn(NO)₂(P(OMe)₂Ph)₃]⁺ (13: 3) show TBP geometry with equivalent nitrosyl ligands in the equatorial plane. For Mn(NO)₂Cl(P(OMe)₂Ph)₂ it is clear that the observed Mn—N—O angles of 163° to 166° are due to electronic

Table 13 Structural Data for Metal Dinitrosyl Complexes: $\{M(NO)_2\}^8$

Compound	M–N (Å)	N–O (Å)	M–N–O (deg.)	N–M–N (deg.)	Reference
1. $Mn(NO)_2(P(OMe)_2Ph)_2Cl$, monoclinic form	1.650(10) 1.665(10)	1.18(1) 1.18(1)	163(1) 166(1)	111.5(5)	158, 159
2. $Mn(NO)_2(P(OMe)_2Ph)_2Cl$, triclinic form	1.657(10) 1.644(10) 1.633(10) 1.642(10)	1.19(1) 1.18(1) 1.19(1) 1.19(1)	167(1) 165(1) 166(1) 165(1)	112.1(5) 113.5(5)	160
3. $[Mn(NO)_2(P(OMe)_2Ph)_3][BF_4]$	1.649(10) 1.649(10)	1.18(1) 1.19(1)	168(1) 170(1)	116.5(5)	161

4. $[Ru(NO)_2Cl(PPh_3)_2][PF_6]$
  Axial
  Equatorial
	
	1.853(19)
	1.743(20)
		
		1.166(20)
		1.158(19)
			
			138(2)
			178(2)
				
				102(1)
					194, 198

5. $[Os(NO)_2(OH)(PPh_3)_2][PF_6]$
  Axial
  Equatorial
	
	1.86(1)
	1.63(1)
		
		1.17(2)
		1.24(2)
			
			133.6(12)
			177.6(12)
				
				99.3(6)
					61, 244

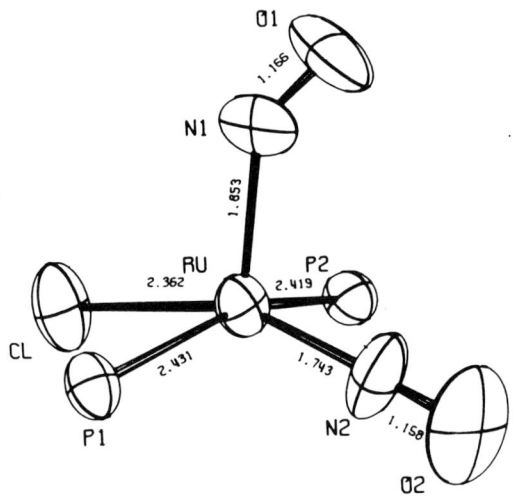

Fig. 16. An example of a $\{M(NO)_2\}^8$ complex with both a linear and a bent MNO group: $[Ru(NO)_2Cl(PPh_3)_2]^+$. (Reproduced from ref. 194.)

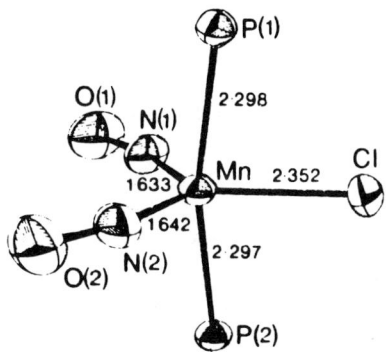

Fig. 17. The molecular geometry of $Mn(NO)_2Cl(P(OMe)_2Ph)_2$, an $\{M(NO)_2\}^8$ complex with equivalent NO groups in the equatorial plane of a TBP. (Reproduced from ref. 160.)

factors, since the same stereochemistry is found for three independent modifications (Fig. 17).

It should be noted that in a previous review (91) five-coordinate $\{M(NO)_2\}^8$ complexes were predicted to exist in two different forms: TP complexes with two nonequivalent nitrosyl groups, which were known; and the then-unknown TBP complexes with two equivalent nitrosyl groups in the equatorial plane. It was also predicted that the latter structure was most likely to occur with elements of the first transition series. The structures found for $Mn(NO)_2Cl(P(OMe)_2Ph)_2$ and $[Mn(NO)_2(P(OMe)_2Ph)_3]^+$ confirm those predictions. It has been generally assumed (194) that the two nitrosyl groups in $Ru(NO)_2Cl(PPh_3)_2^+$ (13:4) equilibrate in solution, but this has never been demonstrated explicitly.

Very recently the five-coordinate $\{Fe(NO)_2\}^9$ complex $Fe(NO)_2(GAP)$ (14:5) has been prepared and its structure determined. The molecule has approximate TBP geometry with the NO groups in the equatorial plane. The slight bending of the nitrosyl ligands (Fe–N–O = 159°) is in the FeN_2 plane, so as to move the oxygen atoms closer together. The same manner of bending occurs in $Mn(NO)_2Cl(P(OMe)_2Ph)_2$ (Fig. 17).

C. Four-Coordination

There is one example, $Fe(NO)_2Cl(PPh_3)$ (14:2), of a monomeric four-coordinate complex having the $\{M(NO)_2\}^9$ electron configuration. Other four-coordinate complexes involving the $\{Fe(NO)_2\}^9$ moiety are binuclear and diamagnetic (Table 14).

Structural studies of four-coordinate $\{M(NO)_2\}^{10}$ complexes have been stimulated in part by the observation that certain of these complexes react with CO to produce CO_2 and N_2O (e.g. [1]) (88). The structures of

$$Ir(NO)_2(PPh_3)_2Br + 3CO \longrightarrow Ir(CO)_2(PPh_3)_2Br + CO_2 + N_2O \quad [1]$$

four-coordinate $\{M(NO)_2\}^{10}$ complexes are listed in Table 15. All the complexes in Tables 14 and 15 exhibit pseudotetrahedral to severely flattened tetrahedral geometry with two equivalent nitrosyl groups. However several of these complexes have distinctly nonlinear M–N–O groups. The geniculation (i.e., relative manner of bending) of the two nitrosyl groups in four-coordinate $M(NO)_2$ complexes has been discussed previously by Enemark and Feltham (91), Summerville and Hoffman (230), and Martin and Taylor (169), using molecular orbital theory. The correlation of N–M–N and O–M–O angles noted by Martin and Taylor is shown in Figure 18. Complexes with N–M–N angles less than ~130° are geniculated so that the two oxygen atoms bend toward each other (Fig. 19). Complexes with N–M–N angles greater than ~130° are geniculated to move the oxygen atoms farther apart (Fig. 20).

D. Polynitrosyl Complexes

The syntheses of $Mn(NO)_3CO$ and $Cr(NO)_4$ have been reported (13,232). The stereochemistry of these molecules has been predicted from spectroscopic and theoretical arguments, but the metrical details await structure determinations.

Table 14 Structural Data for Metal Dinitrosyl Complexes: $\{M(NO)_2\}^9$

Compound	M–N (Å)	N–O (Å)	M–N–O (deg.)	N–M–N (deg.)	Reference
1. $[Fe(NO)_2]_2(\mu_2\text{-}P(CF_3)_2)_2$	1.661(3) 1.655(3)	1.158(4) 1.156(4)	175.2(3) 174.0(3)	122.4(1)	64
2. $Fe(NO)_2Cl(PPh_3)$	1.679(5) 1.681(5)	1.136(7) 1.163(7)	166.4(5) 165.5(5)	115.6(3)	156
3. $[Fe(NO)_2]_2(\mu_2\text{-}I)_2$	1.73(3) 1.66(2) 1.64(2) 1.65(3)	1.06(3) 1.16(3) 1.21(3) 1.17(4)	157(3) 165(3) 163(2) 161(3)	117.8(12) 115.0(13)	73
4. $[Fe(NO)_2]_2(\mu_2\text{-}SEt)_2$	1.66(8) 1.67(4)	1.18(1) 1.16(1)	167.7(35) 167.2(35)	117.4(2)	234
5. $Fe(NO)_2(GAP)$	1.706(4) 1.700(4)	1.158(5) 1.148(5)	157.9(4) 159.3(5)	108.1(2)	54

Table 15 Structural Data for Metal Dinitrosyl Complexes: $\{M(NO)_2\}^{10}$

Compound	M–N (Å)	N–O (Å)	M–N–O (deg.)	N–M–N (deg.)	Reference
1. $Fe(NO)_2(CO)_2$ [a]	1.77(2)	1.12(3)	180 [b]	109 [b]	33
2. $Fe(NO)_2(PPh_3)_2$	1.650(7)	1.190(10)	178.2(7)	123.8(4)	3
3. $Fe(NO)_2(CO)(PPh_3)$ [c]	1.709(8)	1.15(1)	177.9(9)	114.4(4)	3
4. $Fe(NO)_2(f_6fos)$	1.661(7) 1.645(7)	1.184(10) 1.177(10)	177.8(7) 176.9(7)	125.4(4)	122
5. $Ru(NO)_2(PPh_3)_2$	1.763(6) 1.776(6)	1.190(7) 1.194(7)	177.7(6) 170.6(5)	139.2(3)	108
6. $Ru(NO)_2(PPh_3)_2$	1.748(20) 1.688(20)	1.215(18) 1.229(18)	174.7(17) 168.0(16)	139.9(8)	17
7. $Os(NO)_2(PPh_3)_2$	1.776(7) 1.771(6)	1.195(8) 1.211(7)	178.7(7) 174.1(6)	139.1(3)	124
8. $[Co(NO)_2I]_n$	1.61(4) 1.69(4)	1.16(5) 1.19(5)	168(5) 173(4)	118.3(18)	73
9. $[Co(NO)_2Cl]_n$ [c]	1.73(3)	1.12(5)	166(3)	110(2)	138
10. $[Co(NO)_2(DPPE)][PF_6]$	1.656(10) 1.671(11)	1.130(14) 1.142(13)	176.6(14) 172.3(11)	131.7(5)	146
11. $[Co(NO)_2(PPh_3)_2][PF_6]$	1.645(6)	1.174(6)	171.0(5)	136.7(4)	205

12. Co(NO)$_2$(DPPEO)I	1.677(20)	1.052(17)	120.5(19) 166.3(26) 154.6(43) 149.0(57)	119.9(11)	103
13. Co(NO)$_2$(sacsac)	1.659(6)	1.120(5)	168.9(5)	115.5(3)	169
14. [Co(NO)$_2$]$_4$(NO$_2$)$_2$(N$_2$O$_2$)	1.656^b 1.644^b	1.151^b 1.137^b	162.1^b 164.0^b	113.3^b	14
15. [Co(NO)$_2$(PPh$_3$)$_2$][ClO$_4$]	1.663(3) 1.665(3)	1.155(4) 1.155(4)	171.2(3) 171.1(3)	132.4(1)	211
16. [Co(NO)$_2$(NO$_2$)]$_n$	1.67^b 1.67^b	1.13^b 1.11^b	n.r.^{d}	112^b	229
17. [Co(NO)$_2$(PEPEOA)][BPh$_4$]	1.622(1) 1.61(1)	1.19(1) 1.17(1)	175.8(10) 175.3(10)	129.5(6)	109
18. [Rh(NO)$_2$(PPh$_3$)$_2$][ClO$_4$]	1.818(4)	1.158(6)	158.9(4)	157.5(3)	145
19. [Ir(NO$_2$)(PPh$_3$)]$_2$	1.787^b	1.19^b	167^b	156^b	9
20. [Ir(NO)$_2$(PPh$_3$)$_2$][ClO$_4$]	1.771(12)	1.213(13)	163.5(10)	154.2(7)	179

^{a} Electron diffraction data.
^{b} Estimated standard deviation not reported.
^{c} Disordered.
^{d} Data not reported.

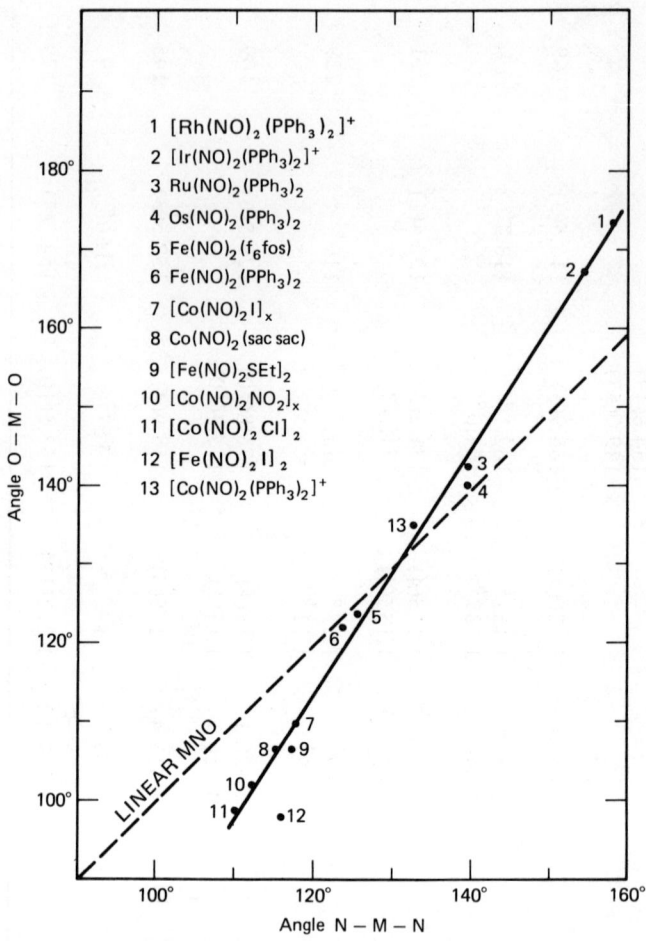

Fig. 18. The correlation of N–M–N angles and O–M–O angles in four-coordinate $\{M(NO)_2\}^{10}$ complexes. (Reproduced from ref. 169.)

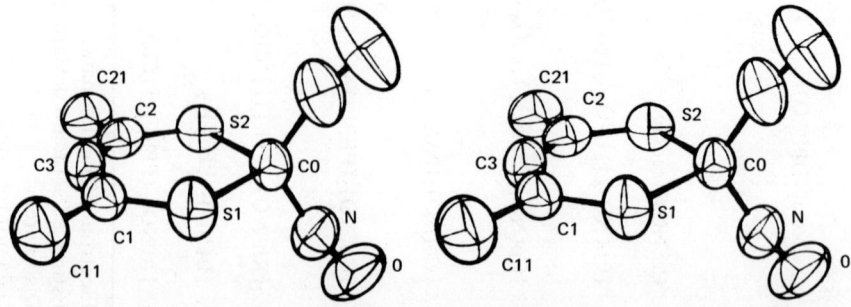

Fig. 19. Stereoview of the molecular geometry of $Co(NO)_2$(sacsac). The oxygen atoms are bent toward each other. (Reproduced from ref. 169.)

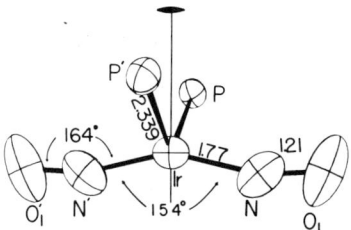

Fig. 20. Molecular geometry of $[Ir(NO)_2(PPh_3)_2]^+$. The oxygen atoms are bent away from each other. (Reproduced from ref. 179.)

IV. BRIDGING AND POLYNUCLEAR NITROSYLS

The structural parameters for the compounds with bridging nitrosyl groups are set out in Table 16. Compounds containing bridging NO ligands are less common than their carbonyl counterparts. However the data show that NO can bridge between two metal atoms (μ_2) and three metal atoms (μ_3). In its μ_2 configuration, examples of both singly bridged and doubly bridged metal atoms have been characterized, but thus far no examples of compounds containing three μ_2-NO ligands for two metal atoms have been described. Although some qualitative molecular orbital descriptions of complexes with bridging nitrosyl groups have been suggested, (16), detailed analysis of the bonding in these complexes has not yet appeared.

Nitrosyl cluster compounds exhibit MNO structural features similar to those that have already been discussed. However in most cases the electrons are delocalized over the several metal atoms of the cluster and the assignment of values for n can be ambiguous. Consequently, the structural data for these polynuclear species have been collected separately (Table 18, Sect. IV-D).

A. $M_2(\mu_2$-NO) Complexes

There are six examples of compounds with a single bridging NO group, (Table 16). Since each of these compounds has one additional bridging ligand, in many respects this category is closely related to the $M_2(\mu_2$-NO$)_2$ species discussed in Sect. IV-B. No examples of compounds in which NO bridges two different metal atoms have been structurally characterized, but the structure of $Os(HNO)(CO)Cl_2(PPh_3)_2$ (16:5) is known.

The important structural features for this class of complexes include the M—N distances, the N—O distance, and the coplanarity, or lack thereof, of the M_2NO group. Comparison of the M_1—N and M_2—N distances shows that generally these compounds can be described as having symmetrical NO bridges. The observed M—N distances are somewhat longer than the corresponding linear

Table 16 Structural Data for Complexes with μ_2-NO Bridges

Compound	M_1-N_1 M_1-N_2 (Å)	M_2-N_1 M_2-N_2 (Å)	N_1-O_1 N_2-O_2 (Å)	$M_1-N_1-O_1$ $M_1-N_2-O_2$ (deg.)	$M_2-N_1-O_1$ $M_2-N_2-O_2$ (deg.)	$M-N_1-M$ $M-N_2-M$ (deg.)	Reference
1. [Mn(Cp)]$_3$(μ_2-NO)$_3$(μ_3-NO) $\begin{array}{c} O \\ \| \\ N \\ / \backslash \\ M_1 \quad M_2 \end{array}$	1.855(4) 1.840(4) 1.841(3)	1.857(3) 1.851(4) 1.842(4)	1.210(5) 1.207(5) 1.218(5)	137.6(3) 137.3(3) 136.9(3)	137.4(3) 136.4(3) 137.1(3)	85.0(1) 85.6(2) 85.6(2)	85
2. [Mn(NO)(Cp)](μ_2-NO)(μ_2-CO)-[Mn(CO)(Cp)][a]	1.911(4)	1.901(3)	1.194(4)	137.1(3)	138.1(3)	84.8(1)	154
3. [Cr(NO)(Cp)]$_2$(μ_2-NO)(μ_2-NH$_2$)[a]	1.936(12)	1.936(12)	1.121(22)	135.1(17)	135.4(17)	86.4(5)	47
4. [{Fe(MEEN)}$_2$(μ_2-NO)][BF$_4$]	1.818(5)	1.818(5)	1.193(8)	137.2(1)	137.2(1)	85.5(3)	148,201
5. Os(HNO)(CO)Cl$_2$(PPh$_3$)$_2$	1.915(6)	0.94(11)	1.193(7)	136.9(6)	99(7)	123(7)	250
6. [{Co(MEEN)}$_2$(μ_2-NO)][BF$_4$]	1.800(6)	1.817(6)	1.211(7)	130.8(5)	129.2(5)	99.9(3)	148,201
7. [Co(Cp)]$_2$(μ_2-NO)(μ_2-CO)[a]	1.829(4)	1.831(4)	1.200(5)	139.5(3)	139.8(3)	99.3(2)	16
8. [Et$_4$N][(PtCl$_2$)(μ_2-NO)(μ_2-Cl)(Pt(NO)Cl$_3$)] $\begin{array}{c} O_1 \\ \| \\ N_1 \\ / \backslash \\ M_1 \quad M_2 \\ \backslash / \\ N_2 \\ \| \\ O_2 \end{array}$	1.89(4)	1.93(3)	1.27(5)	123(3)	118(3)	119(2)	98
9. [Cr(NO)(Cp)]$_2$(μ_2-NO)$_2$	1.960(3)	1.969(3)	1.193(4)	136.4(3)	136.4(3)	86.7(1)	39

10. [Mn(NO)(Cp)](μ_2-NO)$_2$[Mn(Cp)(η^1-Cp)]	1.944(3) 1.944(3)	1.752(3) 1.752(3)	1.223(4)	n.r.^{b}	n.r.^{b}	n.r.^{b}	41
11. [Mn(NO)(Cp)](μ_2-NO)$_2$[Mn(Cp)(NO$_2$)]	1.943^c 1.943^c	1.775^c 1.775^c	1.205^b 1.205^b	n.r.^{b} n.r.^{b}	n.r.^{b} n.r.^{b}	n.r.^{b} n.r.^{b}	37
12. [Fe(Cp)]$_2$(μ_2-NO)$_2$	1.768(9)	1.768(9)	1.254(12)	138.8(8)	138.8(8)	82.3(4)	40
13. Ru$_3$(CO)$_{10}$(μ_2-NO)$_2$	2.05(1) 2.03(1)	1.99(1) 2.04(1)	1.20(1) 1.24(1)	129.4(6) 127.2(6)	129.4(6) 130.2(6)	101.3(3) 102.5(4)	190
14. [Co(Cp)]$_2$(μ_2-NO)$_2$	1.827(7)	1.824(7)	1.187(10)	139.4(5)	139.6(5)	99.0(3)	16
15. [Pt$_2$(μ_2-OAc)$_2$]$_2$(μ_2-NO)$_2$(μ_2-OAc)$_2$	1.869(21)	1.955(21)	1.22(3)	120.0(17)	120.0(17)	119.9(12)	77,78

^{a} Disordered.
^{b} Data not reported.
^{c} Estimated standard deviation not given.

terminal MNO species (see Table 6), but approximately the same as for the strongly bent terminal MNO species, (see Table 8, e.g.). The average NO distance (1.204 Å) is somewhat longer than for the terminal MNO complexes (1.159 Å) but is still well within the range of distances found for the terminal nitrosyl species. In contrast with bridging carbonyls, the M—N—M angles of [{Co(MEEN)}$_2$-(μ_2-NO)] [BF$_4$], [CoCp]$_2$(μ_2-NO)(μ_2-CO) and [(PtCl$_2$)(μ_2-NO)(μ_2-Cl)(Pt-(NO)Cl$_3$)]$^-$, (Table 16, compounds 6—8) are significantly greater than 90°, as are those of some of the doubly bridged complexes discussed below. There appears to be only one example (67) of a bridging CO group with a bond angle of 120°; but there are two examples of bridging nitrosyl with M—N—M angles of ~120°, and two others near 100°. The underlying source of these differences remains to be elucidated. Since each of these compounds has coplanar M$_2$NO moieties, the M—N—O angles are usually larger than 120°.

B. M$_2$(μ_2-NO)$_2$ Complexes

The general structural features found for the M$_2$(μ_2-NO)$_2$ group of compounds closely resemble those for the singly bridged complexes discussed above. The average NO distance is 1.215 Å, although the M—N—O angles have a somewhat wider range of values (120°—140°). However [Co(Cp)]$_2$(μ_2-NO)$_2$ and [Fe(Cp)]$_2$(μ_2-NO)$_2$ (Table 16, compounds 14 and 12) offer the opportunity for comparing the possible effects which electron count has on the structural features of the M$\overset{N}{\underset{N}{\diamond}}$M moiety. These complexes contain 20 and 22 valence electrons ($d + \pi^*$), respectively. Although the M—N—O angles and NO distances differ only marginally, the M—N distances and M—N—M angles differ significantly. Moreover, [Fe(Cp)]$_2$(μ_2-NO)$_2$ (16: 12) has planar M—(NO)—M groups, whereas in [Co(Cp)]$_2$(μ_2-NO)$_2$ (16: 14), the M—(NO)—M groups are nonplanar. These structural features suggest that the two additional electrons in the cobalt complex reside in an orbital that is antibonding with respect to the two metal atoms. The Fe—Fe distance (2.326(4)Å) and the Co—Co distance (2.372(1) Å) as well as those of the [Co(Cp)]$_2$(μ_2-CO)$_2$ complexes (112) support this conclusion, as does the ESR spectrum of compound 7 in Table 16, which shows that the unpaired electron is in an orbital that is distributed equally between the two cobalt nuclei.

C. Triply Bridging NO Complexes M$_3$(μ_3-NO)

The metrical details of the only complex of the M$_3$(μ_3-NO) type that has been structurally characterized are set out in Table 17. The NO distance does not differ significantly from the values found for the μ_2-NO complexes described

Table 17 Structural Data for Complexes with μ_3-NO Bridges

Compound	M_1-N M_2-N M_3-N (Å)	$N-O$ (Å)	M_1-N-M_2 M_2-N-M_3 M_3-N-M_1 (deg.)	M_1-N-O M_2-N-O M_3-N-O (deg.)	Reference
[Mn(Cp)]$_3$(μ_2-NO)$_3$(μ_3-NO)	1.932(3) 1.917(4) 1.938(4)	1.247(5)	81.1(1) 81.1(2) 80.8(2)	131.4(3) 132.3(3) 130.5(3)	85

above. Moreover, the M—N—O angles are also within the range found for the μ_2-NO complexes, although the M—N—M angles are somewhat smaller, a feature that is probably attributable to the triangular structure of the metal cluster. Comparison with triply bridging CO ligands shows that the M—N—M angles are marginally larger than those of the M—C—M species (127).

D. Polynuclear Metal Nitrosyls

Although relatively few polynuclear metal nitrosyls have been prepared, Roussin's black salt, $[Fe_4(NO)_7S_3]^-$ was one of the earliest nitrosyl complexes to be structurally characterized (Table 18, compound 3). This anion, whose structure was first elucidated by Johansson and Lipscomb (141) and later by Chu and Dahl (55), exhibits most of the features common to compounds of this type. Although the extraordinary stability of $[Fe_4(NO)_7S_3]^-$ remains to be explained, it has no structural features not exhibited by the mononuclear species. The apical iron of the four-iron tetrahedral cluster has one terminal NO ligand with an Fe—N distance of 1.65(1) Å and an Fe—N—O angle of 176.3(9)°. Each of the other three iron atoms that lie at the base of the distorted tetrahedron is coordinated to two terminal NO ligands. Although it is tempting to compare the FeNO bond distances and angles with those found for the $\{FeNO\}^6$ and $\{Fe(NO)_2\}^9$ species (Tables 6 and 14), a simple electron count coupled with the fact that apical iron atom is only four-coordinate shows that such a simple analogy does not apply. Although these complications in electron accounting have prevented assignment of electron configurations to the individual $\{M(NO)_m\}$ groups in these clusters, examination of the data in Table 18 fails to show any structural features that are not encountered in the simple monomeric nitrosyl complexes. It is also worth noting that CO, sulfur, and halogen ligands seem particularly adept at stabilizing metal nitrosyl clusters. The current interest in nitrosyl clusters by several research groups should elicit a clearer understanding of these species. It is worth noting that if the NO ligand in these complexes is counted as a three-electron donor, the application of Wade's rules to $Fe_4(NO)_4S_4$ (18: 4) and $Co_4(NO)_4(\mu_3\text{-NCMe}_3)_4$ (18: 10) (241) gives a total of 12 valence electrons, and predicts a tetrahedral array of the four metal atoms, as is observed experimentally.

V. NITROSYL COMPLEXES CONTAINING SO_2 AND C_3H_5 LIGANDS

In addition to NO, other small molecules that also exhibit more than one mode of bonding to transition metals include O_2, SO_2, and C_3H_5 (134,178). Although a complete review of the structures of such complexes is beyond the scope of this chapter, several compounds are known in which either the SO_2

Table 18 Structural Data for Polynuclear Metal Nitrosyls

Compound	M–N (Å)	N–O (Å)	M–N–O (deg.)	μ_2-M–N (Å)	μ_2-N–O (Å)	μ_2-M–N–O (deg.)	M–N–M (deg.)	Reference
1. $[NH_4]_4[Mo_4(NO)_4S_{13}]$	1.742(16)	1.219(22)	175(2)	–	–	–	–	187
2. $[\{Mo_3(NO)_3(CO)_6(OMe)_3(O)\}_2Na][Ph_3PNPPh_3]$	1.77(2) 1.84(2) 1.83(2)	1.21(3) 1.21(2) 1.17(3)	179(3) 178(2) 177(2)	–	–	–	–	163
3. $[AsPh_4][Fe_4(NO)_7(\mu_3\text{-}S)_3]$	1.65(1)	1.17(1)	176.3(9) 166.8(9)	–	–	–	–	55,141
4. $Fe_4(NO)_4(\mu_3\text{-}S)_4$	1.663(5)	1.152(6)	177(5)	–	–	–	–	106
5. $Fe_4(NO)_4(\mu_3\text{-}S)_2(\mu_3\text{-}NCMe_3)_2$	1.661(5)	1.166(7)	178.6(5)	–	–	–	–	106
6. $Ru_3(CO)_{10}(\mu_2\text{-}NO)_2$	–	–	–	2.03(1) 2.04(1) 2.05(1) 1.99(1)	1.20(1) 1.24(1)	129.1(6) 129.4(6) 130.2(6) 127.2(6)	101.3(3) 102.5(4)	190
7. $([Ru(NO)]_2(\mu_2\text{-}Cl)_2)_2(\mu_2\text{-}PPh_2)_2$	1.779(7)	1.160(10)	160.3(8)	–	–	–	–	87
8. $Os_3(NO)_2(CO)_8\{P(OMe)_3\}$	1.77(4) 1.65(4)	1.16(5) 1.21(5)	171(4) 165(4)	–	–	–	–	208
9. $Ru_3(\mu_2\text{-}NO)(\mu_2\text{-}H)(CO)_7\{P(OMe)_3\}_3$	–	–	–	1.989(8) 1.973(5)	1.230(9)	134.1(5) 135.2(6)	90.6(3)	143
10. $Co_4(NO)_4(\mu_3\text{-}NCMe_3)_4$	1.645(9) 1.662(10)	1.16(1) 1.15(1)	174(1) 168(1)	–	–	–	–	107

Table 19 Nitrosyl Complexes Containing SO_2 and C_3H_5 Ligands

Compound	Type	M–N (Å)	N–O (Å)	M–N–O (deg.)	Geometry of SO_2 or C_3H_5	Reference
1. $Ru(NO)Cl(SO_2)(PPh_3)_2$	$\{MB_2\}^8$	1.740(6)	1.122(8)	177.8(5)	$\eta^2(S,O)$	249
2. $Co(NO)(SO_2)(PPh_3)_2$	$\{MB_2\}^{10}$	1.68(1)	1.13(2)	169(2)	η^1-planar	186
3. $Rh(NO)(SO_2)(PPh_3)_2$	$\{MB_2\}^{10}$	1.802(6)	1.195(7)	140.4(6)	$\eta^2(S,O)$	184,185
4. $Mo(NO)(C_3H_5)(Cp)(S_2CNMe_2)$	$\{MB_2\}^6$	n.r.[a]	n.r.[a]	178[b]	η^3	12
5. $Mo(NO)(C_3H_5)(Cp)I$	$\{MB_2\}^6$	1.783(2)	1.178(2)	172.6(2)	η^3	100
6. $Ru(NO)(C_3H_5)(PPh_3)_2$	$\{MB_2\}^{10}$	1.751(6)	1.188(8)	173.8(6)	η^3	218,220
7. $[Ir(NO)(C_3H_5)(PPh_3)_2][BF_4]$	$\{MB_2\}^{10}$	1.95(1)	1.11(1)	129(1)	η^3	219

[a] Data not reported.
[b] Estimated standard deviation not reported.

ligand or the allyl ligand is attached to the same transition metal atom as the nitrosyl group. The structural details for these complexes are listed in Table 19. In describing the complexes containing two ambidentate ligands, we adopt the suggestion of Mingos (178) that the $\{M(NO)_m\}^n$ convention be extended. Thus we classify the complexes of NO, SO_2, and C_3H_5 as derivatives of $\{MB_m\}^n$.

A. $M(NO)(SO_2)$

The binding of SO_2 to transition metals is structurally more complicated than that of the nitrosyl ligand. To date, three different modes of coordination of SO_2 to transition metals have been identified and structurally characterized. These include η^1-planar, η^1-pyramidal, and η^2-(S,O) in which one S and one O atom of the SO_2 ligand are attached to a transition metal. Others (178,210) have previously pointed out that the conversion of a η^1-planar SO_2 group to η^1-pyramidal geometry is similar to the bending of the nitrosyl group. These three modes of bonding are shown schematically in Fig. 21.

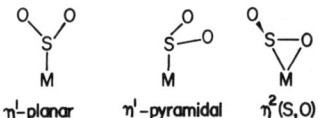

Fig. 21. Three alternative modes of SO_2 bonding: η^1-planar, η^1-pyramidal, and η^2-(S,O).

Both η^1-planar and η^2-(S,O) coordination of the SO_2 ligand have been found for complexes that also contain the nitrosyl group (Table 19). Of particular interest is the homologous pair of $\{MB_2\}^{10}$ complexes, $Co(NO)(SO_2)(PPh_3)_2$ (19: 2) and $Rh(NO)(SO_2)(PPh_3)_2$ (19: 3). The Co compound has a Co–N–O angle of 169° and η^1-planar coordination of the SO_2 ligand (Fig. 22), whereas the Rh compound has an Rh–N–O angle of 140° and η^2-coordination of the SO_2 ligand (Fig. 23). Since differences in both the M–N–O angle and the mode of SO_2 coordination occur for these two homologous compounds, it seems likely that the energetics of the interaction of NO and SO_2 with transition metals are very similar.

B. $M(NO)(C_3H_5)$

The formal similarities in the electronic structure of the allyl ligand (C_3H_5) and the nitrosyl ligand have been noted by Eisenberg (220). The coordinated allyl ligand also has two limiting modes of attachment to a transition metal, namely, η^1 and η^3 (Fig. 24), which are sometimes called σ and π complexes, respectively, in the older literature. The structures of four molecules that contain both the allyl group and the nitrosyl group have been reported. Two

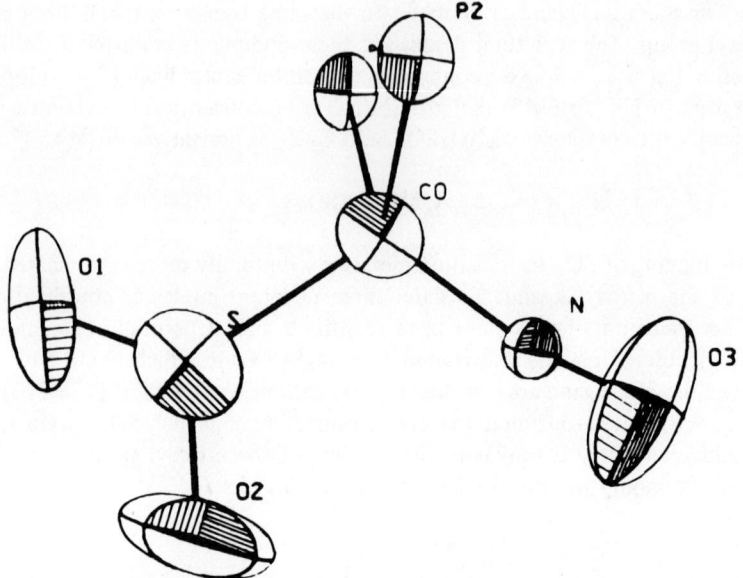

Fig. 22. Coordination geometry of Co(NO)(SO$_2$)(PPh$_3$)$_2$. (Reproduced from ref. 186.)

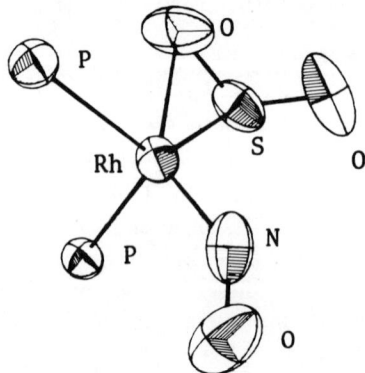

Fig. 23. Coordination geometry of Rh(NO)(SO$_2$)(PPh$_3$)$_2$. (Reproduced from ref. 185.)

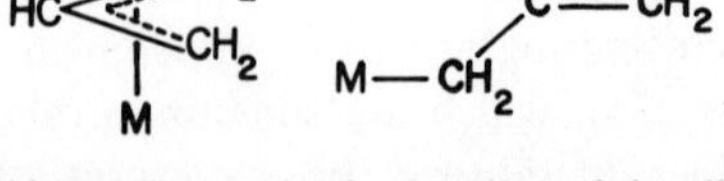

Fig. 24. Alternative modes for bonding of the allyl ligand.

are $\{MB_2\}^6$ complexes and two are $\{MB_2\}^{10}$ complexes. All four complexes contain an η^3-allyl ligand. Both the $\{MB_2\}^6$ species have nearly linear nitrosyl groups. However the $\{MB_2\}^{10}$ complexes are more interesting. The two examples for which structural data are available show that $Ru(NO)(C_3H_5)(PPh_3)_2$ (19:6) has an η^3-allyl ligand and a linear nitrosyl group (Fig. 25), whereas $[Ir(NO)(C_3H_5)(PPh_3)_2][BF_4]$ (19:7) has an η^3-allyl ligand but a bent nitrosyl group (Fig. 26). Stereospecific reactions of allyl complexes containing nitrosyl ligands

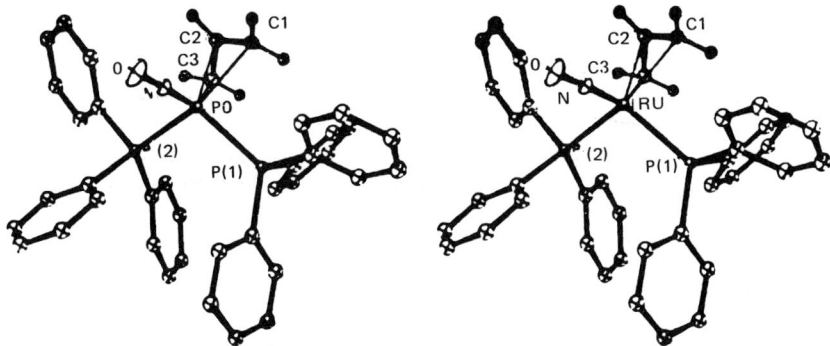

Fig. 25. Stereoview of the molecular geometry of $Ru(NO)(C_3H_5)(PPh_3)_2$. (Reproduced from ref. 220.)

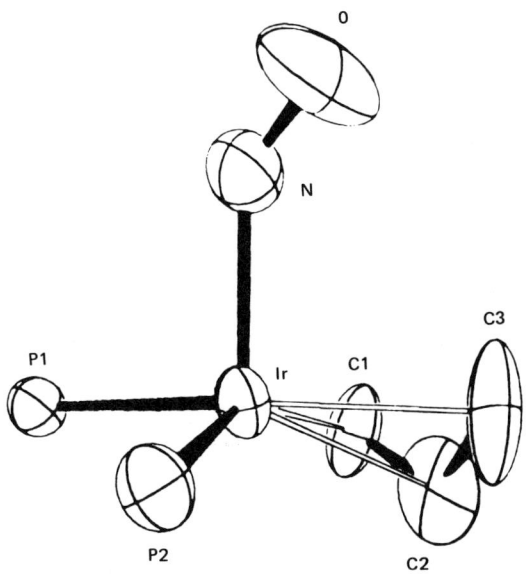

Fig. 26. Coordination geometry of $[Ir(NO)(C_3H_5)(PPh_3)_2][BF_4]$. (Reproduced from ref. 219.)

are also of current interest (1), as is the use of NO complexes for direct formation of N–C bonds. Although there has been a marked increase in activity in the area during the past two years, the organometallic chemistry of NO has been rather neglected and consequently should provide a fruitful area for future research.

VI. SUMMARY

The nearly 200 molecules whose structural features have been summarized in this chapter provide an important data base for understanding the chemical and physical properties of NO complexes. The structural data have had particular impact on the development of the current theories of bonding for metal nitrosyls. This combination of structure and theory can now account for many properties of NO complexes that formerly were difficult to reconcile: the existence of both linear and strongly bent MNO groups, the amphoteric properties of coordinated NO, and the large variations in electron density on coordinated NO. The general structural relationships apparent in the tables underscore the utility of the $\{M(NO)_m\}^n$ framework for understanding and interpreting the properties of nitrosyl complexes. However the discovery of structural features that have not been considered previously, such as the nonlinear RhNO group in $Rh(NO)(PPh_3)_3$, continue to provide stringent tests for any bonding model.

The vast majority of the data tabulated here come from crystallographic studies of the static structures of metal nitrosyl complexes in the solid state. Much less is known about the stereochemical dynamics of metal nitrosyl complexes in the gas phase or in liquid solutions. The existence of linear, bent, and bridging nitrosyl groups (sometimes in the same molecule) raises the possibility of dynamic interconversions among various nitrosyl geometries. It has been difficult to investigate the stereochemical integrity of metal nitrosyl complexes in solution because of the lack of a suitable direct probe of the nitrosyl group. NMR is often the best technique for studying structural dynamics in solution. The increasing availability of ^{15}N NMR facilities should stimulate studies of the solution structures and stereochemical dynamics of metal nitrosyl complexes. Such experiments should also provide additional information about the structure and bonding relationships in metal nitrosyl complexes.

ACKNOWLEDGMENTS

We thank the National Science Foundation and the National Institutes of Health for support of our contributions to this field, and each of our colleagues who so generously supplied preprints, reprints, and helpful discussions. We especially thank Ms. Leslie Boyer for collecting and organizing the references and Ms. Theresa Sequence for typing the manuscript.

[Note Added June 1980] Several additional structures of metal nitrosyl complexes have been reported since the original tabulation of the data for this chapter. These include: [Mo(NO)(η_2-H$_2$NO)(o-phen)$_2$]I$_2$ · o-phen · H$_2$O) (K. Wieghardt, W. Holzbach, B. Nuber, and J. Weiss, *Chem. Ber.*, 113, 629 (1980)); Cr(NO)$_2$Cl(Cp) and W(NO)$_2$Cl(Cp) (T. J. Greenhough, B. W. S. Kolthammer, P. Legzdins, and J. W. Trotter, *Acta Crystallogr., Sect. B*, 36, 795 (1980)); [Ir(NO)(NCMe)$_3$(PPh$_3$)$_2$][PF$_6$]$_2$ (M. Lanfranchi, A. Tiripicchio, G. Dolcetti, and M. Ghedini, *Trans. Met. Chem.*, 5, 21 (1980)); [Fe(NO)$_2$(Me$_2$Pz)]$_2$ and [Co(NO)$_2$(Me$_2$Pz)]$_2$ (K. S. Chong, S. J. Rettig, A. Storr, and J. Trotter, *Can. J. Chem.*, 57, 3119 (1979));Fe(NO)L' and Fe(NO)$_2$(L'H), where L' is $^-$S(CH$_2$)$_2$-NMe(CH$_2$)$_3$NMe(CH$_2$)$_2$S$^-$ (L. M. Baltusis, K. D. Karlin, H. N. Rabinowitz, J. C. Dewan, and S. J. Lippard, *Inorg. Chem.*, 19, 2627 (1980)); Mn(NO)$_3$(PPh$_3$) (R. D. Wilson and R. Bau, *J. Organomet. Chem.* 191, 123 (1980)); [Os$_3$(μ_2-NO)$_2$-(CO)$_9$(NMe$_3$)] (B. F. G. Johnson, J. Lewis, P. R. Raithby, and C. Zuccaro, *J. Chem. Soc., Chem. Commun.*, 1979, 916).

ABBREVIATIONS

acacen	N,N'-ethylenebis(acetylacetoneiminato)-
benacen	N,N'-ethylenebis(benzoylacetoneiminato)-
Bu^n	n-butyl
Bu^t	t-butyl
Cp	η^5-cyclopentadienyl
η^1-Cp	η^1-cyclopentadienyl
DAS	o-phenylenebis(dimethylarsine)
DIPIC	dipicolinato
DPPE	1,2-bis(diphenylphosphino)ethane
DPPEO	1,2-bis(diphenylphosphino)ethane monoxide
DPPP	1,3-bis(diphenylphosphino)propane
en	ethylenediamine
Fl	η^5-fluorenyl
GAP	(CH$_3$)$_2$Ga(N$_2$C$_5$H$_7$)$_2$(OCH$_2$CH$_2$N(CH$_3$)$_2$)
f$_6$fos	1,2-bis(diphenylphosphino)hexafluorocyclopropene
Me	methyl
MeCp	η^5-methylcyclopentadienyl

MEEN	N,N'-dimethyl-N,N'-bis(2-mercaptoethyl)-ethylenediamine
Me$_6$[14]4,11-diene-N$_4$	5,7,7,12,14,14-hexamethyl-1,4,8,11-tetraaza-cyclotetradeca-4,11-diene
1-MeIm	1-methylimidazole
4-MePi	4-methylpiperidine
MEPN	N,N'-dimethyl-N,N'-bis(2-mercaptoethyl)-1,3-propanediamine
MNT	maleonitriledithiolate
MPADP	methyl(1-phenylethyl)aminodiphenylphosphine
NP$_3$	tris(2-diphenylphosphinoethyl)amine
OAc	$CH_3CO_2^-$
P$_3$	bis(1-diphenylphosphinopropyl)phenylphosphine
PBP	pentagonal bipyramid
P$_3$C	$CH_3C(CH_2P(C_6H_5)_2)_3$
PEPEOA	bis(2-diphenylphosphinoethyl)2-diphenyl-phosphineoxideethylamine
Ph	phenyl
POC	$CH_3C(CH_2O)_3P$
PP$_3$	tris(2-diphenylphosphinoethyl)phosphine
Pr^i	isopropyl
PT	trigonal planar
Py	pyridine
Pz	pyrazoyl
QH	quinolinium
sacsac	dithioacetylacetonato
salen	N,N'-ethylenebis(salicylideniminato)
SP	square planar
T	tetrahedral
TBP	trigonal bipyramid
TMC	1,4,8,11-tetramethyl-1,4,8,11-tetraaza-cyclotetradecane
TP	tetragonal pyramid
TPP	tetraphenylporphine dianion
TTP	tetratolylporphine dianion

REFERENCES

1. R. D. Adams, D. F. Chodosh, J. W. Faller, and A. M. Rosan, *J. Am. Chem. Soc.*, **101**, 2570 (1979).
2. C. C. Addison and J. Lewis, *Q. Rev.*, **9**, 115 (1955).
3. V. G. Albano, A. Araneo, P. L. Bellon, G. Ciani, and M. Manassero, *J. Organomet. Chem.*, **67**, 413 (1974).
4. V. G. Albano, P. L. Bellon, and G. Ciani, *J. Organomet. Chem.*, **38**, 155 (1972).
5. V. G. Albano, P. Bellon, and M. Sansoni, *J. Chem. Soc. A.*, **1971**, 2420.
6. P. R. H. Alderman and P. G. Owston, *Nature (London)*, **178**, 1071 (1956).
7. P. R. H. Alderman, P. G. Owston, and J. M. Rowe, *J. Chem. Soc.*, **1962**, 668.
8. M. Andrews, D. L. Tipton, S. W. Kirtley, and R. Bau, *J. Chem. Soc., Chem. Commun.*, **1973**, 181.
9. M. Angoletta, G. Ciani, M. Manassero, and M. Sansoni, *J. Chem. Soc., Chem. Commun.*, **1973**, 789.
10. J. L. Atwood, R. Shakir, J. T. Malito, M. Herberhold, W. Kremnitz, W. P. E. Bernhagen, and H. G. Alt, *J. Organomet. Chem.*, **165**, 65 (1979).
11. N. A. Bailey, P. Frisch, J. A. McCleverty, N. W. Walker, and J. Williams, *J. Chem. Soc., Chem. Commun.*, **1975**, 350.
12. N. A. Bailey, W. G. Kita, J. A. McCleverty, A. J. Murray, B. E. Mann, and N. W. J. Walker, *J. Chem. Soc., Chem. Commun.*, **1974**, 592.
13. C. G. Barraclough and J. Lewis, *J. Chem. Soc.*, **1960**, 4842.
14. R. Bau, I. H. Sabherwal, and A. B. Burg, *J. Am. Chem. Soc.*, **93**, 4926 (1971).
15. D. Berglund and D. W. Meek, *Inorg. Chem.*, **11**, 1493 (1972).
16. I. Bernal, J. D. Korp, G. M. Reisner, and W. A. Herrmann, *J. Organomet. Chem.*, **139**, 321 (1977).
17. S. Bhaduri and G. M. Sheldrick, *Acta Crystallogr., Sect. B*, **31**, 897 (1975).
18. G. Bombieri, E. Forsellini, R. Graziani, and G. Zotti, *Transition Met. Chem.*, **2**, 264 (1977).
19. F. Bottomley, *J. Chem. Soc., Dalton Trans.*, **1974**, 1600.
20. F. Bottomley, *J. Chem. Soc., Dalton Trans.*, **1975**, 2538.
21. F. Bottomley, *Acc. Chem. Res.*, **11**, 158 (1978).
22. F. Bottomley, *Coord. Chem. Rev*, **26**, 7 (1978).
23. F. Bottomley, W. V. F. Brooks, S. G. Clarkson, and S.-B. Tong, *J. Chem. Soc., Chem. Commun.*, **1973**, 919.
24. F. Bottomley, S. G. Clarkson, and S.-B. Tong, *J. Chem. Soc., Dalton Trans.*, **1974**, 2344.
25. F. Bottomley and P. S. White, *Acta Crystallogr. Sect. B*, **35**, 2193 (1979).
26. D. C. Bradley and M. H. Chisholm, *Acc. Chem. Res.*, **9**, 273 (1976).
27. D. C. Bradley, M. B. Hursthouse, C. W. Newing, and A. J. Welch, *J. Chem. Soc., Chem. Commun.*, **1972**, 567.
28. T. F. Brennan and I. Bernal, *J. Chem. Soc., Chem. Commun.*, **1970**, 138.
29. T. F. Brennan and I. Bernal, *Inorg. Chim. Acta*, **7**, 283 (1973).
30. D. M. Bridges, D. W. H. Rankin, D. A. Clement, and J. F. Nixon, *Acta Crystallogr., Sect. B*, **28**, 1130 (1972).
31. C. P. Brock, J. P. Collman, G. Dolcetti, P. H. Farnham, J. A. Ibers, J. E. Lester, and C. A. Reed, *Inorg. Chem.*, **12**, 1304 (1973).
32. C. P. Brock and J. A. Ibers, *Inorg. Chem.*, **11**, 2812 (1972).
33. L. O. Brockway and J. S. Anderson, *Trans. Faraday Soc.*, **33**, 1233 (1937).
34. M. A. Bush, G. A. Sim, G. R. Knox, M. Ahmad, and C. G. Robertson, *J. Chem. Soc., Chem, Commun.*, **1969**, 74.

35. M. A. Bush and G. A. Sim, *J. Chem. Soc. A*, **1970**, 605.
36. M. A. Bush and G. A. Sim, *J. Chem. Soc. A*, **1970**, 611.
37. J. L. Calderon, F. A. Cotton, B. G. DeBoer, and N. Martinex, *J. Chem. Soc., Chem., Commun.*, **1971**, 1476.
38. J. L. Calderon, F. A. Cotton, and P. Legzdins, *J. Am. Chem. Soc.*, **91**, 2528 (1969).
39. J. L. Calderon, S. Fontana, E. Frauendorfer, and V. W. Day, *J. Organomet. Chem.*, **64**, C10 (1974).
40. J. L. Calderon, S. Fontana, E. Frauendorfer, V. W. Day, and S. D. A. Iske, *J. Organomet. Chem.*, **64**, C16 (1974).
41. J. L. Calderon, S. Fontana, E. Frauendorfer, V. W. Day, and B. R. Stults, *Inorg. Chim. Acta*, **17**, L31 (1976).
42. M. Carrondo, P. R. Rudolf, A. C. Skapski, J. R. Thornback, and G. Wilkinson, *Inorg. Chim. Acta*, **24**, L95 (1977).
43. O. L. Carter, A. T. McPhail, and G. A. Sim, *J. Chem. Soc. A*, **1966**, 1095.
44. P. Carty, A. Walker, M. Mathew, and G. J. Palenik, *J. Chem. Soc., Chem. Commun.*, **1969**, 1374.
45. E. E. Castellano, O. E. Piro, A. D. Podjarny, B. E. Rivero, P. J. Aymonino, J. H. Lesk, and E. L. Varetti, *Acta Crystallogr., Sect. B*, **34**, 2673 (1978).
46. K. G. Caulton, *Coord. Chem. Rev.*, **14**, 317 (1975).
47. L. Y. Y. Chan and F. W. B. Einstein, *Acta Crystallogr., Sect. B*, **26**, 1899 (1970).
48. P.-T. Cheng and S. C. Nyburg, *Inorg. Chem.*, **14**, 327 (1975).
49. M. H. Chisholm, F. A. Cotton, M. W. Extine, and R. L. Kelly, *J. Am. Chem. Soc.*, **100**, 3354 (1978).
50. M. H. Chisholm, F. A. Cotton, M. W. Extine, and R. L. Kelly, *Inorg. Chem.*, **18**, 116 (1979).
51. K. S. Chong, S. J. Rettig, A. Storr, and J. Trotter, *Can. J. Chem.*, **57**, 3090 (1979).
52. K. S. Chong, S. J. Rettig, A. Storr, and J. Trotter, *Can. J. Chem.*, **57**, 3099 (1979).
53. K. S. Chong, S. J. Rettig, A. Storr, and J. Trotter, *Can. J. Chem.*, **57**, 3107 (1979).
54. K. S. Chong, S. J. Rettig, A. Storr, and J. Trotter, *Can. J. Chem.*, **57**, 3113 (1979).
55. C. T.-W. Chu and L. F. Dahl, *Inorg. Chem.*, **16**, 3245 (1977).
56. G. Ciani, D. Giusto, M. Manassero, and M. Sansoni, *J. Chem. Soc., Dalton Trans.*, **1975**, 2156.
57. G. Ciani, D. Giusto, M. Manassero, and M. Sansoni, *J. Chem. Soc., Dalton Trans.*, **1978**, 798.
58. G. Ciani, D. Giusto, M. Manassero, and A. Albinati, *J. Chem. Soc., Dalton Trans.*, **1976**, 1943.
59. G. R. Clark, K. R. Grundy, W. R. Roper, J. M. Waters, and K. R. Whittle, *J. Chem. Soc., Chem. Commun.*, **1972**, 119.
60. G. R. Clark, J. M. Waters, and K. R. Whittle, *Inorg. Chem.*, **13**, 1628 (1974).
61. G. R. Clark, J. M. Waters, and K. R. Whittle, *J. Chem. Soc., Dalton Trans.*, **1975**, 463.
62. G. R. Clark, J. M. Waters, and K. R. Whittle, *J. Chem. Soc., Dalton Trans.*, **1975**, 2233.
63. G. R. Clark, J. M. Waters, and K. R. Whittle, *J. Chem. Soc., Dalton Trans.*, **1976**, 2029.
64. W. Clegg, *Inorg. Chem.*, **15**, 2928 (1976).
65. J. Clemens, M. Green, M.-C. Kuo, C. J. Fritchie, J. T. Mague, and F. G. A. Stone, *J. Chem. Soc., Chem. Commun.*, **1972**, 53.
66. M. Colapietro, A. Domenicano, L. Scaramuzza, A. Vaciago, and L. Zambonelli, *J. Chem. Soc., Chem, Commun.*, **1967**, 583.
67. R. Colton, M. J. McCormick, and C. D. Pannan, *Aust. J. Chem.*, **31**, 1425 (1978).
68. N. G. Connelly, *Inorg. Chim. Acta Rev.*, **6**, 47 (1972).

69. N. G. Connelly, B. A. Kelly, R. L. Kelly, and P. Woodward, *J. Chem. Soc., Dalton Trans.*, **1976**, 699.
70. F. A. Cotton and G. A. Rusholme, *J. Am. Chem. Soc.*, **94**, 402 (1972).
71. A. P. Cox and A. H. Brittain, *Trans. Faraday Soc.*, **66**, 557 (1970).
72. W. V. Cruz, P. C. W. Leung, and K. Seff, *Inorg. Chem.*, **18**, 1692 (1979).
73. L. F. Dahl, E. R. de Gil, and R. D. Feltham, *J. Am. Chem. Soc.*, **91**, 1653 (1969).
74. G. R. Davies, J. A. J. Jarvis, B. T. Kilbourn, R. H. B. Mais, and P. G. Owston, *J. Chem. Soc. A*, **1970**, 1275.
75. G. R. Davies, R. H. B. Mais, and P. G. Owston, *J. Chem. Soc., Chem. Commun.*, **1968**, 81.
76. E. R. de Gil, A. V. Rivera, and H. Noguera, *Acta Crystallogr., Sect. B*, **33**, 2653 (1977).
77. P. De Meester and A. C. Skapski, *J. Chem. Soc., Dalton Trans.*, **1973**, 1194.
78. P. De Meester, A. C. Skapski, and J. P. Heffer, *J. Chem. Soc., Chem. Commun.*, **1972**, 1039.
79. T. Diebold, M. Schappacher, B. Chevrier, and R. Weiss, *J. Chem. Soc., Chem. Commun.*, **1979**, 693.
80. M. Di Vaira, C. A. Ghilardi, and L. Sacconi, *Inorg. Chem.*, **15**, 1555 (1976).
81. M. Di Vaira, A. Tarli, P. Stoppioni, and L. Sacconi, *Cryst. Struct. Commun.*, **4**, 653 (1975).
82. A. Domenicano, A. Vaciago, L. Zambonelli, P. L. Loader, and L. M. Venanzi, *J. Chem. Soc., Chem. Commun.*, **1966**, 476.
83. W. L. Dorn and J. Schmidt, *Inorg. Chim. Acta*, **16**, 223 (1976).
84. M. G. B. Drew and C. F. Pygall, *Acta Crystallogr., Sect. B*, **33**, 2838 (1977).
85. R. C. Elder, *Inorg. Chem.*, **13**, 1037 (1974).
86. F. W. B. Einstein, D. Sutton, and P. L. Vogel, *Inorg. Nucl. Chem. Lett.*, **12**, 671 (1976).
87. R. Eisenberg, A. P. Gaughan, C. G. Pierpont, J. Reed, and A. J. Schultz, *J. Am. Chem. Soc.*, **94**, 6240 (1972).
88. R. Eisenberg and C. D. Meyer, *Acc. Chem. Res.*, **8**, 26 (1975).
89. J. H. Enemark, *Inorg. Chem.*, **10**, 1952 (1971).
90. J. H. Enemark and R. D. Feltham, *J. Chem. Soc., Dalton Trans.*, **1972**, 718.
91. J. H. Enemark and R. D. Feltham, *Coord. Chem. Rev.*, **13**, 339 (1974).
92. J. H. Enemark, R. D. Feltham, B. T. Huie, P. L. Johnson, and K. B. Swedo, *J. Am. Chem. Soc.*, **99**, 3285 (1977).
93. J. H. Enemark, R. D. Feltham, J. Riker-Nappier, and K. F. Bizot, *Inorg. Chem.*, **14**, 624 (1975).
94. J. H. Enemark and J. A. Ibers, *Inorg. Chem.*, **6**, 1575 (1967).
95. J. H. Enemark and J. A. Ibers, *Inorg. Chem.*, **7**, 2339 (1968).
96. J. H. Enemark, M. S. Quinby, L. L. Reed, M. J. Steuck, and K. K. Walthers, *Inorg. Chem.*, **9**, 2397 (1970).
97. R. B. English, L. R. Nassimbeni, and R. J. Haines, *Acta Crystallogr., Sect. B*, **32**, 3299 (1976).
98. J. M. Epstein, A. H. White, S. B. Wild, and A. C. Willis, *J. Chem. Soc., Dalton Trans.*, **1974**, 436.
99. G. Evrard, R. Thomas, B. R. Davis, and I. Bernal, *Inorg. Chem.*, **15**, 52 (1976).
100. J. W. Faller, D. F. Chodosh, and D. Katahira, *J. Organomet. Chem.*, **187**, 227 (1980).
101. R. F. Fenske and R. L. DeKock, *Inorg. Chem.*, **11**, 437 (1972).
102. J. E. Fergusson, C. T. Page, and W. T. Robinson, *Inorg. Chem*, **15**, 2270 (1976).
103. J. S. Field, P. J. Wheatley, and S. Bhaduri, *J. Chem. Soc., Dalton Trans.*, **1974**, 74.
104. B. A. Frenz, J. H. Enemark, and J. A. Ibers, *Inorg. Chem.*, **8**, 1288 (1969).

105. B. A. Frenz and J. A. Ibers, *MTP Int. Rev. Sci. Phys. Chem.*, Ser. 1, **11**, 33 (1972).
106. R. S. Gall, C. T.-W. Chu, and L. F. Dahl, *J. Am. Chem. Soc.*, **96**, 4019 (1974).
107. R. S. Gall, N. G. Connelly, and L. F. Dahl, *J. Am. Chem. Soc.*, **96**, 4017 (1974).
108. A. P. Gaughan, B. J. Corden, R. Eisenberg, and J. A. Ibers, *Inorg. Chem.*, **13**, 786 (1974).
109. C. A. Ghilardi and L. Sacconi, *Cryst. Struct. Commun.*, **4**, 687 (1975).
110. G. Gilli, M. Sacerdoti, and P. Domiano, *Acta Crystallogr., Sect. B*, **30**, 1485 (1974).
111. G. Gilli, M. Sacerdoti, and G. Reichenbach, *Acta Crystallogr., Sect. B*, **29**, 2306 (1973).
112. R. E. Ginsburg, L. M. Cirjak, and L. F. Dahl, *J. Chem. Soc., Chem. Commun.*, **1979**, 468.
113. S. Z. Goldberg, C. Kubiak, C. D. Meyer, and R. Eisenberg, *Inorg. Chem.*, **14**, 1650 (1975).
114. S. A. Goldfield and K. N. Raymond, *Inorg. Chem.*, **13**, 770 (1974).
115. M. L. H. Green, *Organometallic Compounds, The Transition Elements*, Methuen & Co., London, 1968.
116. W. P. Griffith, *Adv. Organomet. Chem.*, **7**, 211 (1968).
117. D. Hall and R. B. Williamson, *Cryst. Struct. Commun.*, **3**, 327 (1974).
118. K. J. Haller and J. H. Enemark, *Acta Crystallogr., Sect. B*, **34**, 102 (1978).
119. K. J. Haller and J. H. Enemark, *Inorg. Chem.*, **17**, 3552 (1978).
120. K. J. Haller, P. L. Johnson, R. D. Feltham, J. H. Enemark, J. R. Ferraro, and L. J. Basile, *Inorg. Chim. Acta*, **33**, 119 (1979).
121. A. D. U. Hardy and G. A. Sim, *Acta Crystallogr., Sect. B*, **35**, 1463 (1979).
122. W. Harrison and J. Trotter, *J. Chem. Soc. A*, **1971**, 1542.
123. B. L. Haymore and J. C. Huffman, personal communication, 1979.
124. B. L. Haymore and J. A. Ibers, *Inorg. Chem.*, **14**, 2610 (1975).
125. B. L. Haymore and J. A. Ibers, *Inorg. Chem.*, **14**, 3060 (1975).
126. T. W. Hawkins and M. B. Hall, *Inorg. Chem.*, **19**, 1735 (1980).
127. A. A. Hock and O. S. Mills, *Proc. 6th ICCC*, **1961**, 640.
128. K. D. Hodges, R. G. Wollman, S. L. Kessel, D. N. Hendrickson, D. G. VanDerveer, and E. K. Barefield, *J. Am. Chem. Soc.*, **101**, 906 (1979).
129. D. J. Hodgson and J. A. Ibers, *Inorg. Chem.*, **7**, 2345 (1968).
130. D. J. Hodgson and J. A. Ibers, *Inorg. Chem.*, **8**, 1282 (1969).
131. D. J. Hodgson, N. C. Payne, J. A. McGinnety, R. G. Pearson, and J. A. Ibers, *J. Am. Chem. Soc.*, **90**, 4486 (1968).
132. R. Hoffmann, M. M. L. Chen, M. Elian, A. R. Rossi, and D. M. P. Mingos, *Inorg. Chem.*, **13**, 2666 (1974).
133. E. M. Holt, S. L. Holt, F. Cavalito, and K. J. Watson, *Acta Chem. Scand.*, **A30**, 225 (1976).
134. J. E. Huheey, *Inorganic Chemistry*, 2nd ed., Harper & Row, New York, 1978.
135. J. A. Ibers, *Acta Crystallogr., Sect. B*, **27**, 250 (1971).
136. O. A. Ileperuma and R. D. Feltham, *Inorg. Chem.*, **16**, 1876 (1977).
137. S. Jagner and E. Ljungström, *Acta Crystallogr., Sect. B*, **34**, 653 (1978).
138. S. Jagner and N.-G. Vannerberg, *Acta Chem. Scand.*, **21**, 1183 (1967).
139. S. Jagner and N.-G. Vannerberg, *Acta Chem. Scand.*, **22**, 3330 (1968).
140. S. Jagner and N.-G. Vannerberg, *Acta Chem. Scand.*, **24**, 1988 (1970).
141. G. Johansson and W. N. Lipscomb, *Acta Crystallogr.*, **11**, 594 (1958).
142. B. F. G. Johnson and J. A. McCleverty, *Prog. Inorg. Chem.*, **7**, 277 (1966).
143. B. F. G. Johnson, P. R. Raithby, and C. Zuccaro, *J. Chem. Soc., Dalton Trans.*, **1980**, 99.

144. P. L. Johnson, J. H. Enemark, R. D. Feltham, and K. B. Swedo, *Inorg. Chem.*, **15**, 2989 (1976).
145. J. A. Kaduk and J. A. Ibers, *Inorg. Chem.*, **14**, 3070 (1975).
146. J. A. Kaduk and J. A. Ibers, *Inorg. Chem.*, **16**, 3283 (1977).
147. J. A. Kaduk and J. A. Ibers, *Isr. J. Chem.*, **15**, 143 (1977).
148. K. D. Karlin, D. L. Lewis, H. N. Rabinowitz, and S. J. Lippard, *J. Am. Chem. Soc.*, **96**, 6519 (1974).
149. K. D. Karlin, H. N. Rabinowitz, D. L. Lewis, and S. J. Lippard, *Inorg. Chem.*, **16**, 3262 (1977).
150. B. A. Kelly, A. J. Welch, and P. Woodward, *J. Chem. Soc., Dalton Trans.*, **1977**, 2237.
151. S. F. A. Kettle, *Inorg. Chem.*, **4**, 1661 (1965).
152. T. S. Khodashova, M. A. Porai-Koshits, V. S. Sergienko, N. A. Parniev, and G. B. Bokii, *Zh. Strukt. Khim.*, **13**, 1105 (1972).
153. T. S. Khodashova, V. S. Sergienko, A. N. Stetsenko, M. A. Porai-Koshits, and L. A. Butman, *Zh. Strukt. Khim.*, **15**, 471 (1974).
154. R. M. Kirchner, T. J. Marks, J. S. Kristoff, and J. A. Ibers, *J. Am. Chem. Soc.*, **95**, 6602 (1973).
155. W. Kita, J. A. McCleverty, B. E. Mann, D. Seddon, G. A. Sim, and D. I. Woodhouse, *J. Chem. Soc., Chem. Commun.*, **1974**, 132; P. R. Mallison, G. A. Sim, and D. I. Woodhouse, *Acta Crystallogr., Sect. B*, **36**, 450 (1980).
156. J. Kopf and J. Schmidt, *Z. Naturforsch. B*, **30**, 149 (1975).
157. J. Kopf and J. Schmidt, *Z. Naturforsch. B*, **32**, 275 (1977).
158. M. Laing, R. Reimann, and E. Singleton, *Inorg. Nucl. Chem. Lett.*, **10**, 557 (1974).
159. M. Laing, R. H. Reimann, and E. Singleton, *Inorg. Chem.*, **18**, 324 (1979).
160. M. Laing, R. H. Reimann, and E. Singleton, *Inorg. Chem.*, **18**, 1648 (1979).
161. M. Laing, R. H. Reimann, and E. Singleton, *Inorg. Chem.*, **18**, 2666 (1979).
162. A. H. Lanfranconi, A. G. Alvarez, and E. E. Castellano, *Acta Crystallogr. Sect. B*, **29**, 1733 (1973).
163. R. A. Love, dissertation, University of Southern California, 1975: S. W. Kirtley, J. P. Chanton, R. A. Love, D. L. Tipton, T. N. Sorrell, R. Bau, *J. Am. Chem. Soc.*, **102**, 3451 (1980).
164. R. A. Love, H. B. Chin, T. F. Koetzle, S. W. Kirtley, B. R. Whittlesey, and R. Bau, *J. Am. Chem. Soc.*, **98**, 4491 (1976).
165. B. C. Lucas, D. C. Moody, and R. R. Ryan, *Cryst. Struct. Commun.*, **6**, 57 (1977).
166. C. M. Lukehart and J. M. Troup, *Inorg. Chim. Acta*, **22**, 81 (1977).
167. P. T. Manoharan and H. B. Gray, *Inorg. Chem.*, **5**, 823 (1966).
168. P. T. Manoharan and W. C. Hamilton, *Inorg. Chem.*, **2**, 1043 (1963).
169. R. L. Martin and D. Taylor, *Inorg. Chem.*, **15**, 2970 (1976).
170. J. A. Masek, *Inorg. Chim. Acta Rev.*, **3**, 99 (1969).
171. J. A. McCleverty, *Chem. Rev.*, **79**, 53 (1979).
172. J. A. McCleverty, D. Seddon, N. A. Bailey, N. W. J. Walker, *J. Chem. Soc., Dalton Trans.*, **1976**, 898.
173. J. A. McGinnety, *MTP Int. Rev. Sci. Inorg. Chem., Ser. 1*, **5**, 229 (1972).
174. A. T. McPhail and G. A. Sim, *J. Chem. Soc. A*, **1968**, 1858.
175. J. H. Meiners, C. J. Rix, J. C. Clardy, and J. G. Verkade, *Inorg. Chem.*, **14**, 705 (1975).
176. E. Miki, K. Mizumachi, T. Ishimori, and H. Okuno, *Bull. Chem. Soc. Jap.*, **46**, 3779 (1973).
177. D. M. P. Mingos, *Inorg. Chem.*, **12**, 1209 (1973).
178. D. M. P. Mingos, *Transition Met. Chem.*, **3**, 1 (1978).

179. D. M. P. Mingos and J. A. Ibers, *Inorg. Chem.*, **9**, 1105 (1970).
180. D. M. P. Mingos and J. A. Ibers, *Inorg. Chem.*, **10**, 1035 (1971).
181. D. M. P. Mingos and J. A. Ibers, *Inorg. Chem.*, **10**, 1479 (1971).
182. D. M. P. Mingos, W. T. Robinson, and J. A. Ibers, *Inorg. Chem.*, **10**, 1043 (1971).
183. T. Moeller, *J. Chem. Ed.*, **23**, 441, 542 (1946).
184. D. C. Moody and R. R. Ryan, *J. Chem. Soc., Chem. Commun.*, **1976**, 503.
185. D. C. Moody and R. R. Ryan, *Inorg. Chem.*, **16**, 2473 (1977).
186. D. C. Moody, R. R. Ryan, and A. C. Larson, *Inorg. Chem.*, **18**, 227 (1979).
187. A. Müller, W. Eltzner, and N. Mohan, *Angew. Chem.*, **91**, 158 (1979).
188. A. Müller, U. Seyer, and W. Eltzner, *Inorg. Chim. Acta*, **32**, L65 (1979).
189. T. E. Nappier, D. W. Meek, R. M. Kirchner, and J. A. Ibers, *J. Am. Chem. Soc.*, **95**, 4194 (1973).
190. J. R. Norton, J. P. Collman, G. Dolcetti, and W. T. Robinson, *Inorg. Chem.*, **11**, 382 (1972).
191. J. P. Olsen, T. F. Koetzle, S. W. Kirtley, M. Andrews, D. L. Tipton, and R. Bau, *J. Am. Chem. Soc.*, **96**, 6621 (1974).
192. P. L. Piciulo, G. Rupprecht, and W. R. Scheidt, *J. Am. Chem. Soc.*, **96**, 5293 (1974).
193. C. G. Pierpont and R. Eisenberg, *J. Am. Chem. Soc.*, **93**, 4905 (1971).
194. C. G. Pierpont and R. Eisenberg, *Inorg. Chem.*, **11**, 1088 (1972).
195. C. G. Pierpont and R. Eisenberg, *Inorg. Chem.*, **11**, 1094 (1972).
196. C. G. Pierpont and R. Eisenberg, *Inorg. Chem.*, **12**, 199 (1973).
197. C. G. Pierpont, A. Pucci, and R. Eisenberg, *J. Am. Chem. Soc.*, **93**, 3050 (1971).
198. C. G. Pierpont, D. G. VanDerveer, W. Durland, and R. Eisenberg, *J. Am. Chem. Soc.*, **92**, 4760 (1970).
199. C. S. Pratt, B. A. Coyle, and J. A. Ibers, *J. Chem. Soc. A*, **1971**, 2146.
200. M. Quinby-Hunt and R. D. Feltham, *Inorg. Chem.*, **17**, 2515 (1978).
201. H. N. Rabinowitz, K. D. Karlin, and S. J. Lippard, *J. Am. Chem. Soc.*, **99**, 1420 (1977).
202. A. I. M. Rae, *J. Chem. Soc., Chem. Commun.*, **1967**, 1245.
203. J. Reed, A. J. Schultz, C. G. Pierpont, and R. Eisenberg, *Inorg. Chem.*, **12**, 2949 (1973).
204. J. Reed, S. L. Soled, and R. Eisenberg, *Inorg. Chem.*, **13**, 3001 (1974).
205. B. E. Reichert, *Acta Crystallogr., Sect. B*, **32**, 1934 (1976).
206. M. G. Reisner, I. Bernal, H. Brunner, and J. Doppelberger, *J. Chem. Soc., Dalton Trans.*, **1979**, 1664.
207. R. Renner, *Z. Phys.*, **92**, 172 (1934).
208. A. V. Rivera and G. M. Sheldrick, *Acta Crystallog., Sect. B*, **34**, 3372 (1978).
209. I. A, Ronova, N. V. Alekseeva, N. N. Veniaminov, and M. A. Kravers, *Zh. Strukt. Khim.*, **16**, 476 (1975).
210. R. R. Ryan and P. G. Eller, *Inorg. Chem.*, **15**, 494 (1976).
211. A. P. Sattelberger, Ph.D. thesis, Indiana University, 1975.
212. W. R. Scheidt, A. C. Brinegar, E. B. Ferro, and J. F. Kirner, *J. Am. Chem. Soc.*, **99**, 7315 (1977).
213. W. R. Scheidt and M. E. Frisse, *J. Am. Chem. Soc.*, **97**, 17 (1975).
214. W. R. Scheidt, K. Hatano, G. A. Rupprecht, and P. L. Piciulo, *Inorg. Chem.*, **18**, 292 (1979).
215. W. R. Scheidt and J. L. Hoard, *J. Am. Chem. Soc.*, **95**, 8281 (1973).
216. W. R. Scheidt and P. L. Piciulo, *J. Am. Chem. Soc.*, **98**, 1913 (1976).
217. J. Schmidt, H. Kühr, W. L. Dorn, and J. Kopf, *Inorg. Nucl. Chem. Lett.*, **10**, 55 (1974).

218. M. W. Schoonover and R. Eisenberg, *J. Am. Chem. Soc.*, **99**, 8371 (1977).
219. M. W. Schoonover, E. C. Baker, and R. E. Eisenberg, *J. Am. Chem. Soc.*, **101**, 1880 (1979).
220. M. W. Schoonover, C. P. Kubiak, and R. Eisenberg, *Inorg. Chem.*, **17**, 3050 (1978).
221. A. J. Schultz, R. L. Henry, J. Reed, and R. Eisenberg, *Inorg. Chem.*, **13**, 732 (1974).
222. N. V. Sidgwick and R. W. Bailey, *Proc. R. Soc. London, Ser. A*, **144**, 521 (1934).
223. G. A. Sim, D. I. Woodhouse, and G. R. Knox, *J. Chem. Soc., Dalton Trans.*, **1979**, 83.
224. S. H. Simonsen and M. H. Mueller, *J. Inorg. Nucl. Chem.*, **27**, 309 (1965).
225. D. A. Snyder and D. L. Weaver, *J. Chem. Soc., Chem. Commun.*, **1969**, 1425.
226. D. A. Snyder and D. L. Weaver, *Inorg. Chem.*, **9**, 2760 (1970).
227. T. G. Southern, P. H. Dixneuf, J. Y. LeMarouille, and D. Grandjean, *Inorg. Chim. Acta*, **31**, L415 (1978).
228. F. S. Stephens, *J. Chem. Soc., Dalton Trans.*, **1972**, 2257.
229. C. E. Strouse and B. I. Swanson, *J. Chem. Soc., Chem. Commun.*, **1971**, 55.
230. R. H. Summerville and R. Hoffmann, *J. Am. Chem. Soc.*, **98**, 7240 (1976).
231. D. H. Svedung and N.-G. Vannerberg, *Acta Chem. Scand.*, **22**, 1551 (1968).
232. B. I. Swanson and S. K. Satija, *J. Chem. Soc., Chem. Commun.*, **1973**, 40.
233. J. H. Swinehart, *Coord. Chem. Rev.*, **2**, 385 (1967).
234. J. T. Thomas, J. H. Robertson, and E. G. Cox, *Acta Crystallog.*, **11**, 599 (1958).
235. A. Tullberg and N.-G. Vannerberg, *Acta Chem. Scand.*, **20**, 1180 (1966).
236. A. Tullberg and N.-G. Vannerberg, *Acta Chem. Scand.*, **21**, 1462 (1967).
237. D. G. VanDerveer, A. P. Gaughan, S. L. Soled, and R. Eisenberg, *Abstr. Am. Cryst. Assoc.*, **1**, 190 (1973).
238. J. T. Veal and D. J. Hodgson, *Inorg. Chem.*, **11**, 1420 (1972).
239. J. T. Veal and D. J. Hodgson, *Acta Crystallogr., Sect. B*, **28**, 3525 (1972).
240. M. O. Visscher and K. G. Caulton, *J. Am. Chem. Soc.*, **94**, 5923 (1972).
241. K. Wade, *Adv. Inorg. Chem. Radiochem.*, **18**, 1 (1976).
242. A. D. Walsh, *J. Chem. Soc.*, **1953**, 2266.
243. D. L. Ward, C. N. Caughlan, G. E. Voecks, and P. W. Jennings, *Acta Crystallogr., Sect. B*, **28**, 1949 (1972).
244. J. M. Waters and K. R. Whittle, *J. Chem. Soc., Chem. Commun.*, **1971**, 518.
245. D. Wester, R. C. Edwards, and D. H. Busch, *Inorg. Chem.*, **16**, 1055 (1977).
246. K. Wieghardt, U. Quilitzsch, B. Nuber, and J. Weiss, *Angew. Chem., Int. Ed. Engl.*, **17**, 351 (1978).
247. R. Wiest and R. Weiss, *J. Organomet. Chem.*, **30**, C33 (1971).
248. R. Wiest and R. Weiss, *Rev. Chim. Miner.*, **9**, 655 (1972).
249. R. D. Wilson and J. A. Ibers, *Inorg. Chem.*, **17**, 2134 (1978).
250. R. D. Wilson and J. A. Ibers, *Inorg. Chem.*, **18**, 336 (1979).
251. W.-K. Wong, W. Tam, C. E. Strouse, and J. A. Gladysz, *J. Chem. Soc., Chem. Commun.*, **1979**, 530.
252. J. E. Fergusson and G. A. Rodley, *MTP Int. Rev. Sci. Inorg. Chem., Ser. 2*, **6**, 37 (1975).

The Stereochemistry of Germanium and Tin Compounds

MARCEL GIELEN

Vrije Universiteit Brussel - T. W. - AOSC,
and
*Université Libre de Bruxelles, Collectif de Chimie Organique Physique,
Brussels, Belgium*

I.	The Optical Stability of RR'R''MX Compounds.	218
	A. The Optical Stability of Triorgano-Group(IV) Metal Halides.	218
	B. The Optical Stability of Triorganotin Hydrides	220
	C. The Optical Stability of Other Triorganotin Compounds.	221
II.	The Stereochemistry of Substitution Reactions at the Metal Atom of RR'R''-MX Compounds.	223
	A. The Synthesis of Optically Active Organogermanium Compounds	223
	B. The Synthesis of Optically Active Organotin Compounds	224
	C. The Stereochemistry of Substitution Reactions at the Metal Atom of Triorganogermanium Compounds Compared to That Characterizing Substitutions at the Silicon Atom of Organosilanes and Rationalization Thereof	228
	D. A Stereospecific Substitution Reaction at the Metal Atom of a Chiral Triorganotin Compound.	228
III.	The Intramolecular Rearrangements of Trigonal Bipyramidal Group(IV) Metal Complexes and Their Importance for the Stereochemistry of Substitution Reactions at RR'R''MX Compounds.	228
	A. The Addition of a Nucleophile at a Tetrahedrally Substituted Metal Atom.	229
	B. The Minimum Constraints for Stereospecificity	231
	C. Intramolecular Rearrangements: The Berry Pseudorotation, Belonging to Mode P1.	232
	D. Chemical Constraints for Particular Cases and the Resulting Stereochemistry.	235
IV.	Dynamic Stereochemistry of *cis*-Octahedral Bis(β-diketonato)-group(IV) Metal Complexes	236
	A. Modes of Rearrangement of *cis*-M($\widehat{AA}$)$_2$X$_2$ Complexes	236
	B. Modes of Rearrangement of *cis*-M($\widehat{AB}$)$_2$X$_2$ Complexes	238
	C. Modes of Rearrangement of *cis*-M($\widehat{AB}$)$_2$XY Complexes	239
	D. Graphical Topological Representations of the Modes of Rearrangement of M($\widehat{AB}$)$_2$X$_2$ and M($\widehat{AB}$)$_2$XY Complexes.	242
	E. The Threshold Mode of Rearrangement of *cis*-Dichlorobis(benzoylacetonato)tin(IV)	243

F. The Threshold Mode of Rearrangement of cis-Phenylchlorobis(benzoyl-
acetonato)tin(IV) 245
Acknowledgments 248
References 249

The first three sections of this chapter are interdependent: indeed, *optically stable* RR'R''MX compounds are needed to study the stereochemistry of substitution reactions at the metal atom discussed in Sect. II. Section I is concerned with which compounds can be used to study the stereochemical aspect of the replacement of leaving group X by a reagent Y. In Sect. III a possible interpretation for the experimental results described and discussed in Sect. II is given. Section IV is almost independent of the other three, though its group theoretical background is the same as for Sect. III.

I. THE OPTICAL STABILITY OF RR'R''MX COMPOUNDS

Several RR'R''MX compounds are configurationally stable in nonnucleophilic solvents but are readily transformed into the compound in which the metal atom has been inverted in the presence of small quantities of nucleophiles.

A. The Optical Stability of Triorgano-Group(IV) Metal Halides

Methyl-1-naphthylphenylgermanium chloride is optically stable in hydrocarbons, in CCl_4 or in $CHCl_3$ but racemizes in 8 min in THF (1). Similarly, several RR'R''SnX compounds are configurationally unstable. The inversion at the tin atom of methylneophylphenyltin chloride $CH_3[C_6H_5C(CH_3)_2CH_2]C_6H_5SnCl$ (1) is a second-order process with respect to the nucleophile pyridine, which induces this configurational instability (2). An analogous second-order rate equation has been obtained for the racemization of various chloro- and bromosilanes induced by nucleophiles (HMPT, DMSO, DMF) (3,8). The activation entropy for this process is large ($\Delta S^{\ddagger} \simeq -50$ e.u.). It may be mentioned that the rate of racemization of triorganogermanium chlorides behaves similarly to that of analogous triorganosilicon compounds (6,7). Two possible mechanisms have been proposed to explain the experimental data:

coupled with either

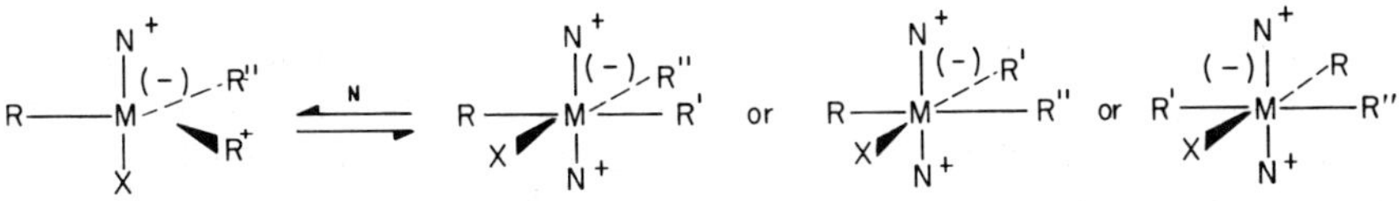

[Structures showing equilibrium between tetracoordinate and pentacoordinate species with N, M, R, R', R'', X substituents]

or

[Structures showing equilibrium between two tetracoordinate species]

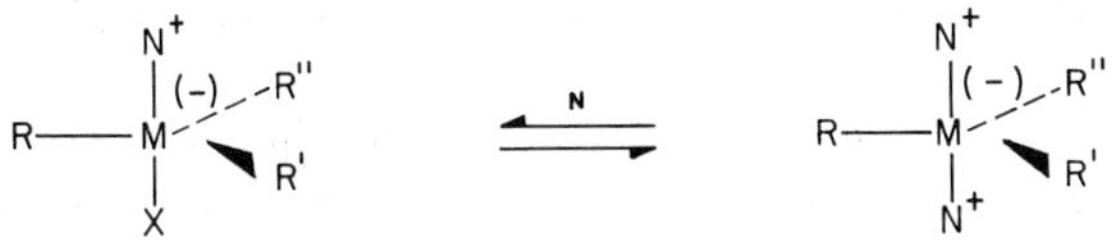

A mechanism with increase of coordination number had already been proposed by Sommer (5) for the racemization of fluorosilanes.

The influence of substituents on the phenyl bound to tin on the rate of racemization of methylneophylphenyltin chloride has been studied. A p-trifluoromethyl group accelerates the racemization by a factor of 5 (9). On the contrary, a p-trifluoromethyl group totally inhibits the racemization of ethyl-1-naphthylphenylsilicon chloride (10).

The presence of bulky groups on the tin atom of triorganotin halides reduces the electrophilicity of the metal atom (11) and increases the optical stability of these compounds: small quantities of DMSO are sufficient to cause the coalescence of the signals of the diastereotopic neophylic methyl groups of methylneophylphenyltin bromide, whereas for methylneophyltrityltin bromide even a $0.5 M$ concentration in DMSO does not cause their coalescence. The last compound is configurationally stable on the NMR time scale up to $150°C$ (12).

Another way to increase the optical stability of triorganotin halides is to use a bidentate ligand that forms a five-membered chelate with the metal atom, like the one studied by van Koten (13):

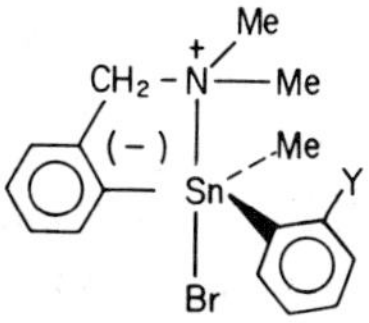

[Structure of tin chelate complex with CH$_2$-N$^+$Me$_2$ group, Sn, Br, and Y substituent]

Y=H, CH$_2$NMe$_2$

(see below).

The optical instability of methylneophylphenyltin chloride (1) has also been studied by optical rotatory dispersion (ORD) (14).

B. The Optical Stability of Triorganotin Hydrides

The optical stability of triorganotin hydrides was first shown by NMR for several compounds like methylneophylphenyltin hydride (2) (15) and methylphenyl-2-phenylpropyltin hydride or deuteride (16), even in nucleophilic solvents like HMPT or DMSO. Subsequently several optically active triorganotin hydrides were prepared: (+)-methyl-1-naphthylphenyltin hydride (3) (17), (+)- or (−)-methylneophylphenyltin hydride (2) (17,19), and (−)methylphenyltertiobutyltin hydride (4) (17).

Triorganotin hydrides are optically stable for long periods (9). In the presence of hydroquinone, compound 2 can be kept for weeks with unchanged optical rotation. In the presence of AIBN, it racemizes in about 12 days at room temperature. A radical mechanism accounts for these observations (17):

$$In \cdot + R\text{---}Sn\text{---}H \rightleftharpoons R\text{---}Sn \cdot + InH$$

$$R\text{---}Sn \cdot \rightleftharpoons \cdot Sn\text{---}R$$

$$RR'R''Sn \cdot + HSnRR'R'' \rightleftharpoons RR'R''SnH + \cdot SnRR'R''$$

$$\begin{array}{c} C_6H_5 \\ | \\ CH_3\text{---}Sn\text{---}X \\ | \\ CH_2C(CH_3)_2C_6H_5 \end{array} \qquad \begin{array}{c} C_6H_5 \\ | \\ CH_3\text{---}Sn\text{---}H \\ | \\ \text{(1-naphthyl)} \end{array} \qquad \begin{array}{c} CH_3 \\ | \\ (CH_3)_3C\text{---}Sn\text{---}H \\ | \\ C_6H_5 \end{array}$$

1 X = Cl
2 X = H   3   4

Compound 3 kept for 4 months neat in the dark at 23°C, for 10 min at 50°C (5 degrees below its melting point), or for 8 min at 60°C does not undergo any racemization. In the presence of polar solvents such as CH_3OD, compound 2 racemizes in less than 1 hr even in the presence of hydroquinone (9). A mechanism analogous to the one explaining the optical instability of triorganotin chlorides may be proposed here (see above).

The absolute configuration of compound 3 could easily be determined by the method of quasi-racemates (20): the absolute configurations of the carbon, silicon, and germanium homologs are indeed known (21) [all three (+)-compounds are R]. It is interesting to mention here that already in 1963 Brook (21) hoped "to report shortly the synthesis and resolution of the related tin compound."

The rapid inversion of configuration of the tin atom of a triorganotin halide observed in 1968 by Peddle and Redl (47) can explain why Belloli (46), extrapolating this optical instability to other organotin compounds, wrote in 1969: "The resolution of the corresponding tin and lead enantiomers would complete the isoconfigurational series, but the finding of rapid inversion for asymmetric tin (and therefore probably also for lead) is a serious obstacle to achieve this objective". Maybe this is why it was only in 1977 that a report appeared on the synthesis of the first optically active triorganotin hydride (18).

C. The Optical Stability of Other Triorganotin Compounds

Like triorganotin halides, triorganotin phenoxides are configurationally unstable on the NMR time scale in the presence of nucleophiles like DMSO (22).

In contrast, triorganostannyliron, -manganese, and -cobalt complexes are configurationally more stable. This has been shown for the methylphenyl(2-phenylpropyl)stannyl transition metal complexes (23). They can exist as a threo and an erythro pair of enantiomers, which together give two anisochronous $^1H_3$C-Sn signals.

The compositions of two different diastereoisomeric mixtures of methylphenyl(2-phenylpropyl)stannylcyclopentadienyldicarbonyliron (5) remain unchanged for weeks *in pyridine*. In contrast, two different diastereomeric mixtures of methylphenyl(2-phenylpropyl)stannylpentacarbonylmanganese (6) are epimerized in a few hours to an equilibrium mixture in the presence of small quantities of pyridine or of DMSO, even though the ratio of erythro to threo isomer remains unaffected for long periods in the absence of nucleophile. The presence of a more electron-donating triphenylphosphine ligand at the manganese atom increases the configurational stability at tin (23). Methylphenyl(2-phenylpropyl)stannyltricarbonyl(triphenylphosphine)cobalt (7) can be obtained as a pure threo or erythro mixture. Its composition does not change after melting under nitrogen (105°C), but the addition of pyridine causes a slow epimerization. The configurational stability of triorganostannyl transition metal complexes is higher when the nucleophilicity of the corresponding transition metalate ion is high; these results show that it is realistic to undertake the synthesis of optically active triorganostannyl transition metal compounds.

Finally, the optical stability of tetraorganotin compounds is very high: (+)-3-(*p*-anisylmethyl-1-naphthylstannyl)-1,1-dimethyl-1-propanol (8) stayed with an unchanged optical rotation for several years (24). Even compounds such as (iodomethyl)methylneophylphenyltin (9) or methylneophylphenyl(phenylethynyl)tin (10) are configurationally stable on the NMR time scale even in the presence of DMSO or pyridine (25).

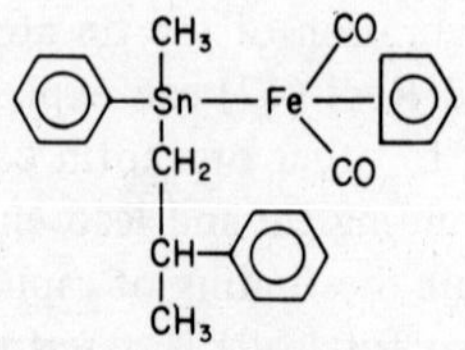

5

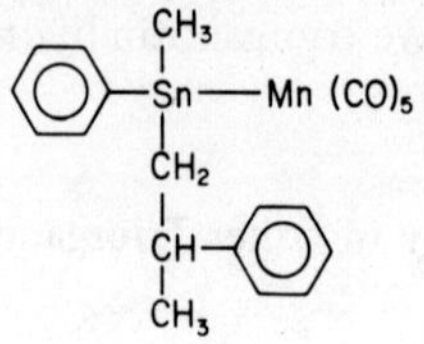

6

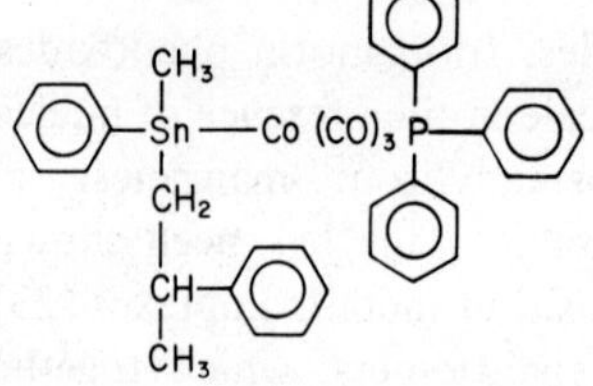

7

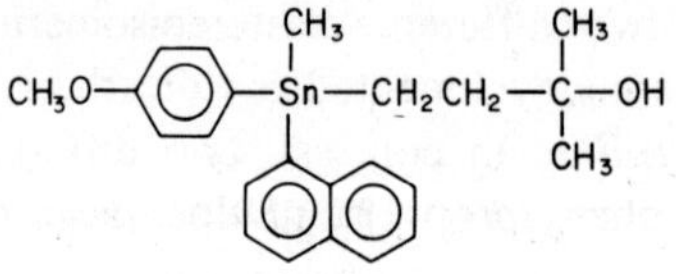

8

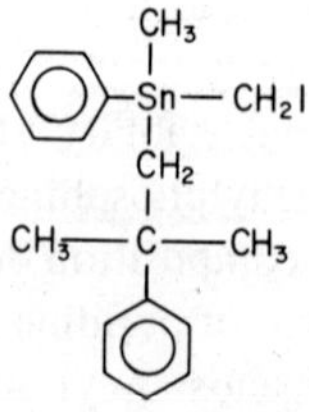

9

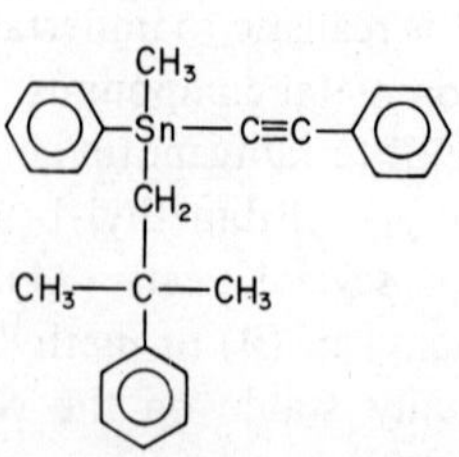

10

II. THE STEREOCHEMISTRY OF SUBSTITUTION REACTIONS AT THE METAL ATOM OF RR'R"MX COMPOUNDS

A. The Synthesis of Optically Active Organogermanium Compounds

The first optically active triorganogermanium compound ethylisopropylphenylgermanium d-α-bromocamphorsulfonate, was prepared in 1931 by Schwartz and Lewinsohn (26), but no attempt was made to replace the chiral resolving agent by an achiral ligand, to make an optically active compound having only one chiral center, viz. the germanium atom.

In 1963 Eaborn (27) and Brook (29) prepared optically active ethyl- and methyl-1-naphthylphenylgermanium hydrides, (+)-11 and (−)-12, respectively, using Sommer's resolution method (28). They described the first Walden cycle at a chiral germanium oxidizing (+)-11 and (−)-12 with Cl_2 to the corresponding chlorides, which were finally reduced back with lithium aluminum hydride to (−)-11 and (+)-12.

More recently, Corriu synthesized a third optically pure triorganogermane: isopropyl-1-naphthylphenylgermanium hydride (13)(32).

Many optically active organogermanium compounds have been used to study the stereochemistry of substitution reactions at their metal atom (see below). The results obtained are generally similar to those obtained for analogous organosilicon compounds, and the mechanisms used to rationalize the experimentally observed stereochemistry are similar too (see below).

B. The Synthesis of Optically Active Organotin Compounds

The first optically active, enantiomerically pure organotin compound containing an asymmetric tin atom as the only chiral center, compound 8, was synthesized in 1971 (33). Several other optically active tetraorganotin compounds have been prepared since (9), either by a classical resolution scheme (34,35) or by asymmetric induction (24,36). An optically active triorganostannylgermane has been made analogously (37): it is optically stable at tin, like the tetraorganotin compounds.

Stereoselective substitution reactions have also been used to prepared optically active tetraorganotin compounds: for instance, the insertion of a carbene into the Sn—H bond of a triorganotin hydride (39) [note that the reaction of ethyl diazoacetate with chiral methyl-1-naphthylphenylsilane proceeds with at least 95% retention of configuration (40)] or the addition of a chiral triorganotin hydride to an activated C=C double bond (39). The last reaction is known to proceed via an intermediate stannyl radical, which must therefore be optically stable like germyl radicals (41). Analogous stereoselective addition reactions of triorganogermanes to double or triple bonds have been described: these proceed with retention of configuration (43,44).

Optically active triorganotin hydrides have also been synthesized either by the reduction of optically unstable triorganotin menthoxides with lithium aluminum hydride or by the reduction of opitcally unstable triorganotin halides with a chiral reducing agent (17,38).

A stereoselective substitution reaction has been used to prepare the first optically active hexaorganoditin compound: methylneophylphenyltin hydride reacts exothermically with Pd/C to give H_2 and the corresponding chiral ditin compound (45).

Another stereoselective substitution reaction transformed methylneophylphenyltin hydride (2) into the corresponding chloride 1, which is optically unstable and racemizes in about 1 hr (14).

Van Koten et al. (61) synthesized, via a C-chiral arylcopper intermediate, a diastereomerically pure triorganotin halide in which the tin atom is pentacoordinated; they obtained a 40:60 mixture of the two possible diastereomers 14 and 15, from which the least soluble compound, **14**, precipitated in pure form owing to the displacement of the equilibrium **14** ⇌ **15** to the left. This epimerization, which is catalyzed by nucleophiles (see above), has been studed experimentally by dynamic NMR.

C. The Stereochemistry of Substitution Reactions at the Metal Atom of Triorganogermanium Compounds, Compared to That Characterizing Substitutions at the Silicon Atom of Organosilanes, and Rationalization Thereof

Table 1 summarizes some of the stereochemical courses observed for substitution reactions at the metal atom of optically active triorganogermanium compounds compared to those observed for similar reactions performed with analogous triorganosilicon compounds. Many reactions of triorganogermyllithium compounds have been studied from the stereochemical point of view.

Table 1 Stereochemical Course of Some Substitution Reactions at the Metal Atom of Triorganogermanium- or Silicon Compounds

Substrate[a]	Reagent	Reaction Product	Stereo-chemistry	Reference
R_3MOCH_3	$LiAlH_4/Et_2O$	R_3MH	Retention	M = Si, 28
				M = Ge, 29
R'_3MOCH_3	$LiAlH_4/Et_2O$	R'_3MH	Retention	M = Si, 48
				M = Ge, 49
$R''_3GeOMenthyl$	$LiAlH_4/Et_2O$	R''_3GeH	Retention	32
R_3MCl	$LiAlH_4 Et_2O$	R_3MH	Inversion	M = Si, 28
				M = Ge, 29
R'_3MCl	$LiAlH_4/Et_2O$	R'_3MH	Inversion	M = Si, 48
				M = Ge, 49
R''_3GeCl	$LiAlH_4/Et_2O$	R''_3GeH (**13**)	Inversion	32
R_3MH (**12**)	Cl_2/CCl_4	R_3MCl	Retention	M = Si, 28
				M = Ge, 29
R'_3MH (**11**)	Cl_2/CCl_4	R'_3MCl	Retention	M = Si, 48
				M = Ge, 49
R''_3GeH (**13**)	Cl_2/CCl_4	R''_3GeCl	Retention	32

[a] $R_3 \equiv Me\phi\text{-}1\text{-}Np$; $R'_3 \equiv Et\phi\text{-}1\text{-}Np$; $R''_3 \equiv i\text{-}Pr\phi\text{-}1\text{-}Np$.

Since triorganogermyllithium compounds, like their triorganosilyllithium analogs (66), invert only slowly, they are optically stable enough in solution (67) to be used as substrates for the study of substitution reactions at the germanium atom. Ethyl-1-naphthylphenylgermyllithium (**16**) reacts with alkyl or allyl bromides or chlorides with retention of configuration, whereas it reacts with inversion of configuration with benzyl bromide, alkyl iodides, and allyl iodide (68). These stereochemical differences have been explained by two distinct mechanisms: a direct coupling four-center process for the first category and a halogen-lithium exchange proceeding with retention of configuration followed by a coupling between the triorganogermanium chloride thus obtained and organolithium reagent in an inverting process for the second category. Methyl-1-naphthylphenylgermyllithium (17) analogously reacts with alkyl chloroacetates with retention (69) and with propargyl bromide with inversion of configuration (70).

[Structures of compounds 16, 17, and 18: 16 is CH₃CH₂—Ge—Li with phenyl and naphthyl; 17 is CH₃—Ge—Li with phenyl and naphthyl; 18 is (CH₃)₂CH—Ge—Li with phenyl and naphthyl]

Analogously, 17 and isopropyl-1-napthylphenylgermyllithium (18) react with retention of configuration with triorganogermanium chlorides or alkoxides, the germanium atoms of which are inverted. This reaction opens the route to *threo*- or *meso*-digermanes (71).

Reactions between compound 16 and carbonyl compounds proceed with retention of configuration at the metal atom and probably go through a four-center intermediate (67). Tetraorganogermanium compounds can also be made by the reaction between triorganogermanium hydrides and Grignard reagents in the presence of a nickel catalyst with retention of configuration (64).

Both the hydrolysis of compound 16 (yielding the corresponding triorganogermane) and the reaction of compound 16 with CO_2 (yielding the triorganogermanecarboxylic acid) proceed with retention of configuration (29,67).

Other triorganogermyl metals have been made: for instance, a bistriorganogermylmercury from the triorganogermyl hydride and di-*t*-butyl mercury, probably with retention of configuration, which reacts with $Hg(CH_2COOCH_3)_2$ to give the corresponding methyl-triorganogermyl acetate, also with retention of configuration (72).

Sommer (50) has rationalized the diverse stereochemistry observed for organosilicon compounds as follows:

1. For poor leaving groups X (those corresponding to very weak acids HX, have a $pK_a > 10$), the most favorable stereochemistry involves retention of configuration.

2. For good leaving groups X (those corresponding to less weak acids, having a $pK_a < 6$), the most favorable stereochemistry involves inversion of configuration.

The study of the stereochemistry of numerous substitutions at an asymmetric silicon atom $RR'R''SiX + R'''-M' = RR'R''SiR''' + XM'$ has been undertaken by Corriu. Three main factors are responsible for the steric course of these reactions: the electronic structure of $R'''-M'$, the structure of the organosilane, and the nature of the leaving group X.

1. The electronic structure of $R'''-M'$ can be a function of the organic group R''', of the metal M' and of the solvent.

The harder the organic anionic group $R'''^{(-)}$ (alkyl), the more the stereochemistry is close to pure retention; the softer the organic group (allyl, benzyl), the more the stereochemistry is close to inversion (52,59).

Also, the harder the metal (Li), the more the stereochemistry is close to retention (53).

Again, a more coordinating solvent polarizes the carbon-metal bond and increases the hardness of the nucleophile, orienting the stereochemistry towards retention of configuration (54): naphthylphenylvinylsilicon fluoride reacts with ethylmagnesium bromide in diethyl ether with inversion and in dimethoxyethane with retention of configuration.

2. Organosilicon chelates are usually more reactive than acyclic compounds and are characterized by more retention of configuration than acyclic compounds (53): 1-phenyl-1-chloro-1-silaacenaphthene reacts with $OH^{(-)}$ and $H^{(-)}$ with retention of configuration (55), whereas inversion is observed with analogous acyclic compounds.

3. The nature of the leaving group is also important. Inversion is generally observed with -Cl, -Br, -F, and -SR, whereas retention is usually characteristic for -OMe and -H (56).

These rules are also valid for bifunctional compounds (57,58) and can be successfully applied to organogermanium compounds (59). They have been rationalized by Nguyen Trong Anh (60).

D. A Stereospecific Substitution Reaction at the Metal Atom of a Chiral Triorganotin Compound

So far, because the optical purity and absolute configuration of the triorganotin hydrides are not known, only the stereoselectivity of some substitution reactions have been evidenced (see above). However there is a substitution reaction for which the stereochemistry can be determined even without knowing the optical purity of the starting material: the replacement of the $^2H^{(-)}$ ligand by an isotopically labeled $^1H^{(-)}$ nucleophile. When chiral methylneophylphenyltin deuteride is left for 9 hr at 40°C, the optical rotation is lowered by 7%. When a mixture of triphenyltin hydride and chiral methylneophylphenyltin deuteride is left for 9 hr at 40°C, the optical rotation is lowered by 14%, and 75% of the initial amount of chiral triorganotin deuteride has been transformed into the corresponding hydride. The H/D exchange is therefore almost stereospecific and proceeds with retention of configuration (17).

Analogous but catalyzed H/D exchanges between two triorganosilanes (62) or between two triorganogermanium hydrides (63) also proceed with retention of configuration at the metal atom.

III. THE INTRAMOLECULAR REARRANGEMENTS OF TRIGONAL BIPYRAMIDAL GROUP (IV) METAL COMPLEXES AND THEIR IMPORTANCE FOR THE STEREOCHEMISTRY OF SUBSTITUTION REACTIONS AT RR'R"MX COMPOUNDS

An interpretation of the stereochemical results can be given on the basis of a mechanism that is generally accepted for such substitution reactions (51) and consists of the following components:

1. The addition of the nucleophilic reagent at the tetrahedrally substituted metal atom, yielding a trigonal bipyramidal intermediate complex.
2. This is (eventually) followed by intramolecular rearrangements of the trigonal bipyramidal complex transforming that intermediate into other accessible trigonal bipyramidal complexes. Finally, the elimination of the nucleophilic leaving group yields the reaction product.

A. The Addition of a Nucleophile at a Tetrahedrally Substituted Metal Atom

Formally, a substitution at a tetrahedrally substituted metal atom can be considered as a game played with five ligands (the four initially attached at the metal atom plus the nucleophilic reagent) and a tetrahedral skeleton. If a tetrahedral species is considered as being formally derived from a pentacoordinate intermediate by the loss of a ligand, it may be named by marking the ligand that has been lost with a "+" if the remaining species is R and a "−" if the enantiomer having the S configuration is produced (73). The parity of one of the possible tetrahedral species can then be defined (74, 75) as the sum of the parity of the absolute configuration (+ is even, − is odd) and of the parity of the ligand that has been lost; thus $\bar{5}$ is even (odd + odd is even), $\overset{+}{4}$ is even (even + even is even), $\overset{+}{5}$ and $\bar{4}$ are odd (even + odd is odd).

A nucleophile can add at the metal atom either by attacking one of the four faces of the tetrahedron (when it is found in the apical position of the trigonal bipyramid: apical attack) or by attacking one of the six edges (when it is found in the equatorial position of the trigonal bipyramid: equatorial attack).

If it attacks a face, then the trigonal bipyramid will have, in the apical positions, the two ligands that are not used to define that face. A bipyramidal molecule is named by marking the two apical ligands starting with the smaller one, with a "−" if the absolute configuration of the trigonal bipyramid is S (74). The parity of the trigonal bipyramid is then defined as the sum of the parities of the symbols used to define it, that is, the absolute configuration and the numbers describing the two apical ligands. It is then easy to show (75) that apical attack is characterized by a parity change and that equatorial attack takes place without any parity change. It is then possible to describe all the possible addition reactions at $\overset{+}{4}$, $\bar{5}$, $\bar{4}$, and $\overset{+}{5}$ for instance by an **AE** matrix, each coefficient a_{ij} of which being one if tetrahedron i can be transformed into trigonal bipyramid j and zero otherwise:

$$\mathbf{AE} = \begin{array}{r} {}_{+4} \\ {}_{-5} \\ {}_{-4} \\ {}_{+5} \end{array} \begin{array}{|cccccccccccccccccccc|} \overline{12} & 13 & \overline{14} & 15 & \overline{23} & 24 & \overline{25} & \overline{34} & 35 & \overline{45} & 12 & \overline{13} & 14 & \overline{15} & 23 & \overline{24} & 25 & 34 & \overline{35} & 45 \\ \hline 1 & 1 & 0 & 0 & 0 & 0 & 1 & 0 & 1 & 0 & 0 & 0 & 1 & 0 & 0 & 1 & 0 & 1 & 0 & 1 \\ 1 & 1 & 1 & 1 & 0 & 1 & 0 & 1 & 0 & 0 & 0 & 0 & 0 & 1 & 0 & 0 & 0 & 0 & 1 & 1 \\ 0 & 0 & 1 & 0 & 0 & 0 & 0 & 0 & 0 & 1 & 1 & 1 & 0 & 1 & 1 & 0 & 1 & 0 & 1 & 1 \\ 0 & 0 & 0 & 0 & 1 & 0 & 1 & 0 & 1 & 1 & 1 & 1 & 0 & 0 & 1 & 0 & 0 & 1 & 1 & 0 \\ \end{array}$$

B. The Minimum Constraints for Stereospecificity

It is easy to show (74) that a substitution reaction proceeds with inversion of configuration if the parity does not change, and with retention of configuration if the parity changes. Thus $\overset{+}{4} \to \overset{+}{5}$ proceeds with retention and $\overset{+}{4} \to \overset{-}{5}$, with inversion of configuration. It is very simple to determine which trigonal bipyramids *must* be excluded if one wants a *stereospecific* substitution occurring by an addition-elimination mechanism (76). Indeed, any trigonal bipyramid allowing the transformation of $\overset{+}{4}$ into $\overset{+}{5}$ (or of $\overset{-}{4}$ into $\overset{-}{5}$) must be discarded if one wants a substitution to proceed with inversion. The **AE** matrix shows that 15, $\overline{25}$, 35, 14, $\overline{24}$, and 34 are indeed connecting $\overset{+}{4}$ and $\overset{+}{5}$ and that $\overline{15}$, 25, $\overline{35}$, $\overline{14}$, 24, and $\overline{34}$ are connecting 4 and 5. Therefore the minimum constraints for inversion are those by which all these bipyramids are not allowed. The **AE** matrix valid for the case for which 14, $\overline{14}$, 15, $\overline{15}$, 24, $\overline{24}$, 25, $\overline{25}$, 34, $\overline{34}$, 35, and $\overline{35}$ are excluded is found by erasing the columns corresponding to these trigonal bipyramids. This **AE** matrix valid for this specific case can be rewritten as follows:

$$\mathbf{AE} \text{ (minimum constraints for inversion)} = \begin{array}{c} \\ \overset{+}{4} \\ \overset{-}{5} \\ \overset{+}{4} \\ \overset{+}{5} \end{array} \begin{array}{c} \overline{12} \quad 13 \quad \overline{23} \quad 45 \quad\quad 12 \quad \overline{13} \quad 23 \quad \overline{45} \\ \left| \begin{array}{cccccccc} 1 & 1 & 1 & 1 & 0 & 0 & 0 & 0 \\ 1 & 1 & 1 & 1 & 0 & 0 & 0 & 0 \\ 0 & 0 & 0 & 0 & 1 & 1 & 1 & 1 \\ 0 & 0 & 0 & 0 & 1 & 1 & 1 & 1 \end{array} \right| \end{array}$$

and can be represented graphically as shown in Figure 1 (75).

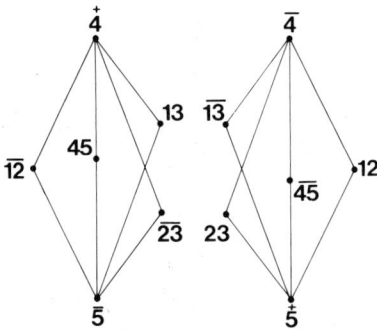

Fig. 1. Addition-elimination reactions allowed when the minimum constraints for inversion are applied.

The graph contains two disjoint subgraphs (because the matrix is block-diagonal), which shows that the substitution reaction is stereospecific.

Of course, the minimum constraints for retention are exactly complementary to those for inversion (i.e., 12, $\overline{12}$, 13, $\overline{13}$, 23, $\overline{23}$, 45, and $\overline{45}$ are excluded). The **AE** matrix valid for this specific case can be rewritten as follows:

$$\mathbf{AE}\text{ (minimum constraints for retention)} = \begin{array}{c} \\ \overset{+}{4} \\ \overset{+}{5} \\ \overset{-}{4} \\ \overset{-}{5} \end{array} \begin{array}{|cccccccccccc|} 14 & 15 & \overline{24} & \overline{25} & 34 & 35 & \overline{14} & \overline{15} & 24 & 25 & \overline{34} & \overline{35} \\ 1 & 1 & 1 & 1 & 1 & 1 & 0 & 0 & 0 & 0 & 0 & 0 \\ 1 & 1 & 1 & 1 & 1 & 1 & 0 & 0 & 0 & 0 & 0 & 0 \\ 0 & 0 & 0 & 0 & 0 & 0 & 1 & 1 & 1 & 1 & 1 & 1 \\ 0 & 0 & 0 & 0 & 0 & 0 & 1 & 1 & 1 & 1 & 1 & 1 \end{array}$$

and can be represented graphically as shown in Figure 2 (76).

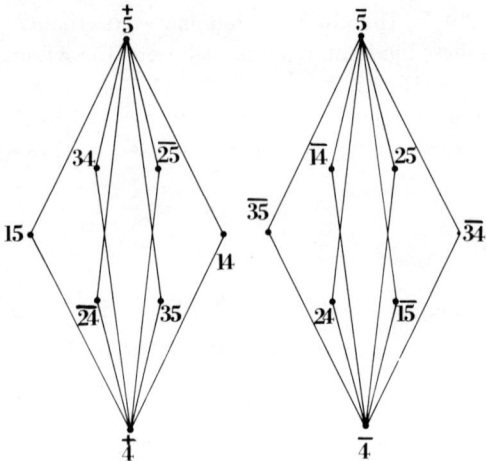

Fig. 2. Addition-elimination reactions allowed when the minimum constraints for retention are applied.

C. Intramolecular Rearrangements: The Berry Pseudorotation Belonging to Mode P1

The intramolecular rearrangement mechanism described by Berry can be visualized as two synchronized bending motions through which the two apical ligands become equatorial and two of the three equatorial ligands become apical. The ligand staying in the equatorial position is named pivot (73). It may be shown that the Berry mechanism [and any mechanism belonging to mode *P1* (*for a definition of mode of rearrangement, see ref. 75 and below*)] is characterized by a parity change.

By definition, the pivots that may be used to transform a given trigonal bipyramid are the equatorial ligands, that is, the ligands that do not appear in the two-digit symbol of the trigonal bipyramid. Furthermore, the isomers formed by a Berry pseudorotation have, as apical substituents, those that are present neither in the symbol of the starting trigonal bipyramid nor in that of the pivot.

All the possible intramolecular rearrangements of trigonal bipyramids can best be described by the **B(P1)** matrix (73, 74), the nonzero elements of which may be considered as the turning points to go from a trigonal bipyramid to another one.

$$\mathbf{B}(P1) = \begin{array}{c|cccccccccc} & \overline{12} & 13 & \overline{14} & 15 & \overline{23} & 24 & \overline{25} & \overline{34} & 35 & \overline{45} \\ \hline 12 & 0 & 0 & 0 & 0 & 0 & 0 & 0 & 1 & 1 & 1 \\ \overline{13} & 0 & 0 & 0 & 0 & 0 & 1 & 1 & 0 & 0 & 1 \\ 14 & 0 & 0 & 0 & 0 & 1 & 0 & 1 & 0 & 1 & 0 \\ \overline{15} & 0 & 0 & 0 & 0 & 1 & 1 & 0 & 1 & 0 & 0 \\ 23 & 0 & 0 & 1 & 1 & 0 & 0 & 0 & 0 & 0 & 1 \\ \overline{24} & 0 & 1 & 0 & 1 & 0 & 0 & 0 & 0 & 1 & 0 \\ 25 & 0 & 1 & 1 & 0 & 0 & 0 & 0 & 1 & 0 & 0 \\ 34 & 1 & 0 & 0 & 1 & 0 & 0 & 1 & 0 & 0 & 0 \\ \overline{35} & 1 & 0 & 1 & 0 & 0 & 1 & 0 & 0 & 0 & 0 \\ 45 & 1 & 1 & 0 & 0 & 1 & 0 & 0 & 0 & 0 & 0 \end{array}$$

Several graphical representations of these intramolecular rearrangements have been described (73–75, 91).

The minimum constraints for inversion can be applied to this **B(P1)** matrix. **B(P1)** (minimum constraints for inversion) can be written as follows:

$$\mathbf{B}(P1)\text{ (minimum constraints for inversion)} = \begin{array}{c|cccc} & \overline{12} & 13 & \overline{23} & \overline{45} \\ \hline 12 & 0 & 0 & 0 & 1 \\ \overline{13} & 0 & 0 & 0 & 1 \\ 23 & 0 & 0 & 0 & 1 \\ 45 & 1 & 1 & 1 & 0 \end{array}$$

The graph corresponding to this **B(P1)** (minimum constraints for inversion) matrix can be combined with the graph corresponding to the **AE** (minimum constraints for inversion) matrix. The resulting graph (Fig. 3) shows that the substitution reaction is still stereospecific if the intramolecular rearrangements of mode *P*1 are allowed (76).

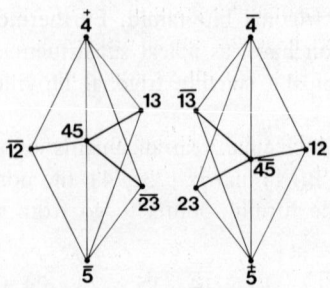

Fig. 3. Addition-*P*1-elimination reactions allowed when the minimum constraints for inversion are applied: the substitution reaction is still stereospecific.

Analogously, the minimum constraints for retention can be applied to $B(P1) \cdot B(P1)$ (minimum constraints for retention) can be written as follows:

$B(P1)$ (minimum constraints for retention) =

	$\overline{14}$	15	24	$\overline{25}$	$\overline{34}$	35
14	0	0	0	1	0	1
$\overline{15}$	0	0	1	0	1	0
$\overline{24}$	0	1	0	0	0	1
25	1	0	0	0	1	0
34	0	1	0	1	0	0
$\overline{35}$	1	0	1	0	0	0

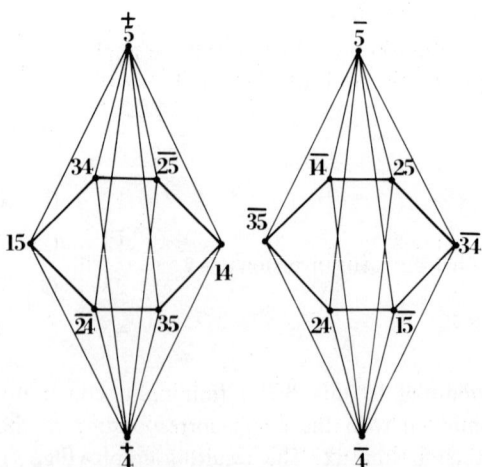

Fig. 4. Addition-*P*1-elimination reactions allowed when the minimum constraints for retention are applied: the substitution reaction is still stereospecific.

The graph corresponding to this **B(P1)** (minimum constraints for retention) matrix can be combined with the graph corresponding to the **AE** (minimum constraints for retention) matrix. The resulting graph (Fig. 4) shows that here too, the substitution reaction is still stereospecific if the intramolecular rearrangements of mode *P*1 are allowed (76).

D. Chemical Constraints for Particular Cases and the Resulting Stereochemistry

Several particular cases can now be examined: the constraints characterizing these cases may be compared with the minimum constraints for stereospecificity, consequently determining the stereochemistry. The most usual chemical constraints encountered in substitution reactions at group IV metals are:

1. *The small cycle constraint*: four-membered rings will prefer to be in apical-equatorial position.
2. *The electronegativity constraint*: electronegative substituents will prefer to be in apical position if possible.

For the case of substitution reactions for which the nucleophile 4 and the leaving group 5 are the most electronegative ligands, the following constraint can be determined: 4 and 5, being electronegative, are allowed only in apical positions and thus all the trigonal bipyramids except 45 and $\overline{45}$ are excluded. This constraint is therefore *stronger* than the minimum constraint for inversion, and the stereochemistry expected for this case is inversion of configuration. This constraint is valid when a good leaving group is replaced by an electronegative nucleophilic reagent. The graph valid for this specific case can be deduced from Figure 3 by erasing 12, $\overline{12}$, 13, $\overline{13}$, 23, and $\overline{23}$ and the lines starting from these forbidden isomers.

When the leaving group is a poor leaving group, then electrophilic assistance is expected

Because of the presence of a small ring containing 4 and 5, the constraints for this specific case are, that all the bipyramids in which 4 and 5 are simultaneously apical (45, $\overline{45}$) or equatorial (12, $\overline{12}$, 13, $\overline{13}$, 23, $\overline{23}$) are not allowed. This is exactly the minimal constraint for retention. The graph of Figure 4 is valid for this specific case, and the expected stereochemistry is retention of configuration.

IV. DYNAMIC STEREOCHEMISTRY OF *CIS*–OCTAHEDRAL BIS(β–DIKETONATO)–GROUP (IV) METAL COMPLEXES

Intramolecular rearrangements of six-coordinate chelate complexes have been reviewed extensively (75,77). However *permutations* have been used in this field only fortuitously, and when they have been used (78,79), the argumentation was more intuitive than inspired by Klemperer's (80) or Ruch's (81) quantitative theories.

Intramolecular rearrangements of octahedral tin chelates have been studied by Serpone (78) and by Finocchiaro (82). The compounds that have provided the most powerful experimental information are *cis*-dichlorobis(benzoylacetonato)tin(IV) and *cis*-phenylchlorobis(benzoylacetonato)tin(IV) (82). We discuss here the permutational approach to the study of the dynamic stereochemistry of such bis chelate complexes. The presentation of the problem is based on the relation (75) existing between the modes of rearrangments of the most symmetrical bis chelate complex $M(\overset{\frown}{AA})_2 X_2$ (09) and those of bis chelate complexes of decreased symmetry, namely, $M(\overset{\frown}{AB})_2 X_2$ (09) and $M(\overset{\frown}{AB})_2 XY$.

A. Modes of Rearrangement of *cis*-$M(\overset{\frown}{AA})_2 X_2$ Complexes

To determine which are the modes of rearrangements of the chiral complexes of the type *cis*-$M(\overset{\frown}{AA})_2 X_2$, two groups have to be defined (75):

1. The restricted group S of allowed permutations for this skeleton.
2. The permutation group A, defining the proper symmetry operations of these complexes.

It has been shown that modes of rearrangements in chiral skeletons correspond to double cosets AxA of S with respect to A (75,81,82) (see below).

For these complexes, (Fig. 5) there is obviously a C_2 symmetry axis, which causes the pairs of sites (1,4), (2,3), and (5,6) to be homotopic. Because the two X ligands we will use later have no influence on the behavior of the dynamic NMR spectra of the complexes, they do not need to be considered here.

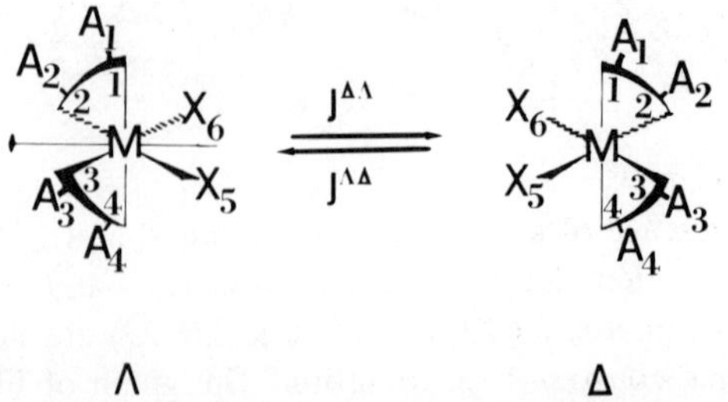

Fig. 5. The inversion of helicity of an octahedral *cis*-$M(\overset{\frown}{AA})_2 X_2$ complex.

Consequently, $A = \{I, (14)(23)\}$, where I denotes the identity and the cyclic permutation $(ijk) = (jki) = (kij)$ means that the ligand on site i replaces the ligand on site j, that the ligand on site j replaces the ligand on site k, and that the ligand on site k replaces the ligand on site i. The restricted group S of allowed permutations is given by the semidirect product $S = (L \wedge K)$ (75, 80), where L is the group of permutations of identical ligands within a given chelate ring, and K is the group of permutations of identical chelate bridges (75); here, $L = S_2^{12} \times S_2^{34} = \{I, (12)\} \times \{I, (34)\} = \{I, (12), (34), (12)(34)\}$ and $K = \{I, (13)(24)\}$.*

Let us determine all the possible modes given by the double cosets AxA. To find a first double coset M_1 (82), we select one of the permutations $x \in S$, for instance $x = (12)$, and we perform the four different AxA products: $I(12)I = (12)$, $I(12)(14)(23) = (1423)$, $(14)(23)(12)I = (1324)$, and $(14)(23)(12)(14)(23) = (34)$. To find the next double coset M_2 (82), we select one of the permutations y belonging to S and not to M_1, for instance, $y = (13)(24)$, and we calculate analogously the four different AyA products, which are $(13)(24)$ (twice) and $(12)(34)$ (twice). Analogously, the last double coset, M_0 (82), can be shown to be the group A itself.

This means that the permutations (12) and (34) of mode M_1 are symmetry equivalent. They are rotationally equivalent to the symmetry equivalent permutations (1423) and (1324) (04)(06); (12)(34) and (13)(24), belonging to mode M_2, are rotationally equivalent. The one-edge exchange corresponds to mode M_1, the two edge exchange, to mode M_2 (and the zero-edge exchange, to mode M_0).

Taking into account the reversal of helicity J (75,82) of the bis(β-diketonato)-tin complex (Fig. 5), which is a feasible (chemically allowed) process (see below), we also need to consider the following modes of rearrangement:

Mode $M_3 = M_0 J = J M_0 = J, (14)(23)J$
Mode $M_4 = M_1 J = J M_1 = (12)J, (1423)J, (1324)J, (34)J$
Mode $M_5 = M_2 J = J M_2 = (13)(24)J, (12)(34)J$

---

*Consequently, S contains the following permutations: I, (12), (34), (12)(34) and $I(13)(24)$, (12)(13)(24), (34)(13)(24), and (12)(34)(13)(24). (Products are read from right to left.) The three last permutation products can be reduced to products of mutually exclusive cycles. For instance, the product $r = (12)(13)(24)$ shows that the ligand on site 1 replaces the ligand on site 3 [permutation (13)]; thus r involves a cycle the two first symbols of which are (13 . . .). We now start from site 3: ligand on site 3 replaces ligand on site 1 [permutation (31) ≡ (13)], but this is followed by the operation: ligand on site 1 replaces ligand on site 2 [permutation (12)], which means that ligand on site 3 replaces ligand on site 2. Hence r involves a cycle starting as (132 . . .). Pursuing this procedure, we get $r = (1324)$. Analogously, $(34)(13)(24) = (1423)$ and $(12)(34)(13)(24) = (14)(23)$. Therefore S can be written as $S = \{I, (12), (34), (12)(34), (13)(24), (1324), (1423), (14)(23)\}$.

B. Modes of Rearrangement of cis-M($\overline{AB}$)$_2$X$_2$ Complexes

We now apply these different modes to the possible isomers of the less symmetrical complex with two identical monodentate ligands but with two nonsymmetrical, identical bidentate ligands.

The six modes defined for the ideal most symmetrical M($\overline{AA}$)$_2$X$_2$ skeleton are no longer *actual* modes for the less symmetrical M($\overline{AB}$)$_2$XY or M($\overline{AB}$)$_2$X$_2$ case (75). They are merely *idealized* modes ($\hat{M}$) (75,83) because rearrangements belonging to them either can lead to different isomers or are no longer symmetry or rotationally equivalent.

Since different isomers are possible for the M($\overline{AB}$)$_2$X$_2$ system, we define the symbols Δij or Λij to characterize them (51,73,84,85), which have the following meaning: Δ or Λ refers to the helicity of the propeller (Fig. 5), i is the label of the chelate moiety trans to ligand X$_5$; j is the label of the chelate moiety trans to ligand X$_6$. The three isomers with Λ helicity are given in Figure 6. Their enantiomers, with Δ helicity, are obtained by taking their mirror images (see Fig. 5).

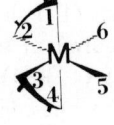

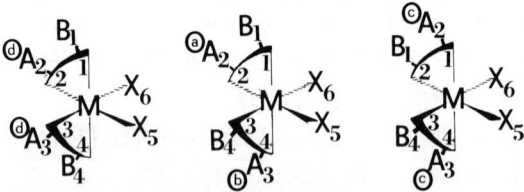

Fig. 6. The skeleton numbering and the three isomers with Λ helicity of cis-M($\overline{AB}$)$_2$X$_2$.
(*left*: ΛA$_2$A$_3$, configuration Λ23 of isomer ΛAA; *center*: ΛA$_2$B$_4$, configuration Λ24 of isomer ΛAB; *right*: ΛB$_1$B$_4$, configuration Λ14 of isomer ΛBB)
The anisochronous magnetic sites corresponding to diastereotopic methyl groups A are denoted a, b, c, and d. These symbols would be barred for the Δ isomers (see text).

From Figure 6 it is obvious that the isomers AA and BB have C_2 symmetry, whereas AB has C_1 symmetry. The two homotopic methyl groups of AA are symbolized d in ΛAA and $\overline{d}$ in ΔAA; the homotopic ones of BB are symbolized c in ΛBB and $\overline{c}$ in ΔBB. Finally, the two diastereotopic methyl groups of AB are symbolized a and b in ΛAB and $\overline{a}$ and $\overline{b}$ in ΔAB.

The application of one of the possible permutations (1324) to one of these isomers (ΛAB) and the resulting magnetization exchanges are described in Figure 7.

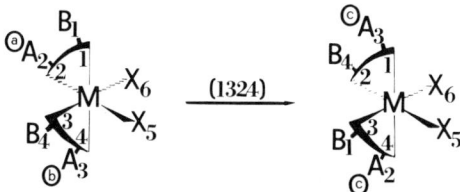

Fig. 7. The application of the permutation (1324) to the configuration Λ24 of isomer ΛAB of cis-M(AB)$_2$X$_2$, which results in the transfer of the magnetizations of a and b to the magnetic sites c (*left*: ΛA$_2$B$_4$, configuration Λ24 of isomer ΛAB; *right*: ΛB$_4$B$_1$, configuration Λ41 of isomer ΛBB).

Analogously, the application of all the permutations belonging to modes $\hat{M}_0$, $\hat{M}_1$, $\hat{M}_2$, $\hat{M}_3$, $\hat{M}_4$, and $\hat{M}_5$ to the six possible isomers of M(AB)$_2$X$_2$ is given in Table 2.

C. Modes of Rearrangement of cis-M(AB)$_2$XY Complexes

Compelling evidence has been provided that in octahedral titanium complexes, stereoisomerizations are accompanied by a reversal of the helicity (86). We make the assumption that this is also the case for the tin complexes. Note that the modes $\hat{M}_0$, $\hat{M}_1$, and $\hat{M}_2$ are indistinguishable by NMR from $\hat{M}_3$, $\hat{M}_4$, and $\hat{M}_5$, respectively, in an achiral environment.

Analogously, the different permutations given in Table 2 can be applied to the eight isomers of the still less symmetrical cis-M(AB)$_2$XY case. The four isomers with Λ helicity are described in Figure 8.

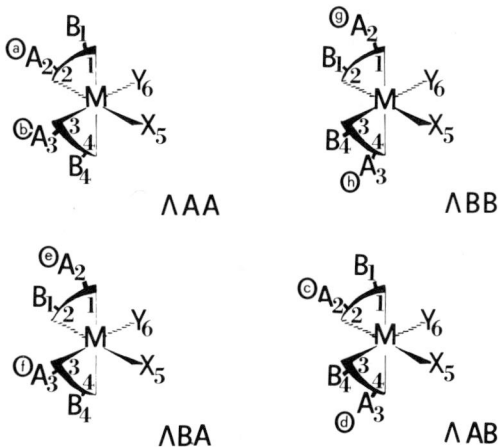

Fig. 8. The four isomers with Λ helicity of cis-M(AB)$_2$XY. The anisochronous magnetic sites corresponding to the diastereotopic methyl groups A are denoted a, b, c, d, e, f, g, and h. The symbols for the Δ isomers would be barred (see text).

Table 2 Application of the Different Permutations (-Inversions) Isomers of cis-M($\overline{AB}$)$_2$X$_2$, the Resulting Magnetization

Idealized Modes	Permutations (-inversions)	Configuration Obtained from 23	Magnetization	
			AA⇌AA	AA⇌BB
$\hat{M}_0, \hat{M}_3$	I (J)	23	[a] [b] [c] [d]	
	(14)(23) (J)	32	(2 +1 +1)	
$\hat{M}_1, \hat{M}_4$	(12) (J)	13		
	(1324) (J)	31		
	(1423) (J)	42		
	(34) (J)	24		
$\hat{M}_2, \hat{M}_5$	(13)(24) (J)	41		[a] [b] [c,d]
	(12)(34) (J)	14		(2 +1)

^{a} The reversibility of all the isomerizations has the consequence that the mutually formations, must be considered together in the NMR analysis of *each* equilibrium having a given intensity.

Table 3 Application of the Different Permutations (-Inversions) of the of cis-M($\overline{AB}$)$_2$XY, the Resulting Magnetization

Idealized Modes	Actual Modes	Permutations (-inversions)	AA⇌AA	AA⇌BB	AA⇌AB	AA⇌BA	BB⇌AB (and number
$\hat{M}_0, \hat{M}_3$	M_{01}, M_{31}	I (J)	[a] [b] (2 +2+2+2)				
	M_{02}, M_{32}	(14)(23) (J)	[a,b] (1 +2+2+2)				
$\hat{M}_1, \hat{M}_4$	M_{11}, M_{41}	(12) (J)				[a,e] [b,f] (2 +2+2)	[e,g] [d,h] (2 +2+2)
	M_{12}, M_{42}	(1324) (J)			[a,d] [b,c] (2 +2+2)	[a,f] [b,e] (2 +2+2)	[c,h] [d,g] (2 +2+2)
	M_{13}, M_{43}	(1423) (J)					
	M_{14}, M_{44}	(34) (J)			[a,c] [b,d] (2 +2+2)		
$\hat{M}_2, \hat{M}_5$	M_{21}, M_{51}	(13)(24) (J)		[a,h] [b,g] (2 +2+2)			
	M_{22}, M_{52}	(12)(34) (J)		[a,g] [b,h] (2 +2+2)			

of the Idealized Modes $\hat{M}_0, \hat{M}_1, \hat{M}_2, \hat{M}_3, \hat{M}_4$, and $\hat{M}_5$ to the Different Exchanges, and the Number of Residual Lines[a]

Exchanges for				Combined Effect of All the Equilibria Associated with the Same Actual Mode (and number of residual lines)
AA⇌AB (and number of residual lines)	BB⇌BB	BB⇌AB	AB⇌AB	
	[a][b][c][d] (2 +1 +1)		[a][b][c][d] (2 +1 +1)	[a][b][c][d] (2 +1 +1)
[a,b,d][c] (1 +1)	[a,b,c][d] (1 +1)			[a,b,c,d] (1)
		[a,b][c][d] (1 +1 +1)		[a,b][c,d] (1 +1)

inverse permutations [(1324) and (1423)], representing microreversible chemical trans-
(10–12). The numbers under the magnetic sites indicate the number of residual lines

Idealized Modes $\hat{M}_0, \hat{M}_1, \hat{M}_2, \hat{M}_3, \hat{M}_4$, and $\hat{M}_5$ to the Different Isomers Exchanges, and the Number of Residual Lines

Exchanges for					Combined Effect of All the Equilibria Associated with the Same Actual Mode (and number of residual lines)
BB⇌BA of residual lines)	BB⇌BB	AB⇌AB	AB⇌BA	BA⇌BA	
	[g][h] (2 +2+2+2)	[c][d] (2 +2+2+2)		[e][f] (2 +2+2+2)	[a][b][c][d][e][f][g][h] (2 + 2 + 2 + 2)
	[g,h] (1 +2+2+2)		[c,f][d,e] (2 +2+2)		[a,b][c,f][d,e][g,h] (1 + 2 +1)
					[a,e][b,f][c,g][d,h] (2 + 2)
					[a,d,f,g][b,c,e,h] (2)
[e,h][f,g] (2 +2+2)					
[e,g][f,h] (2 +2+2)					[a,c][b,d][e,g][f,h] (2 + 2)
		[c,d] (1 +2+2+2)		[e,f] (2 +2+2+2)	[a,h][b,g][c,d][e,f] (2 +1 +1)
			[c,e][d,f] (2 +2+2)		[a,g][b,h][c,e][d,f] (2 + 2)

The application of the permutations corresponding to the idealized modes $\hat{M}_0, \hat{M}_1, \hat{M}_2, \hat{M}_3, \hat{M}_4$, and $\hat{M}_5$ on the eight isomers of cis-$M(\overline{AB})_2XY$ are shown in Table 3. The idealized modes are completely split; that is, every permutation corresponds to an actual mode because all the isomers have C_1 symmetry (83,87–89).

D. Graphical Topological Representations of the Modes of Rearrangement of $M(\overline{AB})_2X_2$ and $M(\overline{AB})_2XY$ Complexes

The topological representations of the modes of rearrangement of $M(\overline{AB})_2X_2$ complexes given in Figure 9 express in a condensed form all the possible isomerization reactions described in Table 2.

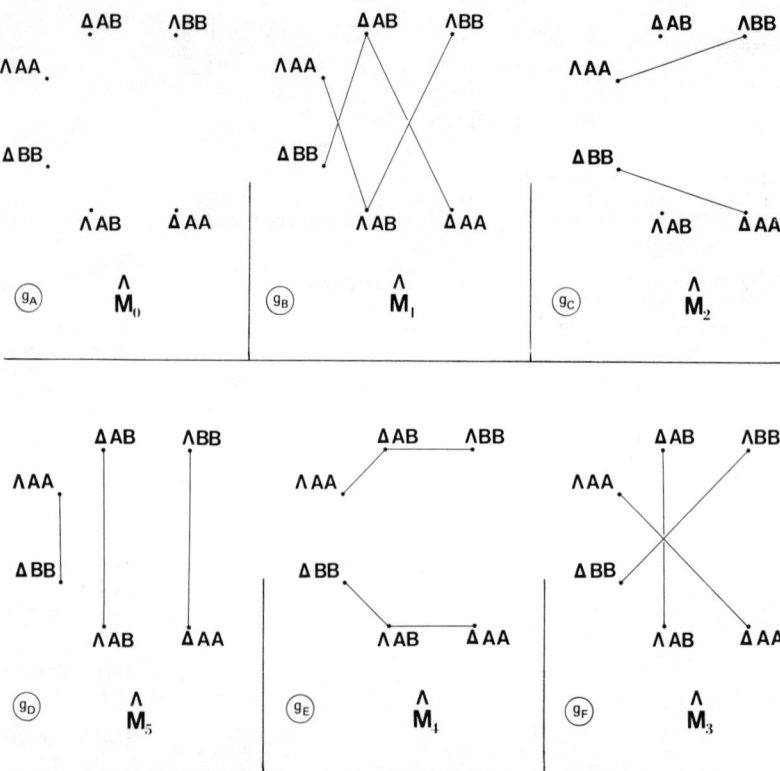

Fig. 9. Graphical topological representations of the six idealized modes of rearrangement applied to the six stereoisomers of cis-$M(\overline{AB})_2X_2$.

Analogously, the graphical representations of the modes of rearrangement characterized by a reversal of the helicity of $M(\overline{AB})_2XY$ complexes can be deduced from Table 3. They are given in Figure 10.

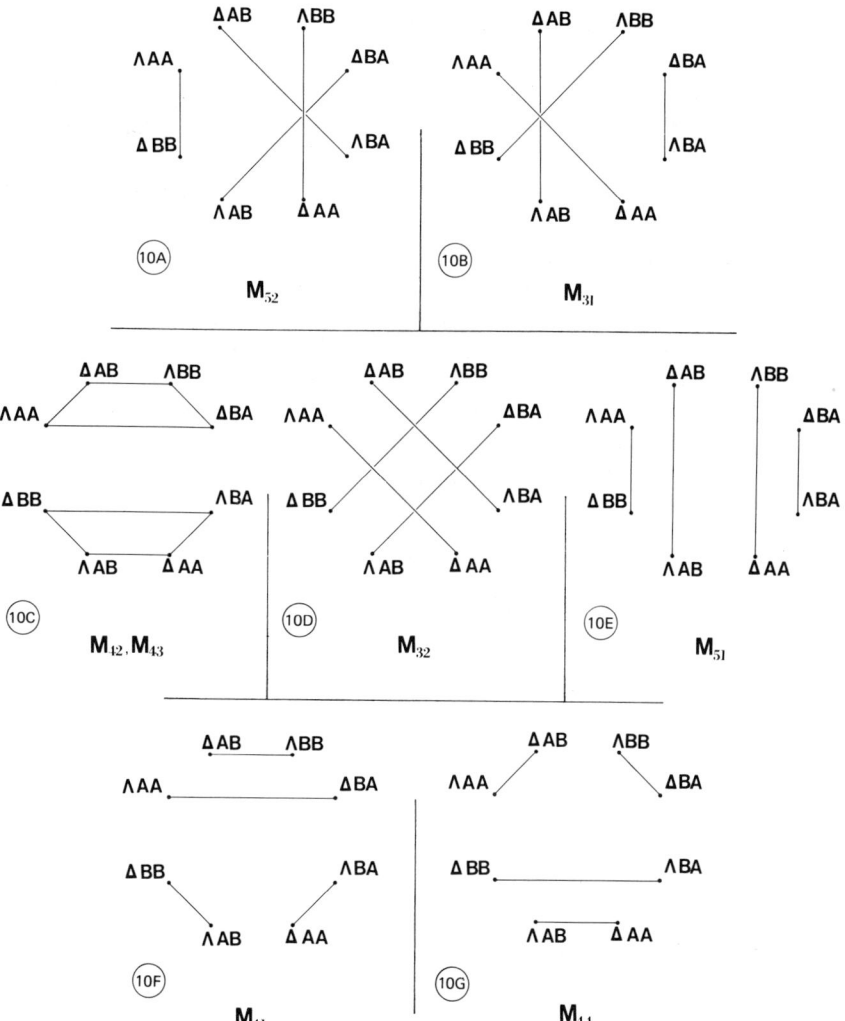

Fig. 10. Graphical topological representations of the possible modes of rearrangement causing an inversion of helicity applied to the eight stereoisomers of cis-$M(A\overrightarrow{B})_2 XY$

E. The Threshold Mode of Rearrangement of cis-Dichlorobis(benzoylacetonato)tin(IV)

The NMR spectrum of a mixture of the three dl pairs of cis-dichlorobis-(benzoylacetonato)tin(IV) in an achiral solvent should show (see above), in the slow exchange region, two methyl signals of equal intensity for the C_1 isomer plus two other methyl singlets for the two C_2 isomers. Table 2 shows:

1. That modes $\hat{M}_0$ and $\hat{M}_3$ are not directly observable by dynamic NMR in an achiral solvent.
2. That if one of the possible exchanges corresponding to the idealized modes $\hat{M}_1$ or $\hat{M}_4$ (AA⇌AB or BB⇌AB) takes place more easily than the other one (BB⇌AB or AA⇌AB), the two lines having initially the same intensity must coalesce with one of the other lines, yielding two singlets of different intensity,
3. That if the exchange AA⇌BB corresponding to the idealized modes $\hat{M}_2$ or $\hat{M}_5$ takes place more easily than the other possible one (AB⇌AB), the two singlets of *different* intensity must coalesce, yielding three lines, two of them having the same intensity, whereas if the exchange AB⇌AB takes place more easily than the other one (AA⇌BB), the two singlets of *equal* intensity must coalesce, yielding three lines of different intensity.

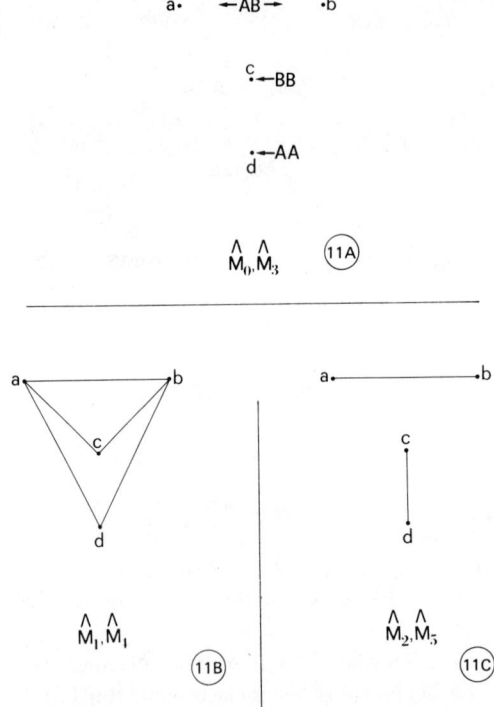

Fig. 11. Magnetization exchange patterns for the idealized modes of rearrangement of cis-M($\overline{AB}$)$_2$X$_2$ complexes.

Experimentally, the 60 MHz ^{1}H NMR spectrum of a chlorobenzene or CDCl$_3$ solution of *cis*-dichlorobis(benzoylacetonato)tin(IV) shows, at room temperature, four lines in the methyl region, two of which are of equal intensity. These four signals coalesce around 80°C (82).

Since this coalescence into one single peak cannot be explained by *one* of the possible exchanges, the different exchanges corresponding to an idealized mode must be considered together.

Figure 11 gives the magnetization exchange patterns corresponding to the combined effect of all the equilibria associated with the six idealized modes of rearrangements of cis-M($\overline{AB}$)$_2$X$_2$. From Figure 11 and Table 2, it is clear that (*a*) modes $\hat{M}_1$ and $\hat{M}_4$, which interconvert all three diastereomers (Fig. 9), are the only ones that should cause the coalescence of the four methyl signals into a single peak, and (*b*) modes $\hat{M}_2$ and $\hat{M}_5$ should cause the coalescence of the two equally intense lines into another peak.

NMR lineshapes based on mode $\hat{M}_4$ were calculated and fit the experimental data satisfactorily (82).

F. The Threshold Mode of Rearrangement of cis-Phenylchlorobis(benzoylacetonato)tin(IV)

The NMR spectrum of a solution of *cis*-phenylchlorobis(benzoylacetonato)-tin(IV) in an achiral solvent should show, in the slow exchange region, two methyl singlets of equal intensity for each of the four possible *dl* pairs. Table 3 shows:

1. That the exchanges corresponding to mode M_{31} are not directly observable by DNMR in an achiral solvent,

2. That two of the three possible exchanges corresponding to mode M_{32} (AA⇌AA or BB⇌BB) and to mode M_{51} (AB⇌AB or BA⇌BA), considered separately, should cause the coalescence of two singlets of equal intensity into a single line, whereas the third one (AB⇌BA for M_{32} or AA⇌BB for M_{51}) should cause the coalescence of two pairs of singlets into one pair of lines having the same intensity.

3. That one of the two possible exchanges corresponding to mode M_{52}, considered separately, or one of the four possible exchanges corresponding to the couple of modes M_{42}/M_{43}, also considered separately, should analogously cause the coalescence of two pairs of singlets into one pair of lines having the same intensity.

Experimentally, the 90 MHz ^{1}H NMR spectrum of a CDCl$_3$ solution of *cis*-phenylchlorobis(benzoylacetonato)tin(IV) cooled to $-25°$C and even at lower temperatures shows six signals in the methyl region that coalesce at 30°C into two singlets of equal intensity, which, in turn, coalesce into a single peak at 44°C (82).

It is obvious that the first coalescence into two singlets of equal intensity cannot be explained by *one* of the possible exchanges. This shows again that the

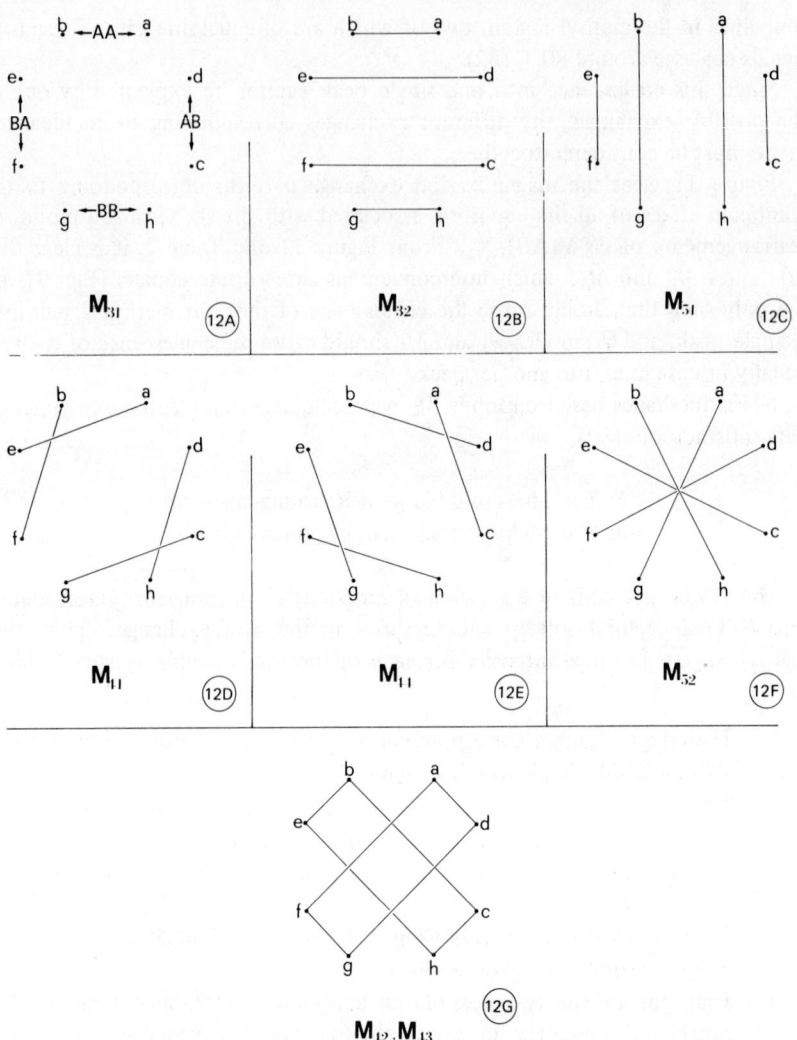

Fig. 12. Magnetization exchange patterns for the modes of rearrangement of cis-M($\overrightarrow{AB}$)$_2$XY complexes

different exchanges corresponding to an actual mode must be considered together and that they have quite close activation parameters.

Figure 12 gives the magnetization exchange patterns corresponding to *the combined effect* of all the equilibria associated to the different rearrangement modes of cis-M($\overrightarrow{AB}$)$_2$XY, which can be deduced from Table 3. They show:

1. That the mutually inverse modes M_{42} and M_{43}, which interconvert the four diastereomers into one another (Fig. 10), have the same magnet-

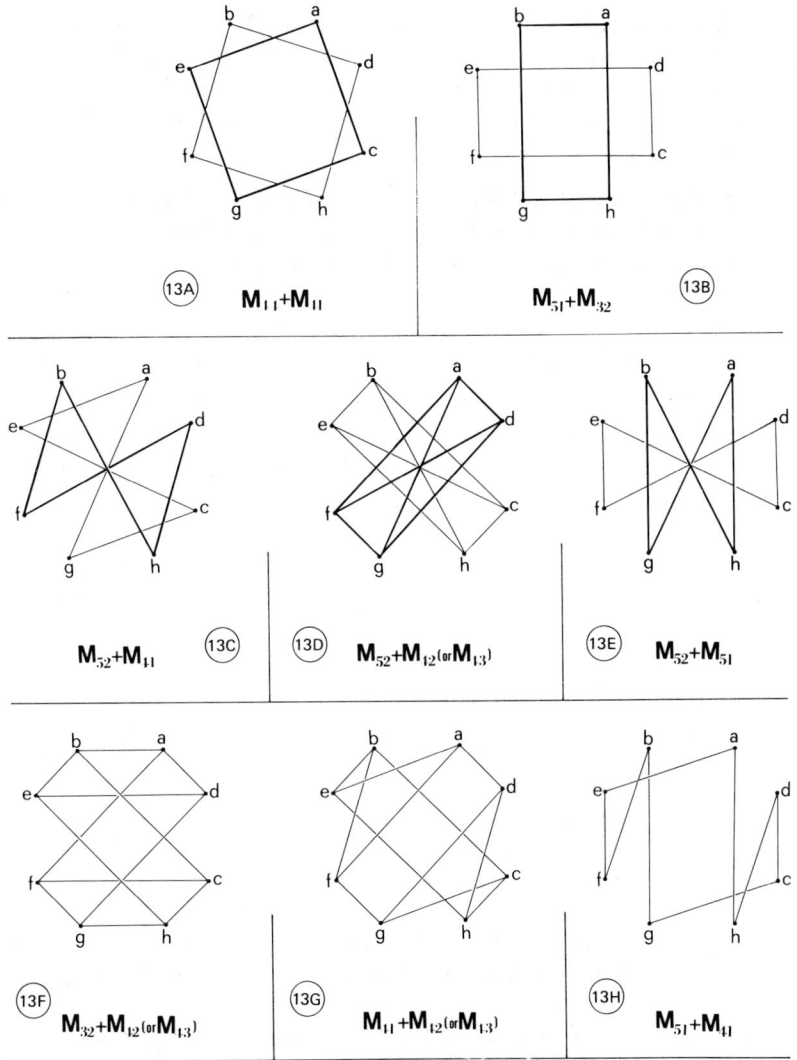

Fig. 13. Magnetization exchange patterns for some combinations of the modes of rearrangement of cis-M(AB)$_2$XY complexes

ization exchange pattern and cause the coalescence of the eight methyl signals into two singlets of equal intensity, corresponding to the diastereotopic methyl groups initially present in each isomer,

2. That modes M_{41}, M_{44}, and M_{52} cause the coalescence of the four pairs of methyl signals into two unequal pairs of equally intense signals,
3. That modes M_{32} and M_{51} cause the coalescence of the eight methyl signals into four lines, two of which have the same intensity.

It is obvious that the second coalescence into a single peak cannot be explained by one of the actual modes. Figure 13 gives the magnetization exchange patterns of *combinations* of modes of rearrangement for cis-$M(\overline{AB})_2XY$, which are obtained by combining the exchange patterns of the corresponding simple modes. They show:

1. That $M_{32} + M_{51}$ and $M_{51} + M_{52}$ (and also $M_{32} + M_{52}$, because M_{32} and M_{51} have analogous magnetization exchange patterns, see Fig. 12) cause the coalescence of the eight methyl signals into two singlets of different intensity.
2. That $M_{41} + M_{44}$, M_{41} (or M_{44}) $+ M_{52}$ and M_{42} (or M_{43}) $+ M_{52}$ cause the coalescence of the four pairs of methyl signals into two singlets of equal intensity (like the mutually inverse modes M_{42} and M_{43}).
3. That M_{42} (or M_{43}) $+ M_{32}$ (or M_{51}), M_{42} (or M_{43}) $+ M_{41}$ (or M_{44}) and $M_{41} + M_{51}$ cause the coalescence of the eight methyl signals into one single line.

We assume that the presence of six signals instead of eight, as expected for the low-temperature region, is due to an accidental isochrony. We do not consider the situations in which one isomer is not observed, or *one* of the equilibria corresponding to one of the actual modes of the idealized mode $\hat{M}_4$ or to mode M_{52} (or AA⇌BB of mode M_{51} or AB⇌BA of mode M_{32}) is already very rapid on the time scale at $-25°C$. With this assumption, the residual diastereotopism observed at $30°C$ can be explained either by the mutually inverse modes M_{42} and M_{43} or by the combination of modes $M_{41} + M_{44}$. The couple of modes M_{42}/M_{43} combined either to mode M_{41} or to mode M_{44} can explain the further coalescence observed at $44°C$. This shows that the idealized mode $\hat{M}_4$ alone can once more account for the experimental results. Note however that M_{42}/M_{43} combined to either M_{32} or M_{51} leads to the same behavior. Moreover, if the combination $M_{41} + M_{44}$ is considered as the threshold rearrangement, the combination $(M_{41} + M_{44})$ with M_{32} or M_{51} can also account for the high-temperature coalescence.

The dynamic stereochemistry of tris(di-o-substituted aryl)germanes (92) can also be described using the same permutational group theoretical concepts (75).

ACKNOWLEDGMENTS

It is a pleasure to thank our co-workers for their important contribution in the field of dynamic organic chemistry. We are indebted to Dr. R. Willem, Prof. Dr. P. Finocchiaro and Prof. Dr. J. Brocas for their critical comments.

We thank the "Fonds voor Kollektief en Fundamenteel Onderzoek" (FKFO), the "Nationaal Fonds voor Wetenschappelijk Onderzoek–Fonds National de la

Recherche Scientifique" (NFWO–FNRS), the "Instituut ter aanmoediging van het Wetenschappelijk Onderzoek in Nijverheid en Landbouw — Institut pour l'Encouragement de la Recherche Scientifique dans l'Industrie et l'Agriculture" (IWONL–IRSIA), and the "Nationale Raad voor Wetenschapsbeleid" for their financial support.

The usual fine secretarial assistance of Mrs. G. Vandendaele–Joris is acknowledged.

REFERENCES

1. F. H. Carré, R. J. P. Corriu, and R. B. Thomassin, *J. Chem. Soc. D*, **1968**, 560.
2. M. Gielen and H. Mokhtar-Jamai, *J. Organomet. Chem.*, **91**, C33 (1975).
3. R. Corriu and M. Henner-Léard, *J. Organomet. Chem.*, **64**, 351 (1974); **65**, C39 (1974).
4. R. Corriu, F. Larcher, and G. Royo, *J. Organomet. Chem.*, **104**, 293 (1976).
5. L. H. Sommer, F. O. Stark, and K. W. Michael, *J. Am. Chem. Soc.*, **85**, 3898 (1963).
6. F. Carré, R. Corriu, and M. Léard, *J. Organomet. Chem.*, **24**, 101 (1970).
7. R. Corriu and M. Henner, *J. Organomet. Chem.*, **74**, 1 (1974).
8. B. G. McKinnie and F. K. Cartledge, *J. Organomet. Chem.*, **104**, 407 (1976); F. K. Cartledge, B. G. McKinnie, and J. M. Wollest, *J. Organomet. Chem.*, **118**, 7 (1976).
9. M. Gielen, "Stereoselective Substitution Reactions at the Metal Atom of Optically Active Organotin Compounds," *IXth International Conference on Organometallic Chemistry*, Pure Appl. Chem., **52**, 657 (1980).
10. R. Corriu and M. Henner, *Bull. Soc. Chim. Fr.*, **1974**, 1447.
11. M. Gielen and M. Declercq, *J. Organomet. Chem.*, **47**, 351 (1973).
12. M. Gielen and H. Mokhtar-Jamai, *Bull. Soc. Chim. Belg.*, **84**, 1037 (1975).
13. J. G. Noltes, G. Van Kooten, and C. A. Schaap, *J. Am. Chem. Soc.*, **98**, 5393 (1976).
14. M. Gielen and Y. Tondeur, *Nouv. J. Chim.*, **2**, 117 (1978).
15. M. Gielen, S. Simon, Y. Tondeur, M. Van de Steen, and C. Hoogzand, *Bull. Soc. Chim. Belg.*, **83**, 337 (1974).
16. M. Gielen and Y. Tondeur, *Bull. Soc. Chim. Belg.*, **84**, 933 (1975).
17. M. Gielen and Y. Tondeur, *J. Organomet. Chem.*, **169**, 265 (1979).
18. M. Gielen and Y. Tondeur, *J. Organomet. Chem.*, **127**, C75 (1977).
19. M. Gielen, S. Simon, Y. Tondeur, M. Van de Steen, C. Hoogzand, and I. Vanden Eynde, *Isr. J. Chem.*, **15**, 74 (1977).
20. A. Fredga, *Tetrahedron*, 8, 126 (1960); *Bull. Soc. Chim. Fr.*, **1973**, 173.
21. A. G. Brook, *J. Am. Chem. Soc.*, **85**, 3051 (1963).
22. U. Folli, D. Iarossi and F. Taddei, *J. Chem. Soc., Perkin Trans. 2*, **1973**, 1284.
23. M. Gielen and I. Vanden Eynde, *Isr. J. Chem.*, **20**, 93 (1980).
24. M. Gielen, "Synthesis and Properties of Optically Active Organotin Compounds," in *Annual Reports in Inorganic and General Syntheses – 1976*, H. Zimmer, Ed., Academic Press, New York, 1977, p. 337.
25. M. Gielen and M. Van de Steen, unpublished results.
26. R. Schwarz and M. Lewinsohn, *Chem. Ber.*, **64**, 2352 (1931).
27. R. W. Bott, C. Eaborn, and I. D. Varma, *Chem. Ind.*, **1963**, 614; C. Eaborn, P. Simpson, and I. D. Varma, *J. Chem. Soc. A*, **1966**, 1133.
28. L. H. Sommer and C. L. Frye, *J. Am. Chem. Soc.*, **81**, 1013 (1959).
29. A. G. Brook and G. J. D. Peddle, *J. Am. Chem. Soc.*, **85**, 1869, 2338 (1963).
30. C. Eaborn and I. D. Varma, *J. Organomet. Chem.*, **9**, 377 (1967).
31. C. Eaborn, R. E. E. Hill, and P. Simpson, *J. Organomet. Chem.*, **15**, 241 (1968).

32. F. Carré and R. Corriu, *J. Organomet. Chem.*, **25**, 395 (1970); see, however, A. Jean and M. Lequan, *Tetrahedron Lett.*, **1970**, 1517.
33. H. Mokhtar-Jamai, C. Dehouck, S. Boué, and M. Gielen, *Proc. 5th Int. Conf. Organomet. Chem.*, **1**, 523 (1971); **2**, 359 (1971); H. Mokhtar-Jamai and M. Gielen, *Bull. Soc. Chim. Fr.*, **9B**, 32 (1972); M. Gielen and H. Mokhtar-Jamai, *Ann. N.Y. Acad. Sci.*, **239**, 208 (1974); M. Gielen and H. Mokhtar-Jamai, *Bull. Soc. Chim. Belg.*, **84**, 197 (1975).
34. M. Lequan and F. Meganem, *J. Organomet. Chem.*, **94**, C1 (1975).
35. M. Gielen and S. Simon, *Bull. Soc. Chim. Belg.*, **86**, 589 (1977).
36. U. Folli, D. Iarossi, and F. Taddei, *Chem. Soc. Perkin Trans. 2*, **1973**, 638.
37. M. Gielen and S. Simon, *Bull. Soc. Chim. Belg.*, **86**, 589 (1977).
38. M. Gielen and Y. Tondeur, *J. Organomet. Chem.*, **127**, C75 (1977).
39. M. Gielen and Y. Tondeur, *J. Organomet. Chem.*, **128**, C25 (1977).
40. A. G. Brook, J. M. Duff, and D. G. Anderson, *J. Am. Chem. Soc.*, **92**, 7567 (1970).
41. H. Sakurai and K. Mochida, *J. Chem. Soc., Chem. Commun.*, **1971**, 1581; H. Sakurai, K. Mochida, A. Hosomi, and F. Mita, *J. Organomet. Chem.*, **38**, 275 (1972); see also ref. 14.
42. C. Eaborn, A. R. Bassingdale, R. A. Jackson, C. L. Turpin, and R. Walsingham, U.S. Clearinghouse, Federal Scientific and Technical Information, AD 1970, No. 712371, from *U.S. Gov. Res. Dev. Rep.*, **70**, 57 (1970); *Chem. Abstr.*, **75**, 19632c (1971).
43. R. J. P. Corriu and J. J. E. Moreau, *J. Chem. Soc. D*, **1971**, 821; *J. Organomet. Chem.*, **40**, 55, 73 (1972).
44. J. Dubac, P. Mazerolles, M. Jolyk and F. Piau, *J. Organomet. Chem.*, **127**, C69 (1977).
45. M. Gielen and Y. Tondeur, *J. Chem. Soc., Commun.*, **1978**, 81.
46. R. Belloli, *J. Chem. Educ.*, **46**, 640 (1969).
47. G. J. D. Peddle and G. Redl, *J. Chem. Soc., Chem. Commun.*, **1968**, 626.
48. R. Corriu and G. Royo, *J. Organomet. Chem.*, **14**, 291 (1968).
49. C. Eaborn, R. E. E. Hill, and P. Simpson, *J. Organomet. Chem.*, **37**, 251 (1972).
50. L. H. Sommer, *Angew. Chem.*, **74**, 176 (1962).
51. M. Gielen, C. Dehouck, H. Mokhtar-Jamai, and J. Topart, *Rev. Si, Ge, Sn, Pb Compd.*, **1**, 9 (1973).
52. R. Corriu and J. Massé, *J. Organomet. Chem.*, **35**, 51 (1972).
53. R. Corriu and J. Massé, *J. Chem. Soc., Chem. Commun.*, **1967**, 1287; *J. Organomet. Chem.*, **34**, 221 (1972).
54. R. Corriu, J. Massé, and G. Royo, *J. Chem. Soc., Chem. Commun.*, **1971**, 252; R. Corriu and G. Royo, *Bull. Soc. Chim. Fr.*, **1972**, 1497.
55. L. H. Sommer and D. N. Roark, *J. Am. Chem. Soc.*, **95**, 969 (1973).
56. L. H. Sommer, G. A. Parker, N. C. Lloyd, C. F. Frye, and K. W. Michael, *J. Am. Chem. Soc.*, **89**, 857 (1967).
57. R. Corriu, G. Lanneau, and G. Royo, *J. Organomet. Chem.*, **35**, 35 (1972).
58. G. Chauviere, R. Corriu, A. Kpoton, and G. Lanneau, *J. Organomet. Chem.*, **73**, 305 (1974).
59. F. Carré and R. Corriu, *J. Organomet. Chem.*, **65**, 343 (1974).
60. Nguyen Trong Anh and C. Minot, to be published.
61. G. Van Koten, J. T. B. H. Jastrzebski, J. G. Noltes, W. M. G. F. Pontenagel, J. Kroon, and A. L. Spek, *J. Am. Chem. Soc.*, **100**, 5021 (1973).
62. L. H. Sommer, J. E. Lyons, and H. Fujimoto, *J. Am. Chem. Soc.*, **91**, 7051 (1969).
63. R. J. P. Corriu and J. J. E. Moreau, *J. Organomet. Chem.*, **40**, 55 (1972).
64. F. H. Carré and R. J. P. Corriu, *J. Organomet. Chem.*, **74**, 49 (1974).
65. F. Carré and R. Corriu, *J. Organomet. Chem.*, **73**, C49 (1974).
66. J. B. Lambert and M. Urdanetta-Perez, *J. Am. Chem. Soc.*, **100**, 157 (1978).

67. C. Eaborn, R. E. E. Hill, and P. Simpson, *J. Organomet. Chem.*, **37**, 267 (1972); **15**, P1 (1968); see also ref. 90.
68. C. Eaborn, R. E. E. Hill, and P. Simpson, *J. Organomet. Chem.*, **37**, 275 (1972).
69. V. M. Vodolazskaya and Yu. I. Baukov, *Zh. Obshch. Khim.*, **43**, 1410 (1973); *Chem. Abstr.*, **79**, 66502c (1973).
70. A. Jean and M. Lequan, *J. Organomet. Chem.*, **42**, C3 (1972).
71. F. Carré and R. Corriu, *J. Organomet. Chem.*, **65**, 349 (1974); see also M. Lequan and Y. Besace, *J. Organomet. Chem.*, **61**, C23 (1973) and F. Carré and R. Corriu, *ibid.*, **73**, C49 (1974).
73. M. Gielen, "Applications of Graph Theory to Organometallic Chemistry," in *Chemical Applications of Graph Theory*, A. T. Balaban, Ed., Academic Press, New York, 1976, pp. 26–298.
74. M. Gielen, *Stéréochimie Dynamique*, Freund Publishing House, Tel Aviv, Israel, 1974.
75. J. Brocas, M. Gielen, and R. Willem, *The Permutational Approach to Dynamic Stereochemistry*, McGraw-Hill, New York (in press).
76. M. Gielen and R. Willem, *Phosphorus Sulfur*, **3**, 339 (1977).
77. (a) F. Basolo and R. G. Pearson, *Mechanisms of Inorganic Reactions*, 2nd ed., Wiley, New York, (1967); (b) J. J. Fortman and R. E. Sievers, *Coord. Chem. Rev.*, **6**, 331 (1971); R. H. Holm, in *Dynamic Nuclear Magnetic Resonance Spectroscopy*, L. M. Jackman and F. A. Cotton, Eds., Academic Press, New York, 1975, Chapter 9; (c) L. H. Pignolet, *Top. Curr. Chem.*, **56**, 91 (1975); (d) L. H. Pignolet and G. N. La Mar, in *NMR of Paramagnetic Molecules, Theory and Applications*, G. N. La Mar, W. D. Horrocks, and R. H. Holm, Eds., Academic Press, New York, 1973, Chapter 8; (e) N. Serpone and D. G. Bickley, *Prog. Inorg. Chem.*, **17**, 391 (1972); (f) N. Serpone and K. A. Hersh, *Inorg. Chem.*, **13**,2901 (1974).
78. D. G. Bickley and N. Serpone, *Inorg. Chem.*, **13**, 2908 (1974); **15**, 948 (1976).
79. S. S. Eaton and G. R. Eaton, *J. Am. Chem. Soc.*, **95**, 1825 (1973); S. S. Eaton, J. R. Hutchinson, R. H. Holm, and E. L. Muetterties, *ibid.*, **96**, 5398 (1974); J. I. Musher, *Inorg. Chem.*, **11**, 2335 (1972).
80. W. G. Klemperer, *J. Am. Chem. Soc.*, **95**, 2105 (1973).
81. W. Hässelbarth and E. Ruch, *Theor. Chim. Acta*, **29**, 259 (1973).
82. P. Finocchiaro, V. Librando, P. Paravigna, and A. Recca, *J. Organomet. Chem.*, **125**, 185 (1977); A. Recca, F. A. Bottino, G. Ronsisvalle, and P. Finocchiaro, *ibid.*, **172**, 397 (1979).
83. J. Brocas, R. Willem, D. Fastenakel, and Y. Buschen, *Bull. Soc. Chim. Belg.*, **84**, 483 (1975).
84. M. Gielen, R. Willem, and J. Brocas, *Bull. Soc. Chim. Belg.*, **82**, 617 (1973).
85. M. Gielen, *Meded. Vlaam. Chem. Ver.*, **31**, 201 (1969).
86. P. Finocchiaro, *J. Am. Chem. Soc.*, **97**, 4443 (1975); R. C. Fay and A. F. Lindmark, *ibid.*, **97**, 5928 (1975).
87. P. Pirlot, "Contribution à l'Étude Théorique des Schémas de Coalescence en Résonance Magnétique Nucléaire de Complexes Bis-chélatés Hexacoordonnés. Une Comparaison de Deux Types d'Approches," Mémoire de Licence, Free University of Brussels U.L.B., 1976.
88. W. G. Klemperer, *J. Am. Chem. Soc.*, **94**, 6940 (1972).
89. D. J. Klein and A. H. Cowley, *J. Am. Chem. Soc.*, **97**, 1633 (1975).
90. E. Colomer and R. J. P. Corriu, *J. Chem. Soc., Chem. Commun.*, **1978**, 435.
91. K. Jurkschat, C. Mügge, A. Tzschach, A. Zschunke, M. F. Larin, V. A. Pestunovich, and M. G. Voronkov, *J. Organomet. Chem.*, **139**, 279 (1977).
92. V. I. Proshutinskii, I. I. Lapkin, and V. A. Dumler, *Konformatsion, Analiza Primenie Sinteze Novykh Organ. Veshchestv.*, **1975**, 120, from *Ref. Zh. Khim.*, **1975**, Abstr. No. 21 Zh. 11, *Chem. Abstr.*, **84**, 73545a (1976).

Stereochemistry of Transition Metal Carbonyl Clusters

BRIAN F. G. JOHNSON and ROBERT E. BENFIELD

*The University Chemical Laboratory, Cambridge University,
Cambridge, England*

I.	Introduction		254
II.	The Central M_m Cluster Unit		263
	A. Cluster Geometries		263
	B. The Nature of the M–M Bond in Clusters: The Main Approaches		263
	1. The 18-Electron Rule		264
	a. $Mn_2(CO)_{10}$		265
	b. $Fe_3(CO)_{12}$		265
	c. $Co_4(CO)_{12}$		265
	d. $Os_5(CO)_{16}$		266
	e. $Os_3(CO)_{12}H_2$		266
	f. $Os_3(CO)_{10}H_2$		266
	2. Skeletal Electron Pair Theory		268
	3. A Modified Approach		273
	C. Metal Polyhedra		278
	1. The M_3 Unit		278
	2. The M_4 Unit		279
	3. The M_5 Unit		282
	4. The M_6 Unit		286
	D. Conclusions		289
III.	The Ligand Envelope		290
	A. CO Bonding Modes		290
	B. Ligand Polyhedra		292
	1. Method		301
	2. Polyhedral Arrangements of $(CO)_n$		302
	C. Summary		331
	Conclusion		332
	Index of Subjects		333
	References		333

I. INTRODUCTION

Transition metal carbonyls have been known since the end of the last century (1). Soon after their initial discovery some of the observed species were shown to be cluster compounds (2). More recent work (3) has led to a considerable increase in our knowledge of these clusters; in particular much information about their molecular structure both in the solid state and in solution has been accumulated. This account has two main objectives; first, to summarize recent progress in this rapidly developing area and, second, to provide a semiquantitative rationalization of the structural properties of such clusters.

Table 1 lists the structures of a range of binary carbonyls of general formula $M_m(CO)_n^l$ (m = number of metal atoms, n = number of carbonyl groups, and l = charge). We shall consider these structures from two points of view: the structure of the central M_m cluster units, and the nature of the surrounding envelope of CO ligands.

Table 1 The Molecular Structures of Some Binary Carbonyls: $M_m(CO)_n^l$.

$m = 2$ $Co_2(CO)_8$ (ref. 7).

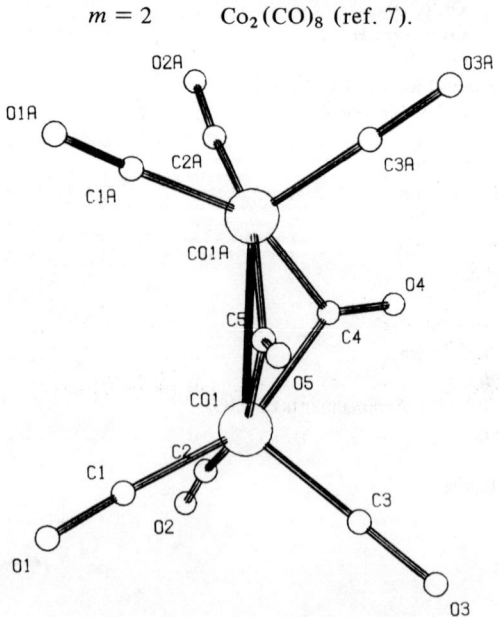

$Fe_2(CO)_9$ (ref. 8).

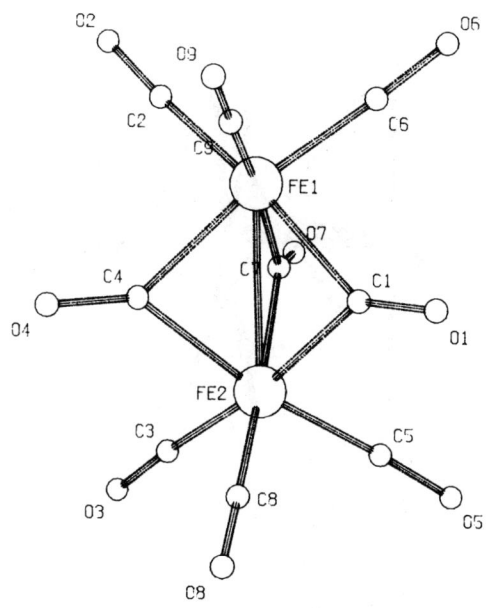

$M_2(CO)_{10}$ (M = Mn, Tc, Re) (ref. 9).

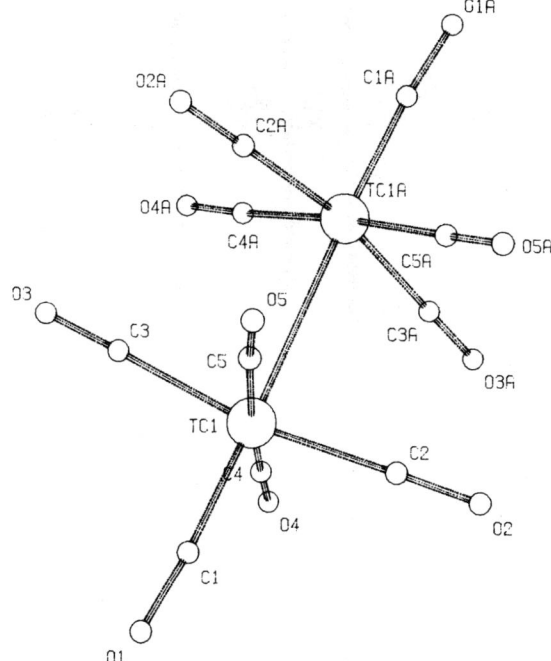

$m = 3$ $Fe_3(CO)_{12}$ (ref. 10).

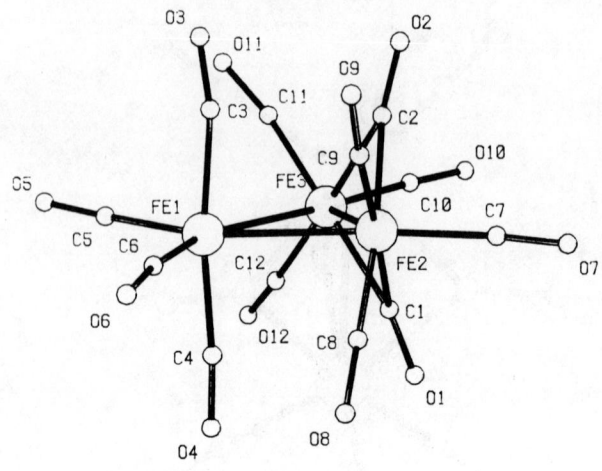

$Os_3(CO)_{12}$ (ref. 11).

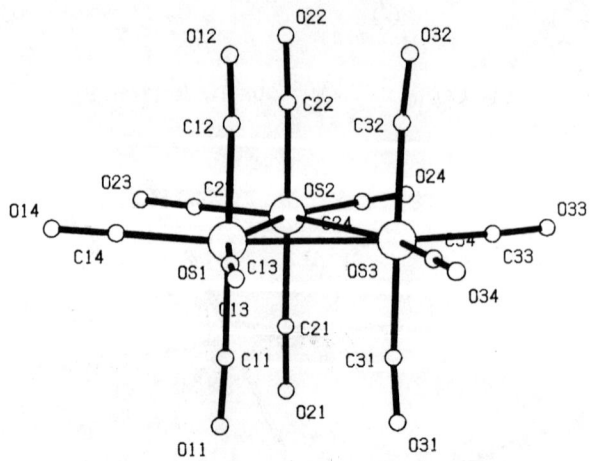

$m = 4$ $Co_4(CO)_{12}$ (ref. 132).

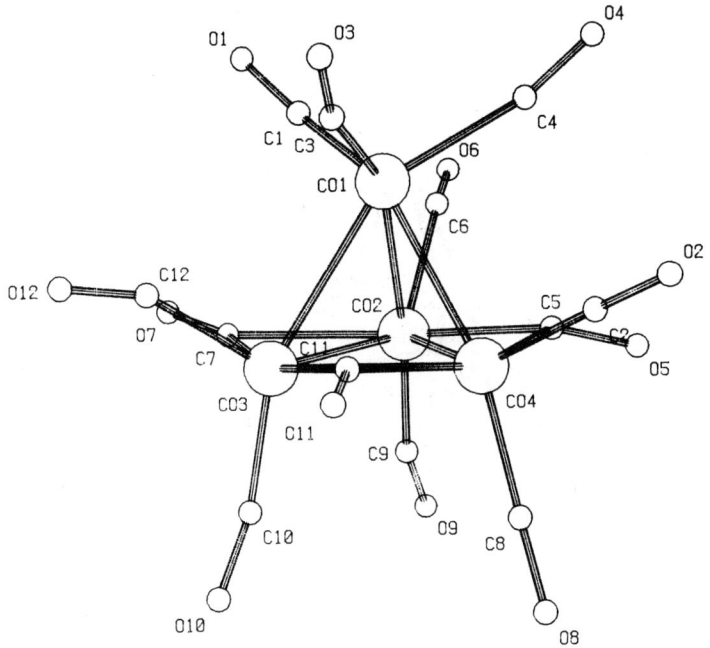

$Ir_4(CO)_{12}$ (ref. 15).

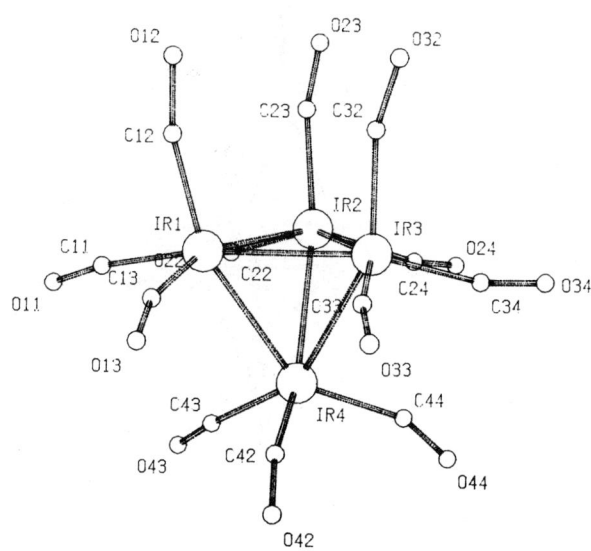

$Fe_4(CO)_{13}^{2-}$ (ref. 16).

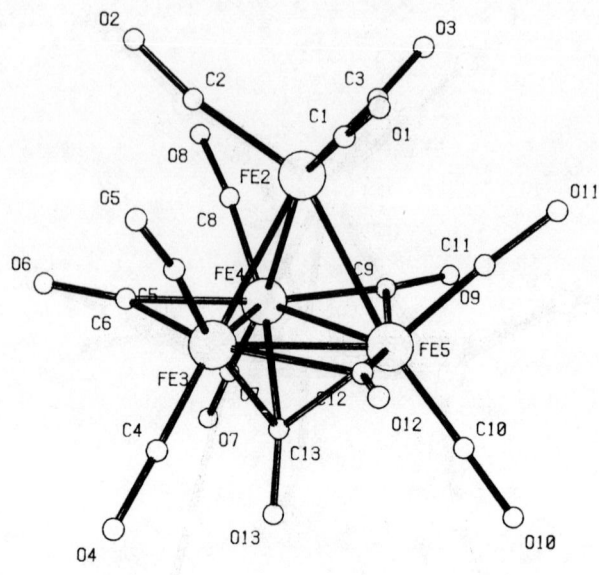

$m = 5$ $Ni_5(CO)_{12}^{2-}$ (ref. 18).

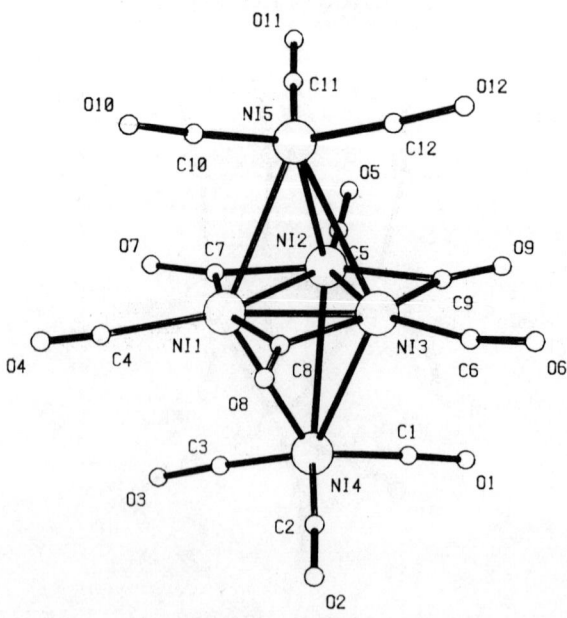

$Os_5(CO)_{16}$ (ref. 19).

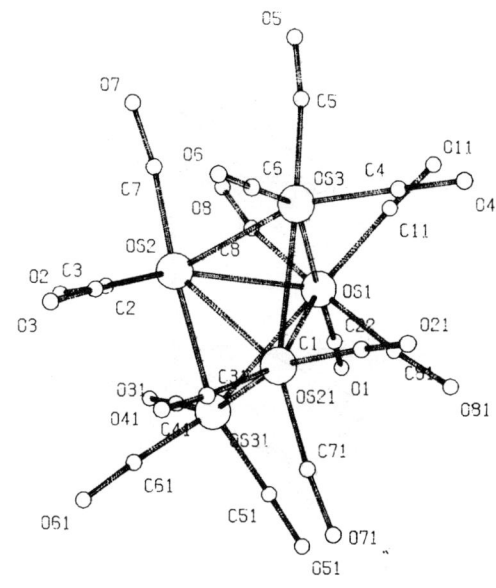

$m = 6$ $Ni_6(CO)_{12}^{2-}$ (ref. 40).

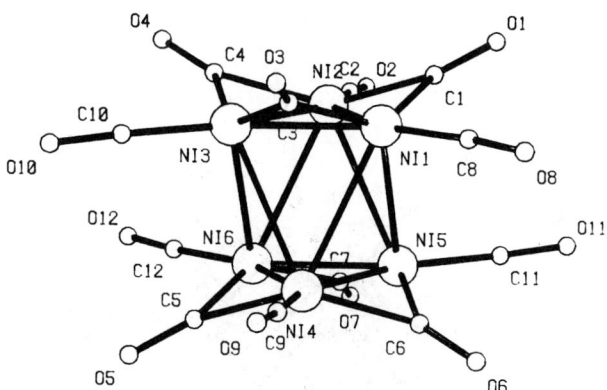

$Pt_6(CO)_{12}^{2-}$ (ref. 12).

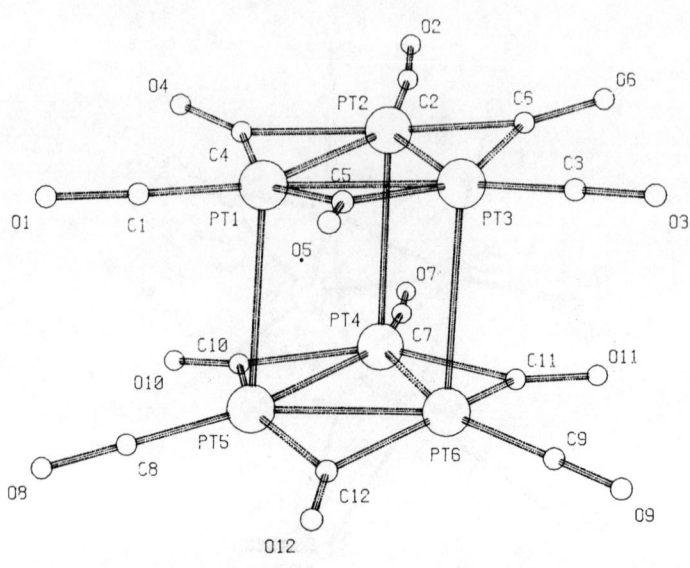

$Co_6(CO)_{14}^{4-}$ (ref. 41).

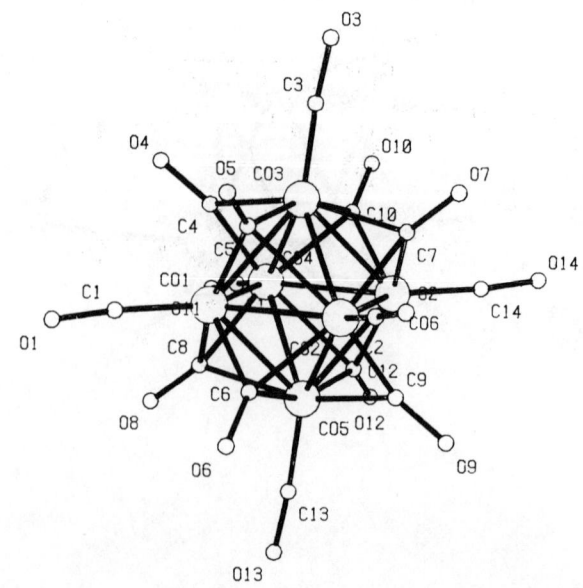

$Co_6(CO)_{15}^{2-}$ (ref. 42).

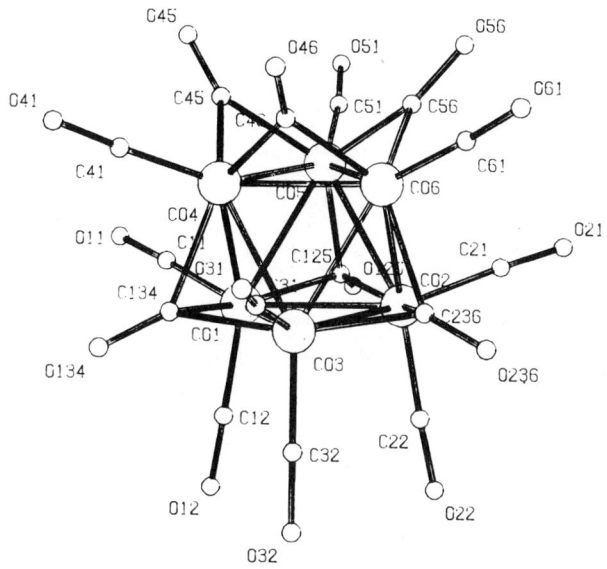

$Rh_6(CO)_{16}$ (ref. 21).

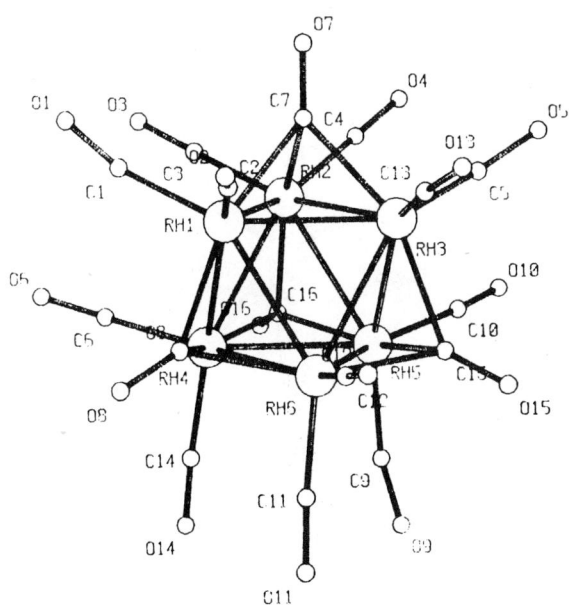

$Os_6(CO)_{18}$ (ref. 22).

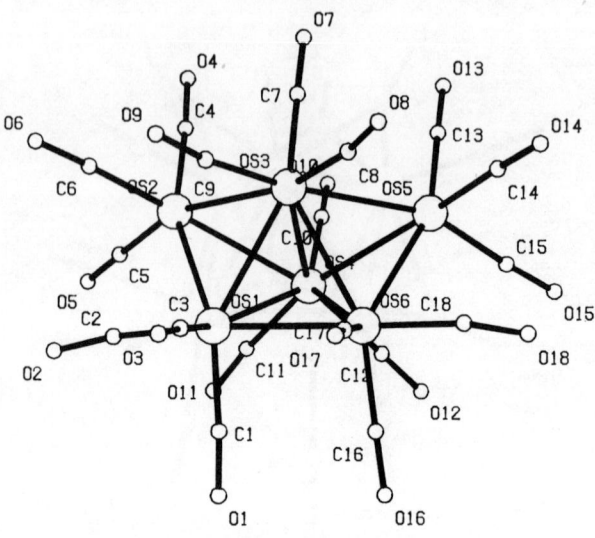

$Os_6(CO)_{18}^{2-}$ (ref. 23).

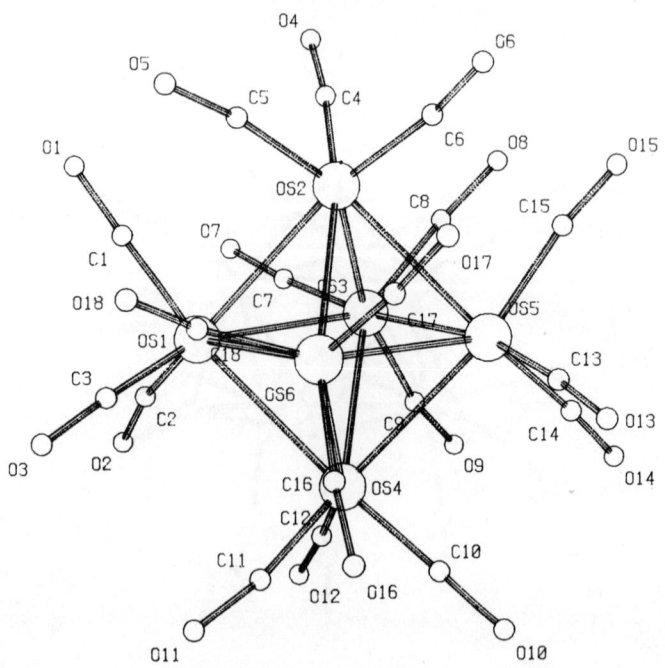

II. THE CENTRAL M_m CLUSTER UNIT

A. Cluster Geometries

It is within the cluster class of binary carbonyls that the largest range of metal M_m polyhedra have been observed. Several general observations can be made:

1. The polyhedra defined by the metal atoms are largely deltahedra, that is, polyhedra with all triangulated faces such as the tetrahedron in $Co_4(CO)_{12}$, the trigonal bipyramid in $[Os_5(CO)_{15}]^{2-}$, the O_h-octahedron in $[Os_6(CO)_{18}]^{2-}$, and the bicapped tetrahedron in $Os_6(CO)_{18}$.
2. Exceptions to observation 1 are relatively few in number but form an ever-increasing class. A number of square planar, trigonal prismatic, and square-based pyramid clusters, for example, are known.
3. The majority of clusters of both groups 1 and 2 represent fragments of close-packed metallic structures, although clusters with pentagonal symmetries are now emerging.
4. Some clusters, although not fragments of hexagonal or cubic close packed structures are nevertheless close packed – a good example is the bicapped tetrahedron found for $Os_6(CO)_{18}$.

Within the M_m unit the number of nearest metal neighbours may range from one to quite high values. The term *connectivity* (c) – the number of nearest metal (or cluster) atoms – is frequently employed under the following circumstances:

1. If $c = 1$, only dimeric species are possible [e.g., $Mn_2(CO)_{10}$, $Fe_2(CO)_9$, or $Co_2(CO)_8$].
2. If $c = 2$, chains or closed rings can be formed [e.g., $OsRe_2(CO)_{14}$ or $Os_3(CO)_{12}$].
3. If $c \geqslant 3$, three-dimensional structures can arise. Strictly speaking, a cluster must fall within this group (by definition); but rings such as $Fe_3(CO)_{12}$ are generally embraced within the general cluster class.

An understanding of the factors responsible for these connectivity values, the cluster geometries, and the nature of M_m cluster bonding is clearly of considerable importance, and several approaches have been adopted.

B. The Nature of the Metal-Metal Bond in Clusters: The Main Approaches

Two methods, both using the molecular orbital approach, have been commonly used for the rationalization of metal-metal bonding in transition metal carbonyl clusters. One (4) has been to examine the isolated metal cluster unit M_m and to characterize the number and type of cluster valence orbitals available for metal-metal and metal-ligand (CO) bonding. This method allows the stoichiometries of carbonyl clusters to be understood. The other (5) considers the individual metal carbonyl fragments that form the cluster vertices. This method, which allocates orbitals to cluster bonding *after* the requirements of the metal-carbonyl bonds have been met, can be used to rationalize and, on occasion, to deduce cluster structures. In general, these two methods are compatible and lead to essentially the same results.

It is to be remembered, however, that although molecular orbital theory is useful in simplifying the calculation of molecular energies, it is at the expense of the simple concept of the chemical bond (6). The idea of localized electron-pair bonds is an important chemical concept, which is especially useful in establishing the structures of complex molecules. The structures of many transition metal clusters are intelligible in such (valence-bond) terms employing the simple concept of *two-*, *three-*, and *multicenter* bonds. Within these terms two additional approaches have been utilized for qualitative discussions of cluster geometries.

1. The 18-Electron Rule

One approach, the 18-electron rule (24), assumes that the skeletal atoms are held together by a network of two-electron pair/two-center ($2e/2$-center) bonds and that each individual cluster atom utilizes its *nine* atomic orbitals to accommodate both metal valence electrons and ligand electron pairs and also to form two electron metal-metal bonds. Thus each transition metal can accommodate a total of 18 electrons — hence the name "18-electron rule." The large majority of low oxidation state, diamagnetic complexes, especially the binary metal carbonyls, obey this rule. Indeed, the rule has been outstandingly successful as a means of both predicting and rationalizing the structures of low oxidation state, transition metal organometallic complexes. Exceptions do occur, particularly with d^8 metal ions, where many examples of 16-electron (square planar) complexes are known. In these complexes the high-lying p_z orbital is found to be nonbonding and empty. This deviation from the rule is considered to be due to the large $s(d) - p$ promotion energies found for the free atoms. As the atomic number increases across a given transition metal series, the energy of the s and d orbitals drops more rapidly than that of the p orbitals, thereby increasing the $s(d) - p$ promotion energies. Exceptions to the 18-electron rule are also found for the group 1B metals. Here again the p orbitals may not be fully utilized and, for example, Au(I) forms primarily two-coordinate, 14-electron complexes.

The useful application of the 18-electron rule to cluster carbonyl systems is, in general, restricted to the smaller clusters containing five or fewer metal atoms.

Three assumptions are made:

1. Metal-metal bonds correspond to polyhedral edges.
2. All metal-metal bonds will be $2e/2$-center bonds.
3. Ligands are considered to serve as a source of electron pairs only, leading to the view that the same metal polyhedron will be derived, irrespective of whether electrons are present as anionic charge or ligand pairs.

The molecules $Mn_2(CO)_{10}$, $Fe_3(CO)_{12}$, $Co_4(CO)_{12}$, and $Os_5(CO)_{16}$ provide good examples of the application of the 18-electron rule.

a. $Mn_2(CO)_{10}$. Each Mn has seven valence electrons and each CO donates two electrons.
   1. Total number of electrons $= 34$
      $(2 \times 7) + (10 \times 2)$
   2. Valence orbitals available $= 18$
      (9×2)
   3. Required number of electrons to fill 18 atomic orbitals $= 36$
      (18×2)
      Difference, item 3 − item 1 $= 2$
      Therefore number of Mn−Mn $2e/2$-center bonds $= 2/2 = 1$

b. $Fe_3(CO)_{12}$. Each Fe has eight valence electrons.
   1. Total number of electrons $= 48$
      $(3 \times 8) + (12 \times 2)$
   2. Valence orbitals available $= 27$
      (9×3)
   3. Required number of electrons to fill 27 atomic orbitals $= 54$
      (27×2)
      Difference, item 3 − item 1 $= 6$
      Therefore number of Fe−Fe $2e/2$-center bonds $= 6/2 = 3$
   Hence triangular Fe_3 unit.

c. $Co_4(CO)_{12}$. Each Co has nine valence electrons.
   1. Total number of electrons $= 60$
      $(9 \times 4) + (12 \times 2)$
   2. Valence orbitals available $= 36$
      (9×4)
   3. Required number of electrons to fill 36 atomic orbitals $= 72$
      (36×2)

Difference, item 3 − item 1	= 12
Therefore number of Co–Co 2e/2-center bonds	= 12/2 = 6

Hence tetrahedral (six edges) Co_4 unit.

d. $Os_5(CO)_{16}$. Each Os has eight valence electrons.
   1. Total number of electrons = 72
      $(8 \times 5) + (16 \times 2)$
   2. Valence orbitals available = 45
      (9×5)
   3. Required number of electrons to fill 45 atomic orbitals = 90
      (45×2)

      Difference, item 3 − item 1 = 18

      Therefore number of Os–Os 2e/2-center bonds = 18/2 = 9

Hence a trigonal bipyramidal (nine edges) Os_5 unit.

The total number of electrons available is also known as the "magic number" for a given polyhedron. Thus species with 34 electrons will be linear dimers [e.g., $Co_2(CO)_8$ and $Fe_2(CO)_9$], species with 48 electrons will be triangular [e.g., $Ru_3(CO)_{12}$ and $H_3Mn_3(CO)_{12}$], species with 60 electrons will be tetrahedral [e.g., $H_4Os_4(CO)_{12}$], and so on. The 18-electron rule may also be used with success to predict the bonding in some electron-rich clusters (i.e. clusters containing more than the magic number of electrons) and electron-poor (not to say electron-deficient) clusters. Again this is best illustrated by example. Consider the two compounds $Os_3(CO)_{12}H_2$ and $Os_3(CO)_{10}H_2$, as examples of electron-rich (50 electrons) and electron-poor (46 electrons) clusters, respectively:

e. $Os_3(CO)_{12}H_2$
   1. Total number of electrons = 50
      $(3 \times 8) + (12 \times 2) + (2 \times 1)$
   2. Valence orbitals available = 27
      (9×3)
   3. Required number of electrons to fill 27 atomic orbitals = 54
      (27×2)

      Difference, item 3 − item 1 = 4

      Therefore number of Os–Os 2e/2-center bonds = 4/2 = 2

Hence linear or open triangle Os_3 unit.

f. $Os_3(CO)_{10}H_2$
   1. Total number of electrons = 46
      $(3 \times 8) + (10 \times 2) + (2 \times 1)$

2. Valence orbitals available = 27
   (9 × 3)
3. Required number of electrons to fill 27 atomic orbitals = 54
   (27 × 2)
   Difference, item 3 − item 1 = 8
   Therefore number of Os–Os 2e/2-center bonds = 8/2 = 4

Thus the Os_3 unit may be represented as triangle with two Os–Os single bonds and one Os=Os double bond.

It follows from the 18-electron rule that as electron pairs are added to an electron-precise polyhedron, edges (bonds) will be broken. Thus addition of one electron pair to a tetrahedral, 60-electron species, will produce a butterfly or planar arrangement:

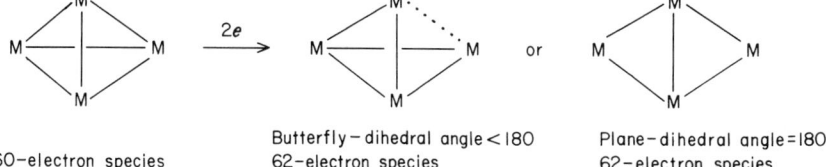

60-electron species ; Butterfly – dihedral angle < 180, 62-electron species ; Plane – dihedral angle = 180, 62-electron species

Put simply, two of the atomic orbitals used in the formation of a metal-metal bond become fully occupied and nonbonding:

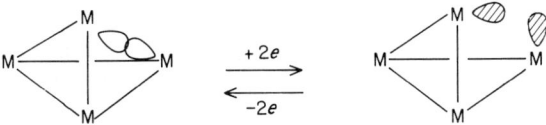

For example,

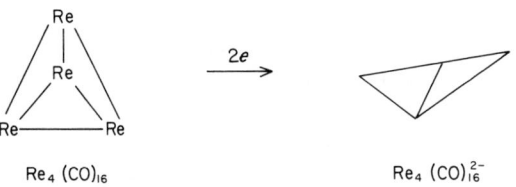

$Re_4(CO)_{16}$; $Re_4(CO)_{16}^{2-}$

Thus on the assumption that atomic orbitals retain their directional properties on the addition (or subtraction) of electron pairs, the idea that electron-rich (or poor) clusters essentially retain the shape of their parent polyhedron is easily understood.

Conversely, the removal of electron pairs leads to the formation of additional electron pair metal-metal bonds:

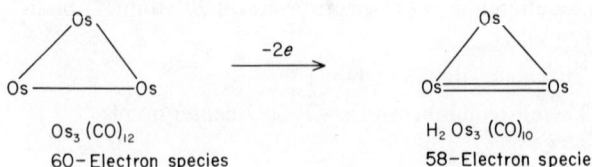

Os$_3$(CO)$_{12}$
60-Electron species

H$_2$Os$_3$(CO)$_{10}$
58-Electron species

This is not general and, according to the rule, subtraction of electron pairs can also lead to a change in polyhedral shapes:

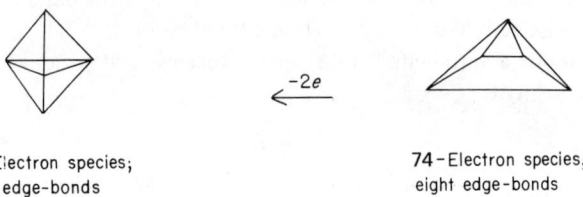

72-Electron species;
nine edge-bonds

74-Electron species;
eight edge-bonds

So far, the examples cited show a correspondence between the number of electron pairs and the number of metal-metal bonds. Most low-nuclearity clusters show this correlation, but the same is by no means true for carbonyl clusters containing five or more metal atoms. Exceptions are also found for certain smaller clusters such as H$_4$Re$_4$(CO)$_{12}$ which, although possessing a tetrahedral metal unit, contains 56 rather than the expected 60 electrons.

The major limitations of the 18-electron rule, however, become apparent in attempts to rationalize the bonding and geometries of larger metal clusters containing five or more metal atoms. There are many 86-electron octahedral clusters known [e.g., Rh$_6$(CO)$_{16}$, [Co$_6$(CO)$_{15}$]$^{2-}$, [Co$_6$(CO)$_{14}$]$^{4-}$, and Ru$_6$C(CO)$_{17}$], and according to the 18-electron rule these clusters should contain 11 electron pair metal-metal bonds, rather than the 12 observed. The effective atomic number (E.A.N.) rule "magic number" for the regular octahedron is 84, and one carbonyl cluster, Os$_6$(CO)$_{18}$, does contain this number of electrons. This cluster, which significantly accepts readily two electrons to produce [Os$_6$(CO)$_{18}$]$^{2-}$, adopts a bicapped tetrahedral structure, for which 12 2e/2-center bonds may be predicted; but it has C_{2v} symmetry rather than the expected O_h symmetry.

Attempts to describe the bonding in clusters solely in terms of the 18-electron rule are thus limited, and other alternative schemes have been offered.

2. Skeletal Electron Pair (S.E.P.) Theory (25)

The skeleton electron pair approach, which retains the condition that each metal atom use its nine valence atomic orbitals, allows the probable cluster shape to be derived from the number of electron pairs available for cluster bonding. It relies on the results of molecular orbital calculations for the boron hydride clusters and assumes that the metal ion behaves as a quasi–main group element capable of contributing three orbitals with predominant s and p character to cluster bonding.

According to this theory the numbers of bonding skeletal electron pairs (or bonding (MOs) for metal polyhedra are as follows:

Number of SEPs (25)	Fundamental polyhedron
6	Trigonal bipyramid
7	O_h-Octahedron
8	Pentagonal bipyramid
9	Dodecahedron
10	Tricapped trigonal prism
11	Bicapped square antiprism
12	Octadecahedron
13	Icosahedron

It follows from this rule that since *three* atomic orbitals per metal are made available for cluster (M_m) bonding, the remaining six will be available for M–CO bond formation and nonbonding pairs. Furthermore, the $6m$ MOs generated from this set of six are regarded as low lying and are filled preferentially.

The three atomic orbitals employed for cluster bonding consist of two tangential (π) orbitals and one σ orbital, which points toward the center of the polyhedron with m vertices:

Since $2m$ orbitals (π) can combine to generate m bonding molecular orbitals (and m antibonding molecular orbitals) and the m σ orbitals can combine to generate *one* strongly bonding molecular orbital, it follows that there will be $m + 1$ bonding molecular orbitals, which will require $m + 1$ electron pairs. This will be true for all spherically symmetrical deltahedra but only those with *five* or more metal atoms. It is important to note that this arrangement does *not* apply to the tetrahedron which almost invariably requires *six* skeletal electron pairs and may be regarded as having localized bonding (see II-B-1).

The application of this method may be simply illustrated by the cluster carbonyl $Rh_6(CO)_{16}$:

$Rh_6(CO)_{16}$

1. Total number of electrons = 86
   $(6 \times 9) + (16 \times 2)$
2. Number of orbitals used for Rh–CO bonds
   and nonbonding pairs = 36
   (6×6)
3. Number of electrons occupying orbitals (2) = 72
   (36×2)
   Difference, item 3 − item 1 = 14
   Therefore number of SEPs $m + 1$ = 7

Hence Rh_6 octahedron. (See above).

It has become customary in discussion of the SEP theory to regard the cluster as constituted of a group of fragments. Thus $Rh_6(CO)_{16}$ *may be regarded* as a combination of six $Rh(CO)_2$ fragments and four additional CO ligands (corresponding to the four three-center bridges). The number of electrons contributed by $M(CO)_x$ fragments for skeletal bonding in the cluster $[M(CO)_x]_n$ is $\nu + 2x - 12$ (where ν = the number of valence electrons of metal M). Thus an alternative calculation for $Rh_6(CO)_{16}$ would be:

1. Each $Rh(CO)_2$ unit contributes
   $(\nu + 2x - 12)$ electrons = 1
2. Therefore contribution from six $Rh(CO)_2$ units = 6
3. Contribution from four CO ligands = 8
   Sum, item 2 + item 3 = 14
   Hence number of SEPs = 7

Table 2 summarizes the number of electrons contributed by a variety of $M(CO)_x$ groups on the assumption that all also contribute three atomic orbitals.

Table 2 The Number of Electrons Contributed for Skeletal Bonding for $M(CO)_x$ Fragments (25)

Cr, Mo, W	Mn, Tc, Re	Fe, Ru, Os	Co, Rh, Ir	Ni, Pd, Pt
–	–	M(CO)	M(CO)	M(CO)
		−2	−1	0
$M(CO)_2$	$M(CO)_2$	$M(CO)_2$	$M(CO)_2$	$M(CO)_2$
−2	−1	0	1	2
$M(CO)_3$	$M(CO)_3$	$M(CO)_3$	$M(CO)_3$	$M(CO)_3$
0	1	2	3	4
$M(CO)_4$	$M(CO)_4$	$M(CO)_4$	$M(CO)_4$	–
2	3	4	5	

It has also been established that certain capped polyhedra require the same number of SEPs as the parent platonic polyhedra. Thus the tetrahedron, the trigonal bipyramid (or monocapped tetrahedron), and the C_{2v}-octahedron (or bicapped tetrahedron) all require *six* electron pairs. Similarly, the O_h-octahedron, and the bicapped octahedron (and presumably the 3-, 4-, 5- and 6-capped octahedra) require *seven* electron pairs. Consider, as examples, $Os_5(CO)_{16}$, $Os_6(CO)_{18}$, $Os_7(CO)_{21}$ and $Os_8(CO)_{23}$:

		$Os_5(CO)_{16}$	$Os_6(CO)_{18}$	$Os_7(CO)_{21}$	$Os_8(CO)_{23}$
1.	Total number of electrons	72	84	98	110
2.	Total number of Os–CO bonding and nonbonding pairs	60	72	84	96
3.	Difference, item 1 − item 2	12	12	14	14
	Number of SEPs	6	6	7	7

Thus the Os_5 trigonal bipyramid, the Os_6 monocapped trigonal bipyramid, the Os_7 monocapped octahedron, and the Os_8 bicapped octahedron contain the same numbers of SEPs as their parent polyhedra.

Skeletal electron pair counting also leads to the conclusion that all species with *six* SEPs will be based on (and be part of) the trigonal bipyramid, and species with *seven* SEPs the O_h-octahedron. Thus $Os_3(CO)_{12}$ with *six* SEPs may be regarded as an arachno-trigonal bipyramid and $Os_5(CO)_{15}C$ with *seven* SEPs as a *nido-O_h*-octahedron.

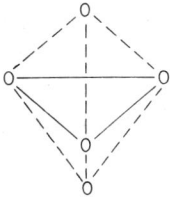 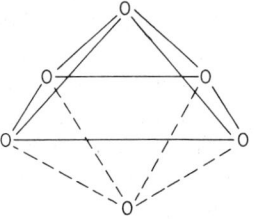

Triangle: *arachno* –trigonal bipyramid (according to SEP theory)   Square–based pyramid: *nido*-octahedron

Since the skeletal atomic orbitals provided by $M(CO)_x$ fragments are similar to (indeed originally derived from) those presented by main group elements, it follows that mixed transition metal–main group element clusters may be treated similarly. The numbers of electrons contributed by common main group elements in clusters are summarized in Table 3.

Clusters containing interstitial atoms (e.g. carbon) are also embraced by this

Table 3 Number of Electrons Contributed for Skeletal Bonding
by Some Main Group Elements or Fragments (25)

B, Al, Ga, In, Tl	C, Si, Ge, Sn	N, P, As, Sb	O, S, Se, Te	F, Cl, Br, I
M	M	M	M	M
1	2	3	4	5
MH	MH	MH	MH	—
2	3	4	5	
MH_2	MH_2	MH_2	MH_2	—
1	2	3	4	

scheme. The interstitial atom donates all its valence electrons to the cluster, but does *not* affect the number of cluster valence molecular orbitals.

The skeletal electron pair theory was originally formulated by Mingos (24) and Wade (25) from empirical observations, and by analogy with the boron hydrides. More recently, King and Rouvray (43) have used graph theory to derive the result that a m-vertex closo-polyhedron requires $(m + 1)$ skeletal electron pairs. This was achieved by treating the closed 3-dimensional graphs of the systems in a way analogous to the 2-dimensional graphs of cyclic hydrocarbons which lead to the $(4n + 2)$ electron Hückel aromaticity rule.

Quantitative calculations by Lauher (4) using the extended Hückel procedure on clusters of rhodium atoms of metallic radius put the skeletal electron pair theory in a new light. Except in the cases of highest symmetry, a complex pattern of orbitals is indicated, and it is not possible to identify a few {i.e. $(m + 1)$} well-defined, low-lying bonding orbitals corresponding to either individual M–M bonds or delocalised face-bridging, etc., bonds. Instead, for a closo-polyhedron of m atoms, $(2m - 1)$ high energy anti-bonding orbitals lie well above the cluster valence molecular orbitals (CVMOs) available for non-bonding, metal-metal bonding and metal-ligand bonding electrons. Thus the skeletal electron pair theory may be more correctly based on the pattern of anti-bonding orbitals. With $(7m + 1)$ orbitals remaining in the CVMO band, the result of $(m + 1)$ skeletal electron pairs follows directly if we assume that six orbitals per metal atom are used for localized bonding. A specific example of this general conclusion was reached by Mingos from calculations on the highly symmetric $[Co_6(CO)_{14}]^{4-}$ (44).

We have seen that the tetrahedron seems to offer a counter-example to SEP theory, requiring 6 rather than 5 electron pairs, because it may be bonded in a localized manner. Lauher's calculations show that a Rh_4 tetrahedron has 6 rather than the usual $(2m - 1) = 7$ inaccessible antibonding molecular orbitals; thus the origin of the "extra" bonding orbital may be understood.

An atom capping a triangular face of an m-vertex metal atom polyhedron has 3 neighbors and may form localized bonds, so that $(m + 1)$ skeletal electron pairs are still required despite the presence of the extra atom, which may donate its surplus electrons towards bonding in the rest of the cluster. Lauher's calculations confirm that capping a closo-polyhedron does not increase the number of CVMOs by more than the three extra "localized" bonds.

The pattern of orbitals actually occupied by electrons in cluster carbonyls may be partly revealed by photoelectron spectroscopy which has recently been applied to these systems in the solid (45) and vapor (46) states.

The S.E.P. bonding scheme works moderately well for carbonyl clusters with relatively few metal atoms. But, as with the 18-electron rule, it is not general and is limited in its application. It does not, for example, provide an explanation for the tetrahedral Re_4 geometry found in $H_4Re_4(CO)_{12}$, nor does it seem to be especially desirable to regard a tetrahedron as a nido-trigonal bipyramid or the triangle as an arachno-trigonal bipyramid. It also fails to explain the trigonal prismatic structure of $[Pt_6(CO)_{12}]^{2-}$, a cluster with seven SEPs, and to distinguish between an O_h-octahedron and a monocapped square-based pyramid, both of which apparently require *seven* SEPs. A more general approach is clearly desirable.

3. A Modified Approach (26)

A fundamental assumption of both the 18-electron and SEP theories concerns the number of atomic orbitals provided for cluster bonding by the individual cluster atoms. According to the 18-electron rule all polyhedra are electron precise; that is, all polyhedral edges are taken to correspond to two-e/2-center bonds. This means that each cluster fragment must contribute a number of atomic orbitals equal to its connectivity. Thus the triangle with a connectivity of 2 requires two atomic orbitals per fragment – the tetrahedron three, the octahedron four, and so on. For transition metal clusters this could be reasonably achieved only for connectivities of 2, 3, or 4. In general, for cluster carbonyls connectivity values will be restricted to values of 2 or 3 on this basis, since few $M(CO)_x$ fragments are able to contribute more than three atomic orbitals to cluster bonding (see below). Electron-precise polyhedra arise when the total number of electrons available for cluster bonding equals the total number of atomic orbitals available. Thus the triangle requires 6 electrons and 6 atomic orbitals, the tetrahedron 12 electrons and 12 atomic orbitals, a trigonal prism, 18 electrons and 18 orbitals, and a cube, 24 electrons and 24 atomic orbitals.

From this point of view a clear analogy with carbon emerges. Less readily available to carbonyl clusters would be the electron-precise octahedron, with a connectivity of 4, or the electron-precise trigonal bipyramid, which requires connectivities of 3 (apical atom) or 4 (equatorial atom). Certain halide clusters (e.g., $[Mo_6Cl_8]^{4+}$) are more able to provide the appropriate number of cluster electrons and the required four atomic orbitals per metal and are, as a consequence, electron precise. It follows that all electron-precise polyhedra obey the

18-electron rule, and the 18-electron rule will be of use only in predicting the structures of electron-precise species.

According to the SEP theory, an $M(CO)_x$ fragment contributes *three* atomic orbitals ($\sigma + 2\pi$) to the cluster bonding scheme. Strictly speaking, this limits the polyhedral forms available to those possessing centrosymmetric structures (see above), but as outlined above, it can be employed to accommodate other less symmetrical arrangements such as capped, nido, and arachno polyhedra.

In this extended approach we reexamine the orbital and electron contributing power of some simple carbonyl fragments $M(CO)_x$, as well as the ways in which such fragments may be combined to produce cluster carbonyls $[M(CO)_x]_y$.

Table 4 lists the numbers of orbitals and electrons that may be donated to cluster bonding by some common fragments. These numbers are, in part, different from those listed in Table 2. According to this scheme, fragments that contribute the same number of orbitals and electrons may form electron-precise (18-electron rule) molecules: for example, $Fe(CO)_4$ (2 orbitals/2e), $Co(CO)_3$ (3 orbitals/3e), $Co(CO)_4$ (1 orbital/1e), and $Mn(CO)_5$ (1 orbital/1e). Apart from the d^{10} metals, which form a special case (see below), all $M(CO)_3$ units are considered to offer three orbitals for cluster bonding and as such are in line with the SEP theory. This is easily understood. Of the nine atomic orbitals per metal, three will be used to form M—CO σ bonds, and three others ($d\pi$) are available for M—CO π-bonding. Interaction of the metal $d\pi$ orbitals with the CO π^*-antibonding orbitals apparently produces low-lying molecular orbitals that will be filled preferentially.

On these grounds $Fe(CO)_3$, $Ru(CO)_3$, and $Os(CO)_3$ are regarded as providing three atomic orbitals and two electrons rather than the possible alternative of four atomic orbitals and four electrons to cluster bonding. This conclusion receives support from the observed structure of $Os_6(CO)_{18}$, which is a bicapped tetrahedron rather than the O_h-octahedron expected for an electron-precise species (24 atomic orbitals/24 electrons). The $M(CO)_2$ fragments (M = Cr, Mo, W, Mn, Tc, or Re), which are capable of contributing four orbitals, function as electron-acceptor units (see Table 4) and, as a consequence, are not expected to form stable, electron-precise $[M(CO)_2]_y$ clusters.

Apart from the $M(CO)_x$ fragments of nickel, palladium, and platinum, the atomic orbitals presented for cluster bonding will be of similar energies and may be readily hybridized. Thus for $Fe(CO)_3$ we may anticipate and employ any one of three possibilities:

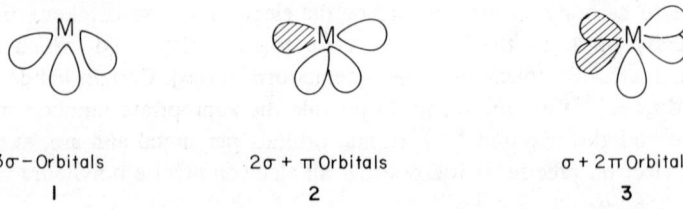

3σ-Orbitals	$2\sigma + \pi$ Orbitals	$\sigma + 2\pi$ Orbitals
1	2	3

Table 4 Numbers of Atomic Orbitals and Electrons (e) Contributed to Cluster Bonding by Some M(CO)$_x$ Fragments

Transition Metal	Number of Valence Electrons	M(CO)$_2$		M(CO)$_2^-$		M(CO)$_3$		M(CO)$_3^-$		M(CO)$_4$		M(CO)$_4^-$		M(CO)$_5$		M(CO)$_5^-$	
		e	a.o	e	a.o	e	a.o	e	a.o	e	a.o	e	a.o	e	a.o	e	a.o
Cr, Mo, W	6	−2	4	−1	4	0	3	1	3	2	3	3	3	2	2	1	2
Mn, Tc, Re	7	−1	4	0	4	1	3	2	3	3	3	2	3	1	1	—	—
Fe, Ru, Os	8	0	3	1	3	2	3	3	3	2	2	1	2	—	—	—	—
Co, Rh, Ir	9	1	3	2	3	3	3	2	3	1	1	—	—	—	—	—	—
Ni, Pd, Pt	10	2	3	3	3	2	2	1	2	—	—	—	—	—	—	—	—

Example 1 will lead to a connectivity of not less than 3, example 2 to a connectivity of not less than four, and example 3 to a connectivity of four or five (or more). SEP theory (see above) only considers example 3. Likewise, $Fe(CO)_4$ may utilize 2σ, or $1\sigma + 1\pi$, or 2π orbitals. These fragments are not necessarily isolobal with either a boron hydride unit (25) or an octahedral fragment (5). The boron hydride unit is necessarily restricted to case 3.

The d^{10} metals nickel, palladium, and platinum form a special class. In general terms, the nickel fragments $Ni(CO)_x$ will be similar to those described above, able to utilize two atomic orbitals ($x = 3$, 2σ, or $\sigma + \pi$) or three atomic orbitals ($x = 2$, 3σ, $2\sigma + \pi$, or $\sigma + 2\pi$). For palladium or platinum, however, the $s(d) - p$ separation becomes an important consideration. There will no longer necessarily be three orbitals available for cluster bonding in, for example, $Pt(CO)_2$. Such a fragment will appear to provide only two orbitals for cluster bonding; the third remaining orbital (p?) apparently being too high in energy to be involved. However, on combination with other $Pt(CO)_x$ fragments to produce $[Pt(CO)_x]_y$ ($y = 3, 4, 5, 6$), we must consider the interactions of these π orbitals. Thus in the formation of the trimer $Pt_3(CO)_6$ the six σ atomic orbitals (two per $Pt(CO)_2$ fragment) interact to generate three σ-bonding molecular orbitals, and the three π atomic orbitals will interact to generate one bonding molecular and two antibonding molecular orbitals. Stability will therefore be achieved for the dianion $[Pt_3(CO)_6]^{2-}$:

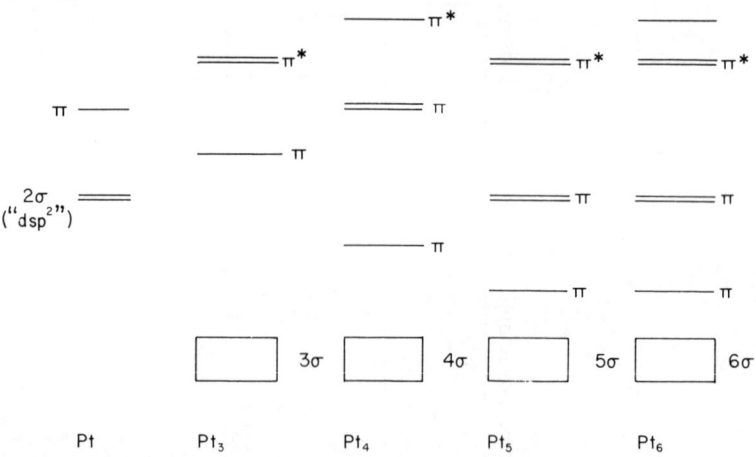

Planar $Pt_4(CO)_8$ might be stabilized either as a 58-electron system $[Pt_4(CO)_8]^{2-}$ ($4\sigma + \pi$) or as a 62-electron system $[Pt_4(CO)_8]^{6-}$ ($4\sigma + 3\pi$). Planar $[Pt_5(CO)_{10}]^{2-}$ and $[Pt_6(CO)_{12}]^{2-}$ will not be stabilized (but see below).

The mixed metal cluster compounds $PtFe_2(CO)_9 PPh_3$ (27) and $Pt_2 Fe(CO)_5$-$(P(OPh)_3)_3$ (59), which have fewer valence electrons than $Fe_3(CO)_{12}$, may be rationalized on a similar basis:

Compound	Number of Valence Electrons
$Fe_3(CO)_{12}$	48
$PtFe_2(CO)_9PPh_3$	46
$Pt_2Fe(CO)_5(P(OPh)_3)_3$	44

A simple bonding scheme for these compounds and that of $[Pt_3(CO)_6]^{2-}$ is outlined below:

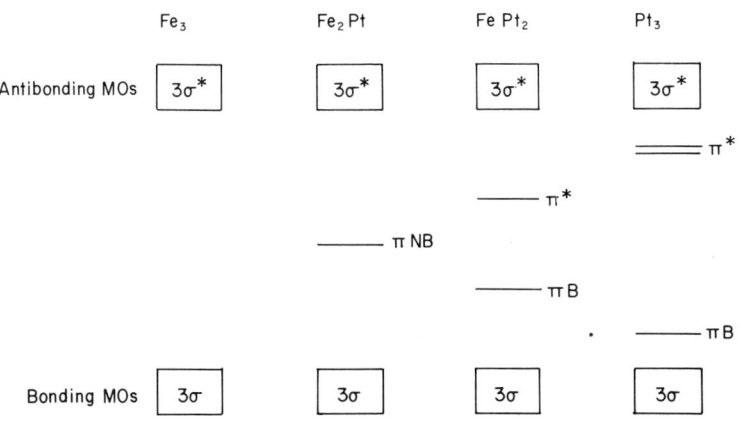

Thus Fe_3, Fe_2Pt, and $FePt_2$ systems may be stabilized by six cluster electrons, whereas Pt_3 requires eight. However it would not be unreasonable, in terms of this simple bonding scheme, to expect the $FePt_2$ system to be reduced to a stable $FePt_2^{2-}$ system.

The butterfly geometry exhibited by $Pt_4(CO)_5(PPhMe_2)_4$ (60), a 58-electron complex that cannot be rationalized by either the 18-electron rule or the SEP theory, may be understood in these terms. As the Pt_4 planar system is folded about the C_2 diagonal axis, two of the four available π orbitals become increasingly less able to interact, leading to a cluster bonding scheme requiring 10 cluster electrons.

Mixed clusters that contain platinum will thus present a slightly more complicated bonding pattern (as will the Pt clusters themselves). As the number of platinum atoms per cluster unit increases, the possibility of forming localized double bonds $(\sigma + \pi)$ will also increase — provided the platinum atoms are adjacent to each other. Thus a cis-Pt_2Rh_4 cluster unit might be stabilized by two electrons more than the corresponding trans system. Simple electron counting theories such as the 18-electron rule and SEP will not work for platinum systems (nor do they necessarily work for other metals). An orbital-electron balance must always be sought.

According to this scheme it is apparent that mixed systems (i.e. systems in which all constituent fragments are not the same) can have electron-acceptor fragments, electron-precise fragments, and finally electron-donor fragments. The analogy between $M(CO)_x$ cluster fragments and main group cluster units is then clear.

(CO)₃Fe "Vacant acceptor orbital"
2 singly occupied + 1 vacant orbital

(CO)₃Co
3 singly occupied orbitals

(CO)₄Fe "Filled donor orbital"
1 filled donor orbital
2 singly occupied orbitals

Main group analog:

H₃B
Electron-pair acceptor

H₃C
Electron-precise fragment

H₃N
Electron-pair donor

Thus in compounds produced by combination of acceptor and donor fragments it is possible to write an alternative view of bonding (4).

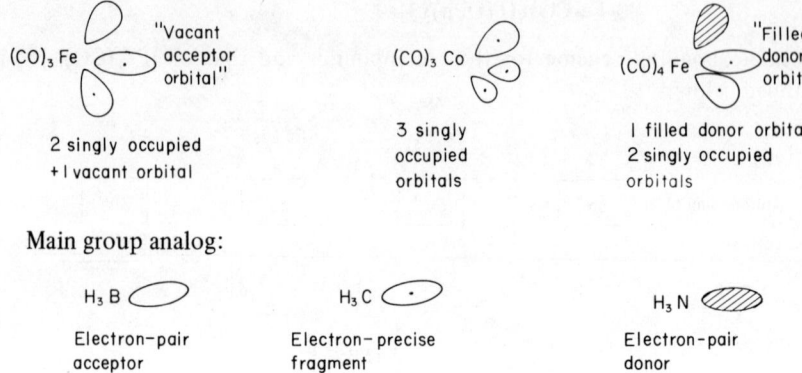

4

Having assessed the various contributions made by the diverse $M(CO)_x$ fragments, we will now consider the construction of polyhedra using the simple ideas of 2e/2-center, 2e/3-center, and multicenter bonds.

C. Metal Polyhedra

1. The M_3 Unit

We shall consider first the simplest structural unit, the triangle, and restrict our attention to the formation of the σ-bonding framework only. Some possible bonding configurations are summarized in Table 5.

Table 5 Possible Bonding Arrangements with an M_3 Triangle

	Bonding description[a]			
	(i) $M_1 \underset{M_2}{\perp} M_3$	(ii) $M_1 \overset{M_2}{\frown} M_3$	(iii) $M_1 \overset{M_2}{\frown} M_3$	(iv) $M_1 \overset{M_2}{\triangle} M_3$
Required number of atomic orbitals per fragment	M_1 1σ M_2 1σ M_3 1σ	M_1 1σ M_2 1π M_3 1σ	M_1 2σ M_2 1π M_3 2σ	M_1 2σ M_2 2σ M_3 2σ
Total number of bonding MOs	1 (3-center)	1 (3-center)	2 (one 2-center, one 3-center)	3 (three 2-center)
Total number of cluster electrons required	2	2	4	6

[a] M_1, M_2, and M_3 represent cluster fragments $M(CO)_x$ rather than individual metal atoms.

It is apparent that the triangle may be constructed in a variety of ways depending on the number and type of orbitals offered by each fragment and the number of electrons available. Theoretically at least, a triangular species may be bound by two, four, or six electrons. Only in examples i and iv will it be possible for $M_1 = M_2 = M_3$. In examples ii and iii it is necessary to invoke π-orbital participation by M_2. For carbonyl clusters, only examples of system iv have been observed. All correspond to the electron-precise arrangement with a total of three bonding electron pairs [e.g., $M_3(CO)_{12}$, M = Fe, Ru, or Os].

2. The M_4 Unit

Within this class fall the tetrahedron, the butterfly, and the square plane. For T_d symmetry there are two bonding alternatives. The tetrahedron requires either six electron pairs (edge bonds: $A_1 + T_2 + E$) (Table 6, example i) or four electron pairs (face bonds: $A_1 + T_2$) (Table 6, example ii). These descriptions differ according to the orientation of the three fragment orbitals (3σ), being aligned above edges in 5 and above faces in 6 leading to six or four bonding MOs respectively.

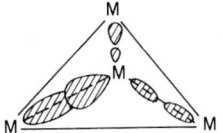

Two-center edge bond
as in $H_4Ru_4(CO)_{12}$, $H_4Os_4(CO)_{12}$
5

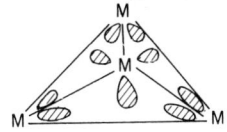

Three-center face bond
as in $H_4Re_4(CO)_{12}$
6

Table 6 Construction of the M_4 Tetrahedron[a]:

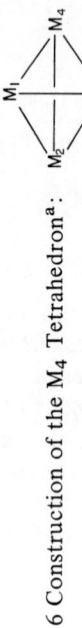

	Bonding Description			
	Six 2e/2-Center Bonds (i)	Four 2e/3-Center Bonds (ii)	Four 2e/2-Center, One 2e/3-Center Bond (iii)	Two 2e/2-Center, Two 2e/3-Center
Required number of atomic orbitals per fragment	M_1 3σ M_2 3σ (Total 12σ) M_3 3σ M_4 3σ	M_1 3σ M_2 3σ (Total 12σ) M_3 3σ M_4 3σ	M_1 σ+π M_2 3σ M_3 3σ (Total 11) M_4 3σ	M_1 σ+π M_2 σ+π M_3 3σ (Total 10) M_4 3σ
Total number of cluster electrons required	3 (Total 12)	2 (Total 8)	(Total 10)	(Total 8)

[a] Only four examples are given. Other possibilities will also exist.

Thus a tetrahedral species may be electron precise (48 electrons) as with $Ir_4(CO)_{12}$, $[H_6Re_4(CO)_{12}]^{2-}$, or $H_4Os_4(CO)_{12}$, or electron deficient as with $H_4Re_4(CO)_{12}$ (44 electrons).

The difference in structure between $H_4Ru_4(CO)_{12}$ (hydrides edge-bridging) and $H_4Re_4(CO)_{12}$ (hydrides face-bridging) may be understood in terms of the different patterns of cluster orbitals to which the hydrides must bond (47).

These structures have also been described in terms of an extension of the styx rules developed by Lipscomb for the boron hydrides (46).

Other methods of constructing quasi-tetrahedral forms using fewer atomic orbitals are also possible. In Table 6, example iii shows the construction of a tetrahedron using four 2e/2-center bonds and one 2e/3-center bond. In this arrangement constituent atom M_1 is required to provide two ($1\sigma + 1\pi$) rather than three atomic orbitals to cluster bonding. An example of this sytem is provided by $H_2Os_3Pt(CO)_{10}PR_3$ (61).

The $Pt(CO)PR_3$ fragment provides two atomic orbitals and two electrons to cluster bonding, and each $Os(CO)_3$ fragment three atomic orbitals and two electrons; the remaining two electrons are provided by the two H-ligands, giving a total count of 10 electrons and 11 orbitals.

In the examples given in Table 6, only i and ii maintain T_d symmetry. Lower symmetries are required for examples iii and iv, and these are to be expected for mixed metal clusters or clusters in which different numbers of CO ligands are associated with the constituent metal atoms.

Some methods of construction of the butterfly are outlined in Table 7. There are a number of methods of construction, but apart from the Pt_4 complex mentioned above, no examples of binary carbonyls possessing this geometry have

Table 7 Some Possible Bonding Arrangements Within the Butterfly

	Bonding Description		
	[diagram]	[diagram]	[diagram]
Required number of atomic orbitals per fragment	M_1 3σ M_2 1π	M_1 2σ M_2 1σ	M_1 3σ M_2 2σ
Total number of bonding MOs	3 (two 3-center, one 2-center)	2 (two 3-center)	5 (five 2-center)
Total number of cluster electrons required	6	4	10

been observed. In general such arrangements are found in systems that also contain organofragments such as RC≡CR and are best viewed as M_4C_2 cluster species (25).

The plane is found for $[Re_4(CO)_{16}]^{2-}$ (17). This cluster may be constructed from two $Re(CO)_4$ (three orbitals and three electrons) and two $Re(CO)_4^-$ (two orbitals and two electrons) giving, as observed, a system with a five σ-bonding framework. As such it is electron precise and may also be derived from an electron-precise tetrahedron (see above).

A more unusual structure, the triangle plus one, is observed with $[H_4Re_4(CO)_{15}]^{2-}$ (14). This molecule may be considered to be constructed from an $Re(CO)_5$ (1σ/1 electron), two $[Re(CO)_4]^-$ (2σ/2 electrons) and one $[Re(CO)_2]^{4-}$ (3σ/3 electrons) fragments [i.e., regard as $Re_4(CO)_{15}^{6-}$, 7].

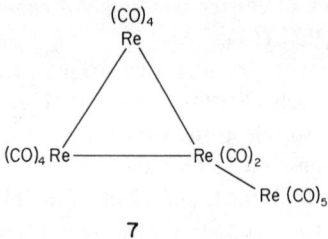

7

3. The M_5 Unit

Two geometries are commonly observed — the trigonal bipyramid and the square-based pyramid. A geometry basically corresponding to the D_{3h}-trigonal bipyramid may be derived from a variety of bonding schemes. Some are listed in Table 8. An electron-precise polyhedron (example i) will be formed, provided each axial fragment contributes three orbitals and three electrons and each equatorial fragment four orbitals and four electrons, a combination not found for any simple M_5 carbonyl species. The osmium cluster $Os_5(CO)_{16}$, although having the correct total electron count (72), possesses only 14, not the required 18 atomic orbitals. The bonding description best suited to this molecule is example v. Each $Os(CO)_3$ unit contributes three orbitals and two electrons, and the unique $Os(CO)_4$ unit two orbitals and two electrons. On this basis, the variation in Os—Os bond distances may be understood. The dianion $[Os_5(CO)_{15}]^{2-}$ (8), with 12 cluster electrons and 15 cluster orbitals is an example of type iv. This corresponds to the SEP approach.

The cluster anions $[Ni_5(CO)_{12}]^{2-}$ and $[Ni_3Mo_2(CO)_{16}]^{2-}$ (48), which to date have not had adequate bonding descriptions, have elongated trigonal bipyramid geometries. The Ni_5 system may be assembled from three $Ni(CO)_2$ fragments (three orbitals and two electrons) and two $Ni(CO)_3$ fragments (two orbitals and

Table 8 Possible Bonding Arrangements Within the Trigonal Bipyramid:

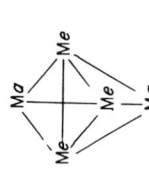

	Bonding Description					
	Nine 2e/2-Center Bonds	Six 2e/3-Center Bonds	Five 2e/2-Center, Two 2e/3-Center	Three 2e/2-Center, Three 2e/3-Center	One 2e/2-Center, Four 2e/3-Center	Three 2e/2-Center, Three 2e/3-Center
	(i)	(ii)	(iii)	(iv)	(v)	(vi)
Required number of atomic orbitals per fragment	M_a 3σ M_e 4σ	M_a 3σ M_e 4σ	M_a 3σ M_e 4σ M_e' 2π	M_a 3σ M_e $2\sigma + \pi$	M_a 3σ M_e $2\sigma + \pi$ M_e' 2π	M_a 2σ M_e $2\sigma + \pi$
Total number of bonding MOs	9	6	7	6	5	6
Total number of cluster electrons required	18	12	14	12	10	12

two electrons), and corresponds to example vi. Of the two atomic orbitals provided by $Ni(CO)_3$, one lies above a face and the other above an edge of the polyhedron. Each may combine with an available π orbital on the $Ni(CO)_2$ fragments ($2\sigma + \pi$) to generate three delocalized molecular orbitals. In the equatorial plane the three $Ni(CO)_2$ fragments combine to generate three $2e/2$-center bonds. The mixed Ni_3Mo_2 cluster may be viewed similarly, since $Ni(CO)_3$ and $Mo(CO)_5$ provide the same numbers of electrons and orbitals to cluster bonding (Table 4). Thus these apparently electron-rich clusters (76 rather than 72 electrons) are easily embraced within this scheme, the four "extra" electrons being assigned to nonbonding orbitals on each of the axial atoms.

There is no example of square pyramidal geometry for a simple binary carbonyl. The only examples contain an additional cluster atom such as C or PR. Table 9 outlines such bonding possibilities and fragment requirements. It is interesting to note that example iv might be expected for a species such as $M_5(CO)_{15}$ (M = Fe, Ru, or Os) [i.e., five $M(CO)_3$ fragments, each contributing three orbitals and two electrons], and according to the views expressed here the $M_5(CO)_{15}C$ (Fig. 1) compounds (49) are best regarded as octahedral M_5C species, that is, the carbon is a surface rather than an interstitial atom.

Table 9 Construction of the M_5 Square-Based Pyramid

	Bonding Description			
	Eight 2e/2-Center Bond (i)	Four 2e/2-Center Two 2e/3-Center (ii)	Four 2e/3-Center Bond (iii)	Four 2e/3-Center One 2e/5-Center (iv)
Required number of atomic orbitals per fragment	M_a 4σ M_e 3σ	M_a 2π (Total 14) M_e 3σ	M_a 2π (Total 10) M_e $\sigma+\pi$	M_a $\sigma+2\pi$ (Total 15) M_e $2\sigma+\pi$
Total number of bonding MOs	8	6	4	5
Total number of cluster electrons required	16	12	8	10

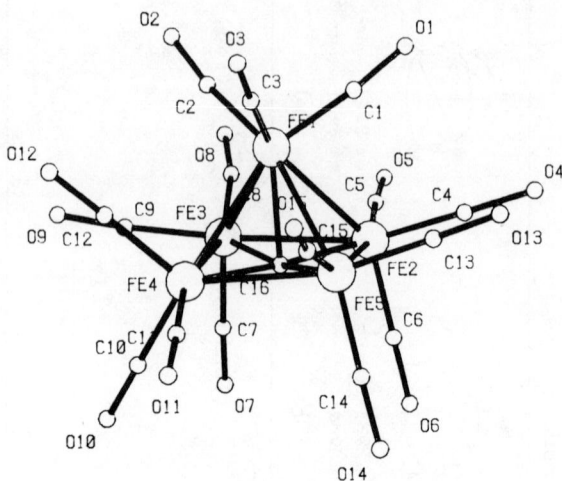

Fig. 1. Molecular structure of $Fe_5C(CO)_{15}$.

4. The M_6 Unit

The O_h octahedron is commonly observed for clusters containing six metal atoms. Thus $Rh_6(CO)_{16}$, $Ru_6C(CO)_{17}$, $[Co_6(CO)_{15}]^{2-}$, and $[Co_6(CO)_{14}]^{4-}$, all with 86 valence electrons, possess an octahedral cluster unit. Other polyhedra are also observed. Both the carbido cluster $[Rh_6C(CO)_{15}]^{2-}$, which has 90 electrons, and the platinum cluster $[Pt_6(CO)_{12}]^{2-}$, with 86 electrons, have a trigonal prism of metal atoms. The compound $Os_6(CO)_{18}$, with 84 valence electrons, has a bicapped tetrahedral arrangement of metal atoms.

For an electron-precise octahedron with 12 M–M bonds, an 84-electron species is required. However each metal fragment must also contribute four atomic orbitals to the overall bonding scheme (Table 10, example i). These criteria are not met by any of the compounds mentioned above (but are satisfied for $[Mo_6Cl_8]^{4+}$). Thus although $Os_6(CO)_{18}$ has the correct magic number of electrons, it can provide a total of only 18 atomic orbitals rather than the required 24.

An alternative bonding pattern involving eight three-center bonds of the type suggested for $[Nb_6Cl_{12}]^{2+}$ is shown in example ii in Table 10. In this arrangement each metal fragment must again provide four atomic orbitals, an unlikely situation for most $M(CO)_x$ fragments.

Example iii corresponds to the SEP viewpoint: a total contribution of 14 electrons and 18 orbitals to cluster bonding is required. This requirement is met by the dianion $[Os_6(CO)_{18}]^{2-}$ and the other 86-electron species mentioned above.

Table 10 Construction of the M_6 Octahedron

	Bonding Description					
	Twelve 2e/2-Center Bonds	Eight 2e/3-Center Bond	Six 2e/3-Center One 2e/6-Center	Six 2e/3-Center Bonds	Four 2e/3-Center One 6-Center	Four 2e/3-Center Bonds
	(i)	(ii)	(iii)	(iv)	(v)	(vi)
Required number of atomic orbitals per fragment	4σ	4σ	$\sigma + 2\pi$	2π	3σ	2σ
Total number of bonding MOs	12	8	7	6	5	4
Total number of cluster electrons required	24	16	14	12	10	8

Other methods of constructing octahedra will exist. Example iv is just one of a number of possibilities. In this case each metal fragment is required to donate two atomic orbitals and two electrons to cluster bonding, leading to six three-center edge bonds. For the criteria to be met, a 98-electron complex is required, a species as yet unobserved.

Closely related to this is the bonding scheme envisaged for the bicapped tetrahedron. This polyhedron may also be constructed from six three-center bonds, but in this case and *in contrast to example iv*, each metal atom is required to donate three (as opposed to two) atomic orbitals to the cluster framework. This is found for $Os_6(CO)_{18}$. Within the bicapped tetrahedron there are three types of metal atom, each employing a different set of atomic orbitals in the bonding scheme. These are indicated below:

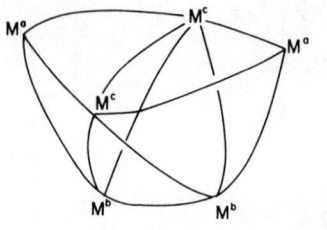

M^a contributes 3σ orbitals

M^b contributes $2\sigma + 1\pi$ orbitals

M^c contributes $\sigma + 2\pi$ orbitals

Bicapped tetrahedron

A total of 18 atomic orbitals and 12 electrons is thus required.

Thus the octahedron in $[Os_6(CO)_{18}]^{2-}$ is constructed from six $Os(CO)_3$ fragments, each contributing three orbitals $(\sigma + 2\pi)$; whereas $Os_6(CO)_{18}$ is constructed of six $Os(CO)_3$ fragments contributing 3σ, $2\sigma + \pi$, or $\sigma + 2\pi$, according to their local connectivities.

The cluster $[Rh_6C(CO)_{15}]^{2-}$ is electron precise; a total of 18 orbitals and 18 electrons is available, leading to a trigonal prismatic geometry.

A trigonal prism is also observed for $[Pt_6(CO)_{12}]^{2-}$. In this case, however, 14 electrons and 18 orbitals are available for cluster bonding, a contribution equivalent to that in $[Os_6(CO)_{18}]^{2-}$. Why, therefore, the structural difference?

The structure of $[Pt_6(CO)_{12}]^{2-}$ may be viewed as being constructed of two planar $Pt_3(CO)_6$ units. As in $[Pt_3(CO)_6]^{2-}$ (see above) three 2e/2-center bonds will be formed from the six available (σ) orbitals (two per Pt); these will be localized along the three triangle edges. In addition the six π orbitals may combine to produce a six-center bonding orbital:

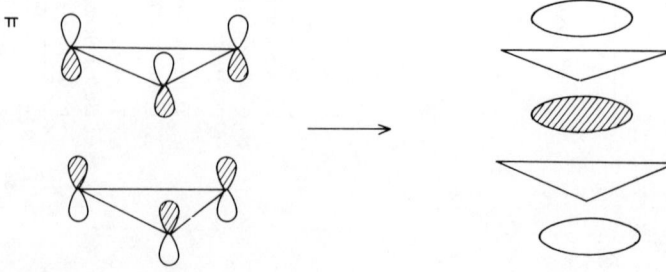

Such a bonding scheme requires three atomic orbitals per Pt ($2\sigma + \pi$) and a total of 14 electrons. Such a bonding scheme would apply equally well to either a trigonal prismatic or a trigonal antiprismatic geometry. A low barrier to octahedron (trigonal antiprism) $\rightleftharpoons$ trigonal prism would, therefore, be expected, a view compatible with the observed rotation of the two Pt triangles. This bonding scheme also falls in line with the established bond length variation found in $[Pt_6(CO)_{12}]^{2-}$:

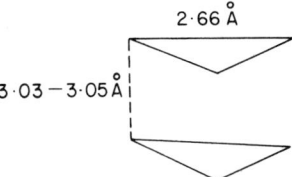

An alternative view is to regard the bond between $[Pt_3(CO)_6]^{2-}$ (filled π-bonding orbital) and $Pt_3(CO)_6$ (empty π-bonding orbital) as dative:

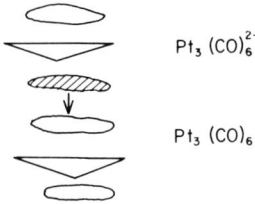

Since the π-bonding extends above and below the Pt_3 triangle, the formation of the linear chains $[Pt_3(CO)_6]_n^{2-}$ is understandable, the π bond extending through in Pt_3 triangles. It follows that there will be a weakening of Pt_3-Pt_3 bonds, hence easy dissociation.

This treatment again emphasizes the differences between the d^{10} and other cluster metals. It also emphasizes the role of the ligand envelope in controlling the ground state geometry (see below).

The bonding arrangements of other polyhedra may be interpreted along similar lines. The number and type of orbitals likely to contribute to the bonding can be inferred from established structures and/or the fragments involved. Experience suggests that the number of orbitals and electrons that $M(CO)_x$ fragments may contribute are as outlined in Table 4.

D. Conclusions

1. The 18-electron rule and the SEP rule are restricted because of the limitations they place on the number and type of atomic orbitals available per cluster fragment. The more general approach (Sect. II-B-3) allows for all possibilities.
2. Simple electron counting schemes will not work for many even simple systems; more important is an *electron-orbital balance* (EOB). Thus an octahedron may be formed with 12, 8, 7, or 6 electron pairs.

3. The isolobal principle may not be generally applied to cluster systems. Outside the nickel and copper triads, any combination of orbital types may be utilized (e.g., 3σ, $2\sigma + \pi$, or $\sigma + 2\pi$).
4. The d^{10} metals form a special case, which although easily embraced within the EOB method, is not within the scope of either the 18-electron rule or the SEP theory.

III. THE LIGAND ENVELOPE

A. CO Bonding Modes

The CO ligand can adopt several modes of coordination. One is as the terminal carbonyl group bonded to a single metal atom and may be linear (**9a**) or significantly bent; another is as a bridging ligand.

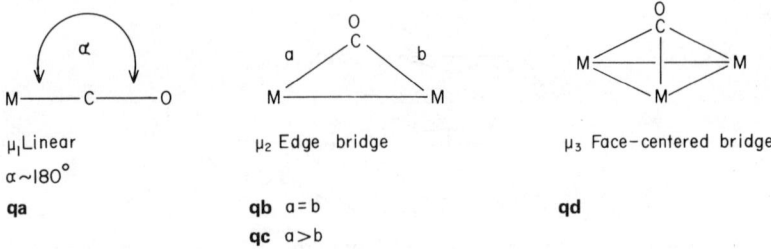

μ_1 Linear
$\alpha \sim 180°$
qa

μ_2 Edge bridge

qb a=b
qc a>b

μ_3 Face-centered bridge

qd

The doubly-bridging or edge-bridging carbonyl group is usually (but *not* always) associated with a metal–metal bond. It may span the M—M bond symmetrically (**9b**) as in $Fe_3(CO)_9(PMe_2Ph)_3$ (28) or with varying degrees of asymmetry (**9c**) (as in $[Fe_4(CO)_{13}]^{2-}$) (16). A metal-metal bond may be spanned by one CO bridge as in $Co_4(CO)_{12}$, two, as in $Co_2(CO)_8$, or three, as in $Fe_2(CO)_9$. Face- (or triply) bridging CO groups (**9d**) are also observed. These span a triangular face of a metal cluster, as in $Rh_6(CO)_{16}$. At present there are no examples of CO *symmetrically* spanning more than three metal atoms, although there are two as yet unique bonding modes, which occur in the compounds $[Fe_4(CO)_{13}H]^-$ (29) and $Mn_2(CO)_5[PPh_2(CH_2)_2PPh_2]_2$ (30):

A recurring problem encountered with the carbonyl clusters is the occurrence of different structural types in closely related compounds. For example, $Ru_3(CO)_{12}$ and $Os_3(CO)_{12}$ have molecular structures of D_{3h} symmetry with no bridging carbonyls, whereas the iron compound $Fe_3(CO)_{12}$ has C_{2v} symmetry with two CO bridges (see Table 11, Sect. III-B). Similarly the compounds $Co_4(CO)_{12}$ and $Rh_4(CO)_{12}$ have molecular structures of C_{3v} symmetry with three edge-bridging and nine terminal CO ligands, but the corresponding compound of iridium, $Ir_4(CO)_{12}$, has T_d symmetry, with all carbonyls terminal.

There have been a number of attempts to account for this variation in the occurrence of terminal, edge-bridging, and face-bridging groups. Cotton (31) has stressed the importance of the M—M bond supporting the CO bridge and has shown for a series of iron compounds that the degree of asymmetry of the edge bridge varies systematically with the M—M distance. He has also drawn attention to the need of heterometallic species, in particular bimetal compounds, to maintain electroneutrality through the M_m cluster unit and has argued that CO bridge formation may provide a mechanism by which neutrality can be achieved. As Cotton (31) points out, however, this cannot be general, since many bimetallic species containing two different metals show no evidence of bridge formation. Related to these ideas are the arguments presented to account for bridge formation in clusters when CO ligands are substituted by more basic ligands such as tertiary phosphines. Thus $Ir_4(CO)_{12}$, for which no bridges are observed, undergoes reaction with triphenylphosphine (PPh_3) to produce $Ir_4(CO)_{10}(PPh_3)_2$, which contains three CO bridges and has a quasi-$Co_4(CO)_{12}$ structure (32). It has been argued that CO bridge formation provides a mechanism by which the additional electron density provided by the more basic PPh_3 ligand (compared to CO) may be passed on to an adjacent metal atom. Certainly these arguments provide a reasonable rationale for the observed rates of substitution across the series $Ir_4(CO)_{12} < Ir_4(CO)_{11}PR_3 < Ir_4(CO)_{10}(PR_3)_2 < Ir_4(CO)_9(PR_3)_3$. However, although this idea is attractive, one might reasonably ask why substituted derivatives of $Os_3(CO)_{12}$ and $Ru_3(CO)_{12}$ show no evidence of CO bridge formation (50).

In $Mn_2(CO)_{10}$ there appears to be a significant bonding interaction between one Mn atom and the equatorial CO groups attached to the adjacent Mn atom (33). Calculations indicate that such interactions are less important for $Re_2(CO)_{10}$ and, although this approach may be used to rationalize the bonding of M—C—O units in the $M_2(CO)_{10}$ species, its application is limited and could not be extended to embrace more complicated cluster species. It has been observed that for certain series of compounds, for example, $Co_6(CO)_{16}$, $[Co_6(CO)_{15}]^{2-}$, and $[Co_6(CO)_{14}]^{4-}$, as the anionic charge increases so does the number of CO bridges (of all descriptions).

Since the bridging CO is considered to be a better π acceptor than the terminally bound CO, this information has been taken to indicate a preference to form bridges as the need to dissipate anionic charge is increased. However it should be recognized that in most cases as the anionic charge increases, the number of CO ligands decreases giving rise to different m/n values. Where the number of CO ligands (n) remains constant as charge is added, [viz. $Os_6(CO)_{18} \rightarrow [Os_6(CO)_{18}]^{2-}$], there is no tendency for CO bridge formation.

In general, CO donates a maximum of two electrons irrespective of its bonding mode. [Examples of four-electron donation are known (29, 30) but are rare and are not considered further here.] The terminal CO bond requires one metal atomic orbital for σ-bonding (and others for π-bonding; see below), whereas a bridging ligand of either type μ_2 or μ_3 does not. Viewed simplistically, the CO ligand in a bridging mode provides orbitals of suitable symmetry for combination with metal cluster bonding orbitals:

$$M\!-\!C\!-\!O \qquad\qquad M\!-\!M\ +\ CO\ \longrightarrow\ M\!-\!(CO)\!-\!M$$
σ Bond $\qquad\qquad\qquad\qquad\qquad$ M_2 CO 3 Center bond

$$\text{or}\quad M_3 \ +\ CO\ \longrightarrow\ M_3(CO)$$
M_3 CO 4-Center bond

Thus the presence of μ_2 or μ_3 CO ligands does not affect the number of atomic orbitals available for cluster bonding. Of course, this very crude picture ignores the more detailed aspects of π interactions.

B. Ligand Polyhedra

It has been noticed by several groups of workers reporting individual crystal structures that the carbonyl ligands in these clusters occupy positions that define (to a fair approximation) the vertices of regular and semiregular polyhedra. In the crystal structure of the anion $[HFe_3(CO)_{11}]^-$ the position of the hydride ligand (undetectable by X-rays in the presence of a large number of heavy atoms) was deduced to be the "missing" vertex of the *nido*-icosahedron defined by the 11 CO ligands (34). It was on the basis of this information that the structure of $Fe_3(CO)_{12}$ was correctly surmised (the twelfth CO completing the icosahedron) at a time when crystal disorder in $Fe_3(CO)_{12}$ itself had prevented its detailed structure determination. The dodecacarbonyls $Os_3(CO)_{12}$ and $Ru_3(CO)_{12}$, which might be expected to have the same structure as $Fe_3(CO)_{12}$, possess the all-CO-terminal structure (see above) in which the CO groups define an anticuboctahedron. Examination of the structures of $Co_2(CO)_8$, $Fe_2(CO)_9$, $M_2(CO)_{10}$ (M = Mn, Tc, or Re), $Co_4(CO)_{12}$, $Rh_4(CO)_{12}$, $Ir_4(CO)_{12}$, and

$Rh_6(CO)_{16}$ reveals that the CO ligands in these species may also be taken to designate regular or semiregular polyhedra (Table 11).

These observations are not unexpected. The idea that the ligands, L, in cluster species $M_m L_n$ adopt recognizable polyhedral arrangements is not new. Several years ago Guggenberger and Muetterties (35) pointed out that there is a surprising consistency of shape not only for simple coordination compounds ML_n but also in clusters $M_m L_n$. Since apparently the very simple binary carbonyls may be taken to describe regular or semiregular CO polyhedra, it would be reasonable to assume that provided we are dealing with a closed-shell system, the more complicated cluster carbonyls will do the same.

In 1976 it was proposed (36) that the number and distribution of bridging and terminal CO groups in cluster carbonyls $M_m(CO)_n^t$ reflected:

1. The polyhedral arrangement of the n CO ligands.
2. The orientation of the M_m unit within this polyhedron.

The problem was then to discover why for a given M_m geometry, which may be established by the methods outlined in Sect. II, one particular CO arrangement was preferred over others. It was argued that the first type of polyhedron listed may be predicted on the basis of relatively simple arguments (see below) and that it will vary (within those available for polyhedra based on n atoms or groups) according to the *size* and *shape* of the encapsulated M_m unit.

Implicit in these arguments was the idea that the CO ligand has a fixed effective radius, estimated to be 3.02 Å.

As an example of the approach consider the binary carbonyls which possess twelve CO ligands:—

$Fe_3(CO)_{12}$, $Ru_3(CO)_{12}$, $Os_3(CO)_{12}$, $Co_4(CO)_{12}$, $Rh_4(CO)_{12}$, $Co_2Ir_2(CO)_{12}$, $Ir_4(CO)_{12}$, $[Ni_5(CO)_{12}]^{2-}$, $[Ni_6(CO)_{12}]^{2-}$ and $[Pt_6(CO)_{12}]^{2-}$. Of the three trinuclear dodecacarbonyls, only $Fe_3(CO)_{12}$ is based on the most favorable icosahedral arrangement of carbonyl groups, $Ru_3(CO)_{12}$ and $Os_3(CO)_{12}$ possessing the slightly less favorable anticuboctahedral disposition. Geometrically, the C_{2v} structure of $Fe_3(CO)_{12}$ is a consequence of placing an Fe_3 triangle within an icosahedron, and the D_{3h} structure of $Ru_3(CO)_{12}$ and $Os_3(CO)_{12}$ that of placing the M_3 triangle within an anticuboctahedron. On the basis of simple radius ratio arguments employing r_{CO} as 3.02 Å and the metallic radii established for 12-coordinate atoms in metallic structures, it was argued that the Fe_3 triangle may be accommodated within the central interstice of the icosahedron, whereas the larger Ru_3 and Os_3 units cannot. By modifying the carbonyl polyhedron from an icosahedron to the less closely packed anticuboctahedron, thereby enlarging the polyhedral interstice, the Ru_3 and Os_3 units could be accommodated.

Of the tetranuclear dodecacarbonyls, $Co_4(CO)_{12}$, $Rh_4(CO)_{12}$ and $Co_2Ir_2(CO)_{12}$, with C_{3v}-type bridged structures, have carbonyl arrangements

Table 11 The Polyhedral Arrangements of CO Ligands Observed in Some Binary Carbonyls $M_m(CO)_n^l$

Connectivity, c	Species	CO Polyhedron	Description: Föppl Notation
0	$Ni(CO)_4$	Tetrahedron	1 : 3
	$M(CO)_5$ (M = Fe, Ru, or Os)	Trigonal bipyramid	1 : 3 : 1
	$M(CO)_6$ (M = Cr, Mo, or W)	O_h-Octahedron	1 : 4 : 1
1	$Co_2(CO)_8$	Bicapped trigonal prism	3 : 2 : 3
	$Fe_2(CO)_9$	Tricapped trigonal prism	3 : (3) : 3
	$M_2(CO)_{10}$ (M = Mn, Tc, or Re)	Bicapped square antiprism	1 : 4 : (4) : 1
2	$Fe_3(CO)_{12}$	Icosahedron	1 : 5 : (5) : 1
	$M_3(CO)_{12}$ (M = Ru or Os)	Anticuboctahedron	3 : 6 : 3
	$Ni_3(CO)_6^{2-}$	Hexagon (chair)	3 : (3)
	$Pt_3(CO)_6^{2-}$	Hexagon (plane)	6
3	$M_4(CO)_{12}$ (M = Co or Rh)	Icosahedron	1 : 5 : (5) : 1
	$Ir_4(CO)_{12}$	Cuboctahedron	3 : 6 : (3)
	$[Pt_6(CO)_{12}]^{2-}$	Hexagonal prism	6 : 6
4	$[Ni_6(CO)_{12}]^{2-}$	Puckered bihexagon	6 : (6)
	$Rh_6(CO)_{16}$	Tetracapped truncated tetrahedron	1 : 6 : (6) : 3

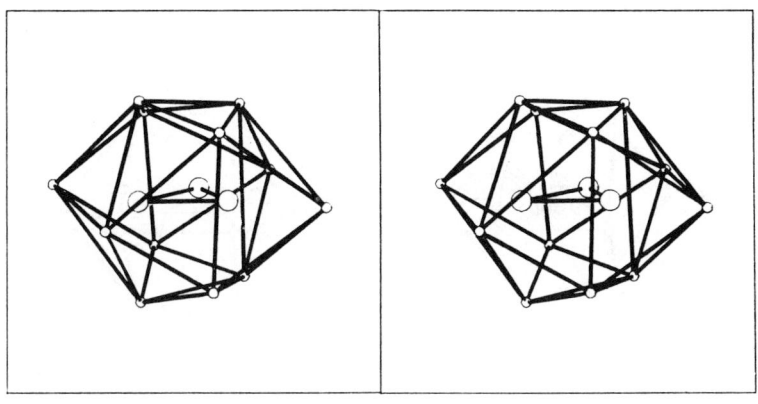

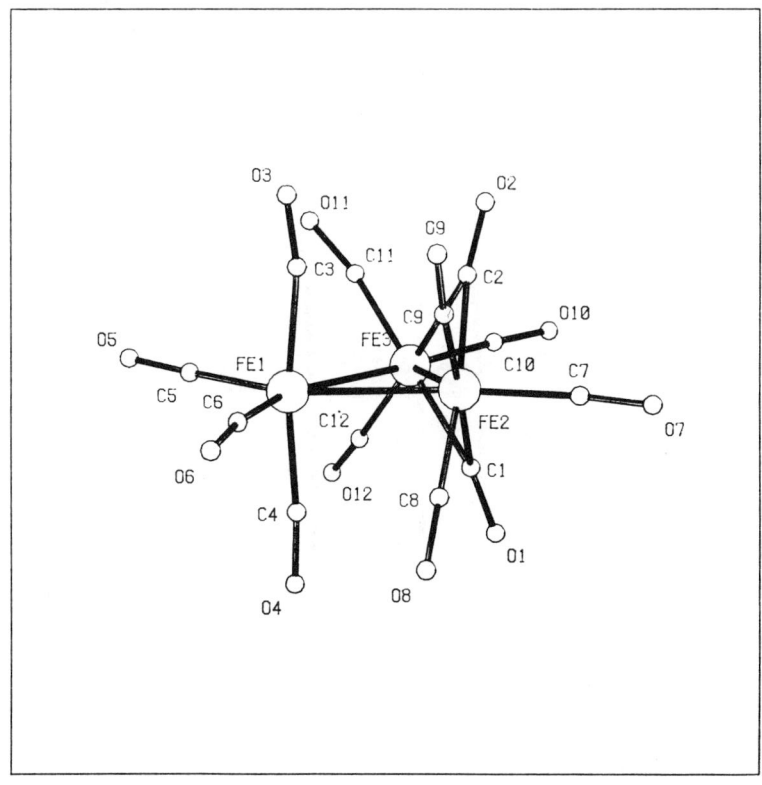

Fig. 2. Molecular structure of $Fe_3(CO)_{12}$.

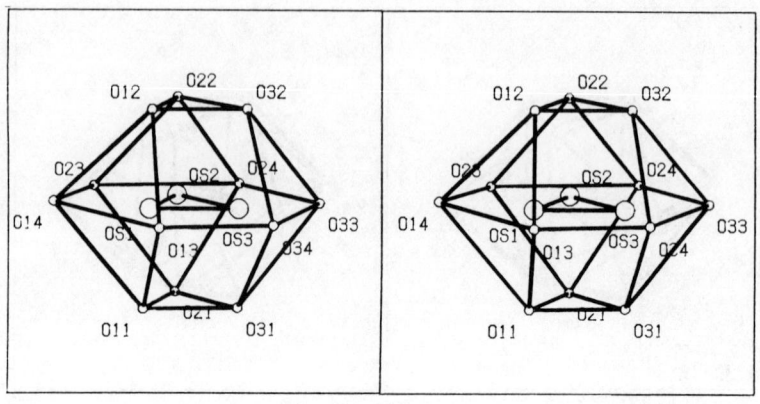

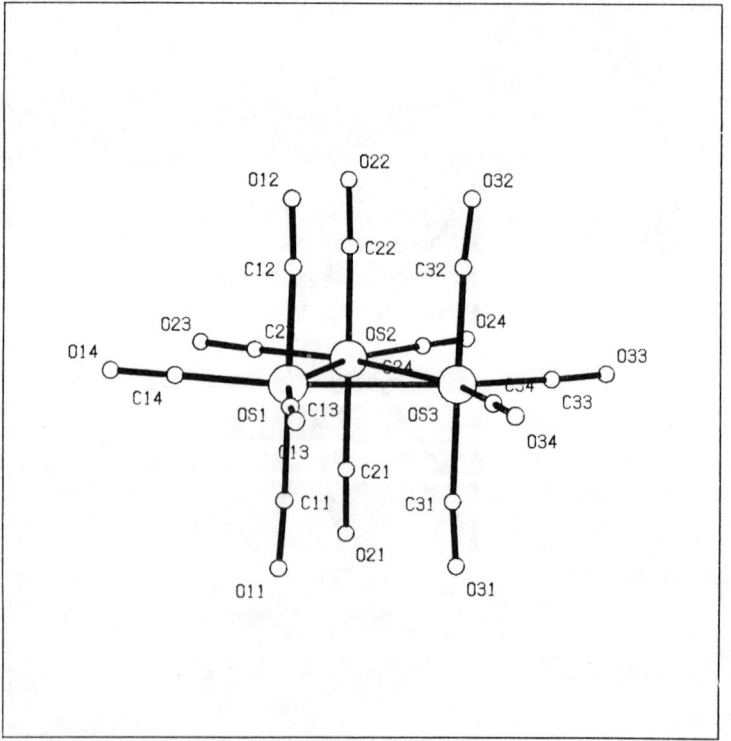

Fig. 3. Molecular structure of $Os_3(CO)_{12}$.

approximating to icosahedra, whereas the unbridged $Ir_4(CO)_{12}$, of T_d symmetry, possesses a cuboctahedral ligand envelope. Thus, for both $M_3(CO)_{12}$ and $M_4(CO)_{12}$ species, the transformation from bridged to nonbridged structures was viewed as a geometric consequence of the increase in size of the M_m unit.

Although the value of 3.02 Å for the radius of a carbonyl ligand worked well for the $M_m(CO)_{12}$ species, it was not possible to view the ligand as a sphere and obtain a value for the radius that could be used irrespective of the number of carbonyls present. The simple radius ratio arguments originally proposed have been replaced by calculations of inter-carbonyl repulsions (38) (see below).

Calculation of the optimum arrangements of twelve points on a sphere confirm that the form minimizing inter-CO repulsions is the icosahedron. Rather less favorable are the cuboctahedron and anticuboctahedron, and the wide range of other polyhedral forms have higher repulsion energies. Thus the implication is that $Fe_3(CO)_{12}$ and $Co_4(CO)_{12}$ adopt icosahedral ligand arrangements because steric interactions of carbonyls on these small clusters are of great importance. The larger clusters, such as $Os_3(CO)_{12}$ and $Ir_4(CO)_{12}$ may, presumably for electronic reasons, possess ligand envelopes of the less sterically favorable cuboctahedral type, because steric interactions are less important when there is more room on the metal cluster surface.

The other dodecacarbonyls illustrate this point further, adopting progressively less sterically favorable (i.e. less close-packed) carbonyl polyhedra as the size of the metal cluster unit increases. $[Ni_5(CO)_{12}]^{2-}$ has the cuboctahedral ligand form, but the octahedral Ni_6^{2-} cluster is enveloped by a puckered bihexagon of carbonyls, and the trigonal prismatic Pt_6^{2-} unit has a hexagonal prismatic carbonyl arrangement. Quantitative calculations of the inter-ligand repulsions in these species confirm strong correlation with the size of the enclosed metal cluster units.

This conclusion is general and may be applied to values of n from 6 to 18. It is a situation commonly found in inorganic systems, namely, that it is the comparatively small, nonbonded ligand-ligand interactions that govern the structures of the binary carbonyls. This view receives support from the observed stereochemical nonrigidity these molecules exhibit in solution (37). At relatively low temperatures the CO groups of many cluster carbonyls have been shown by ^{13}CO NMR spectroscopy to be mobile over part or all of the metal cluster (see below). Activation energies for such processes are commonly about $50\,kJ/mol^{-1}$, indicating the accessibility of additional structural forms. Thus the ground state structure is not governed by strongly directional bonds. It can be reasonably concluded that the adoption of a particular solid state structure (for a molecular rather than an ionic compound) depends on a fine balance of factors. Since the same ground state structure is usually observed in both the solid and solution states, it is not too unreasonable to assume that it is the CO—CO nonbonded interactions that are the most important, at least for neutral systems.

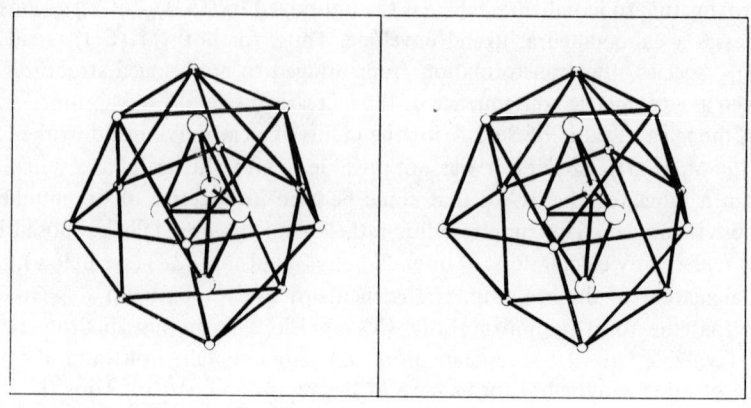

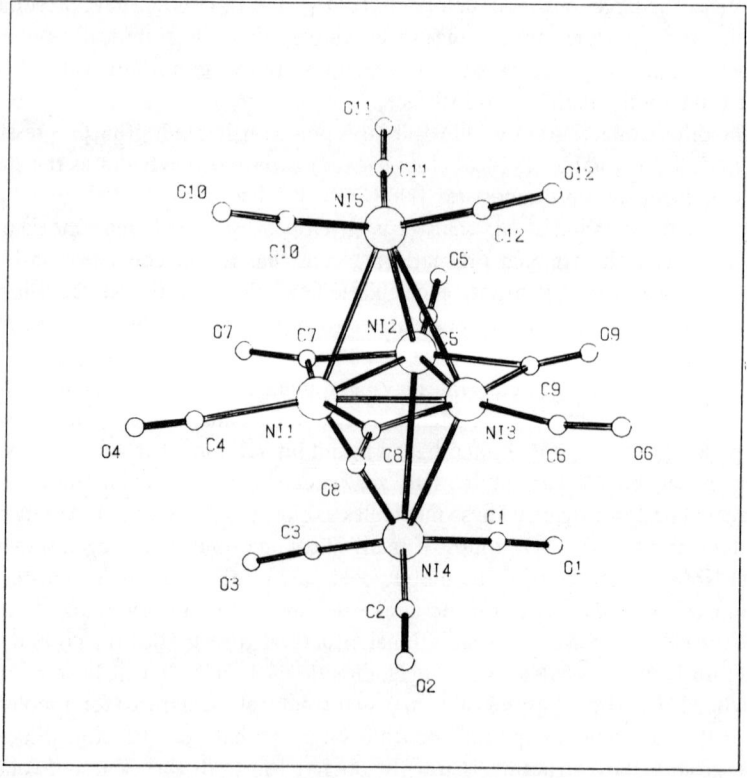

Fig. 4. Molecular structure of $[Ni_5(CO)_{12}]^{2-}$.

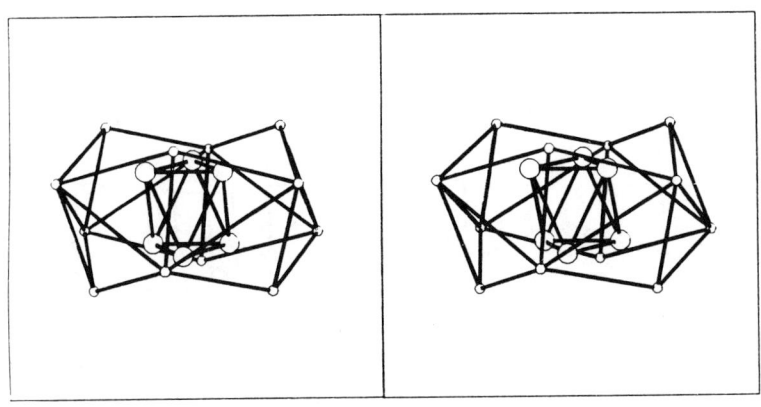

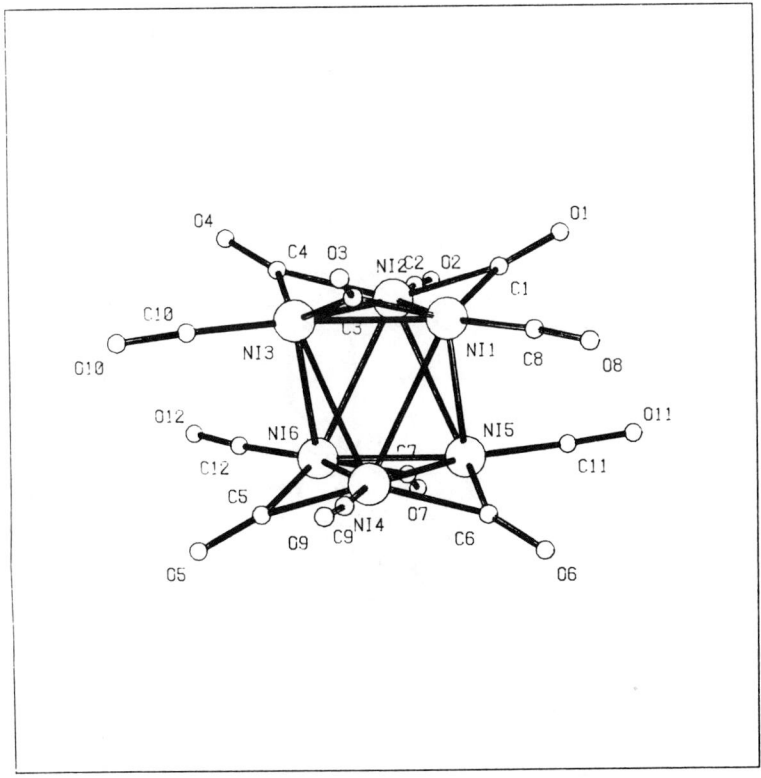

Fig. 5. Molecular structure of $[Ni_6(CO)_{12}]^{2-}$.

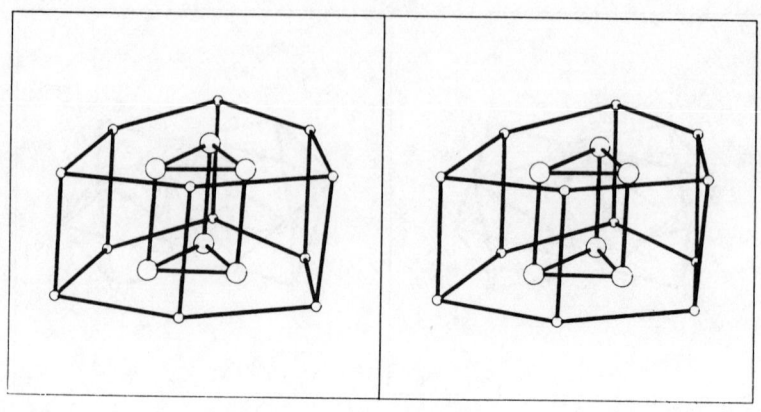

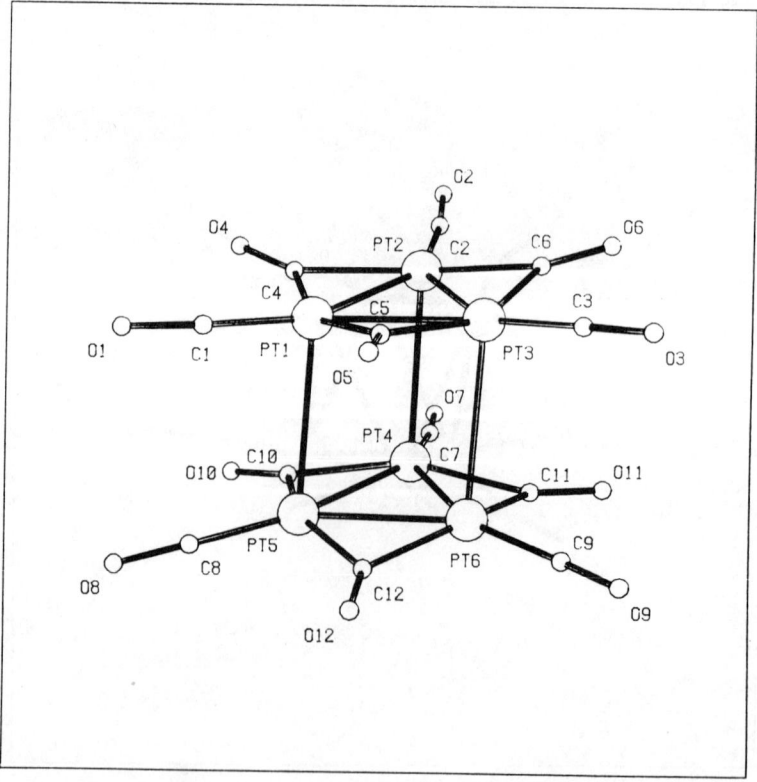

Fig. 6. Molecular structure of $[Pt_6(CO)_{12}]^{2-}$.

The CO polyhedra observed for real structures may show some distortion from the idealized polyhedra. To some extent the nature of these distortions can be predicted. For systems in which the M_m and $(CO)_n$ polyhedra have compatible symmetries (e.g., an M_4 tetrahedron within a cuboctahedron), little or no distortion will be observed. For systems with less compatible symmetries [e.g., the triangle (two dimensional arrangement) within an icosahedron (three-dimensional)] some distortion will be observed. In general terms each polyhedron M_m and $(CO)_n$ will distort so as to produce a common symmetry [e.g., C_{2v} in the case of $Fe_3(CO)_{12}$].

This section shows that the rationalization of structures (CO polyhedra) on the basis of optimization of nonbonded interactions is valid for the cluster carbonyls, provided the M_m unit is reasonably spherical. Where the M_m geometry is extremely anisotropic [e.g., $Re_4(CO)_{16}^{2-}$], the need to maintain reasonable metal-carbonyl bonding distances becomes the more important factor.

Throughout we assume that the metal cluster geometry is predetermined by the number of electrons and orbitals (see above) available for cluster bonding. As noted above, these depend on the number and type of CO ligand present. At first sight this may seem to be a contradictory statement. We have argued that the number and type of cluster orbitals available per cluster fragment will depend on the geometry of that cluster fragment. However, although this is the case, it is possible to generate larger polyhedra (e.g., the icosahedron or cuboctahedron) from the combination of smaller fragments without unduly modifying the local metal geometry. Thus the local CO environment about each metal in the hypothetical example of an $M_2(CO)_6$ octahedral complex is similar to that in an $M_2(CO)_6$ trigonal prismatic complex.

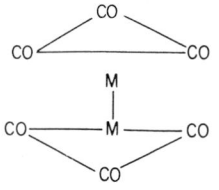

Octahedral CO arrangement

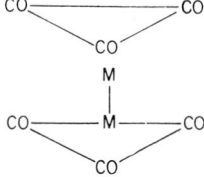

Trigonal prismatic CO arrangement

1. Method

Our method (38) has been to calculate the most favorable arrangements of points on the surface of sphere for numbers of points between 6 and 16. This served as a model for predicting the polyhedral arrangements of carbonyl ligands in cluster carbonyls $M_m(CO)_n^l$. In addition, the relative favorability of different polyhedral forms was calculated, to guide in determining the relative importance of ligand-ligand nonbonded interactions in governing the overall molecular structure. Where polyhedra have similar repulsion energies, nonbonded interactions would be insignificant in comparison with the other energy terms involved.

The ligands are constrained to lie on the surface of a sphere. The nonbonded repulsion between two ligands i and j is assumed to be proportional to the inverse qth power of the distance R_{ij} between them, so that the total repulsion energy $\rangle E \propto \Sigma_{i \neq j} R_{ij}^{-q} \langle$ where the sum is over all pairs of ligands. Since the distance between a pair of points on a sphere is proportional to the angular distance between them, then $E \propto \Sigma_{i \neq j} \Omega_{ij}^{-q}$ for ease of calculation in spherical polar coordinates. The exponent q may have any value, and changing it has little qualitative effect on the result; throughout the work described here it has been taken to have the value of 6. A detailed description of this method has been given by Claxton and Benson (39).

Possible polyhedra for each number n of vertices are listed according to the numbers of vertices in successive latitudinal planes (Föppl notation) so that, for example, the octahedron is described as $1:4:1$ or $3:(3)$ — brackets denoting the staggering of layers with the same number of vertices. This approach is best suited to the larger polyhedra which usually contain planes of 3 or more vertices but results for polyhedra up to ten vertices are consistent with previous work on models for coordination spheres about large transition metal, lanthanide and actinide atoms. The degree to which (CO)-polyhedra in real structures approximate to ideal forms may be quantitatively ascertained by determining their real values of $\Sigma_{i \neq j} \Omega_{ij}^{-6}$. This is done by calculation of the angular distribution of oxygen atoms about the center of gravity of their atomic coordinates.

The various 9-vertex polyhedra have previously been examined to find readily available distortions that may interconvert them (51); this approach is also of importance in cluster carbonyls (52).

The calculations discussed in this review have been reported in detail elsewhere (38).

Tables 12–18 list the polyhedral arrangements predicted for n carbonyl groups for values of $n = 8$ to 16. These arrangements are listed in order of favorability.

2. Polyhedral Arrangements of $(CO)_n$

$n = 6$. This is the simplest case, with few examples. Best known are the octahedral compounds $M(CO)_6$ (M = Cr, Mo, or W). As yet there are no examples of $M_2(CO)_6$ species, for which a trigonal prismatic arrangement of CO ligands might be observed [cf. $Co_2(CO)_8$ below]. The nickel and platinum dianions $[Ni_3(CO)_6]^{2-}$ and $[Pt_3(CO)_6]^{2-}$ provide examples of $M_3(CO)_6$ species. For these dianions the less favorable $3:3$ hexagonal chair and hexagon of CO groups are observed. Thus as with the 12-CO species described above, a transition

Table 12 Favorable Polyhedra for $n = 8$

Order of Favorability	Description:	Föppl Notation
1	4 : (4)	square antiprism
2	2 : (2) : 2 : (2)	dodecahedron
3	1 : 4 : 3	
4	1 : 5 : 2	
5	4 : 4	cube
6	3 : 2 : 3	bicapped trigonal prism

Table 13 Favorable Polyhedra for $n = 9$

Order of Favorability	Description:	Föppl Notation
1	3 : (3) : 3	tricapped trigonal prism
2	1 : 4 : (4)	monocapped square antiprism
3	1 : 5 : 3	
4	1 : 4 : 4	monocapped cube
5	4 : 5	
6	2 : 5 : 2	

Table 14 Favorable Polyhedra for $n = 10$

Order of Favorability	Description:	Föppl Notation
1	1 : 4 : (4) : 1	bicapped square antiprism
2	1 : 3 : (3) : 3	
3	1 : 5 : 4	
4	2 : 4 : (4)	
5	1 : 6 : 3	
6	2 : 5 : 3	

Table 15 Favorable Polyhedra for $n = 12$

Order of Favorability	Description: Föppl Notation	
1	1 : 5 : (5) : 1 or 3 : (3) : 3 : (3)	icosahedron
2	3 : 6 : (3) or 4 : (4) : 4	cuboctahedron
3	3 : 6 : 3	anticuboctahedron
4	2 : 6 : 2 : (2)	

Table 16 Favorable Polyhedra for $n = 13$

Order of Favorability	Description: Föppl Notation
1	1 : 2 : (2) : 2 : (2) : 2 : (2)
2	1 : 3 : (3) : 3 : (3)
3	1 : 5 : 6 : 1
4	3 : (3) : 3 : 4
5	1 : 4 : (4) : 4
6	3 : 6 : 4
7	1 : 5 : (5) : 2

Table 17 Favorable Polyhedra for $n = 14$

Order of Favorability	Description: Föppl Notation
1	1 : 6 : (6) : 1
2	1 : 4 : (4) : 4 : 1
3	1 : 5 : (5) : 3
4	4 : 6 : 4
5	1 : 5 : 6 : 2
6	3 : (3) : 3 : 5

Table 18 Favorable Polyhedra for $n = 15$

Order of Favorability	Description: Föppl Notation
1	3 : (3) : 3 : (3) : 3
2	1 : 6 : (6) : 2
3	1 : 5 : 3 : (3) : 3
4	1 : 5 : 6 : 3
5	2 : (2) : 7 : 2 : (2)
6	1 : 5 : (5) : 2 : (2)

from the most favorable O_h octahedral arrangement to the less favorable hexagon is observed as the size of M_m increases (i.e., for $m = 1$ or 3).

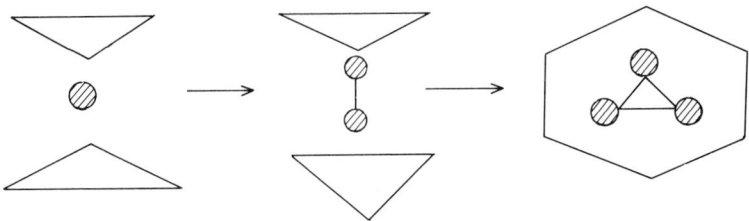

$n = 8$. Common eight-vertex polyhedra are the dodecahedron, the bicapped trigonal prism, the square antiprism, and the square prism (cube). No $M(CO)_8$ species is known. Examples of the $M_2(CO)_8$ class are $Co_2(CO)_8$, $[FeCo(CO)_8]^-$, and $[Fe_2(CO)_8]^{2-}$. Packing two cobalt atoms inside the square antiprism or square prism would lead to "long" Co–Co distances [~3.0 Å; see $Mn_2(CO)_{10}$, below]. The most favorable arrangement of eight CO groups that would accommodate two Co atoms and allow a reasonable Co–Co distance (~2.5 Å) and sensible Co–CO bond lengths is the bicapped trigonal prism. Allowing the two Co atoms to occupy positions in the triangulated faces of the trigonal prism leads to a Co–Co distance of 2.50 Å and two CO bridges. As the size of the M_2 unit is increased in going from Co_2 to $CoFe^-$ to Fe_2^{2-}, the carbonyl polyhedron adjusts from the bicapped trigonal prism to (eventually) a square prism. Consequently the two bridges observed with $Co_2(CO)_8$ (Fig. 7) are not present in the $[Fe_2(CO)_8]^{2-}$ dianion; thus we have another example in which anionic charge does not cause CO-bridge formation.

$n = 9$. The most favorable nine-vertex polyhedron is the tricapped trigonal prism. Other less favorable forms are listed in Table 13. This is the arrangement

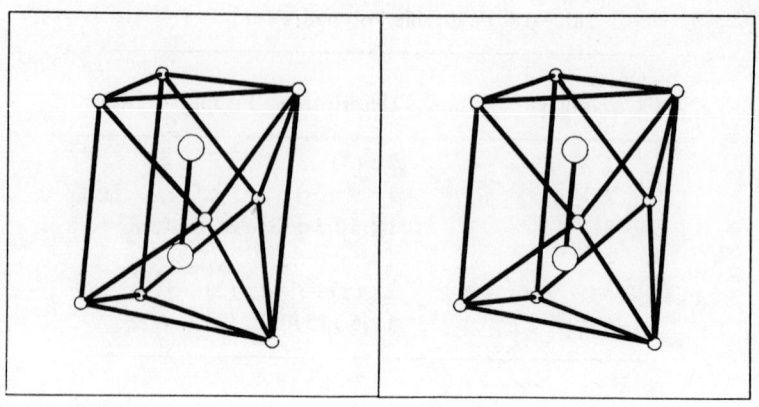
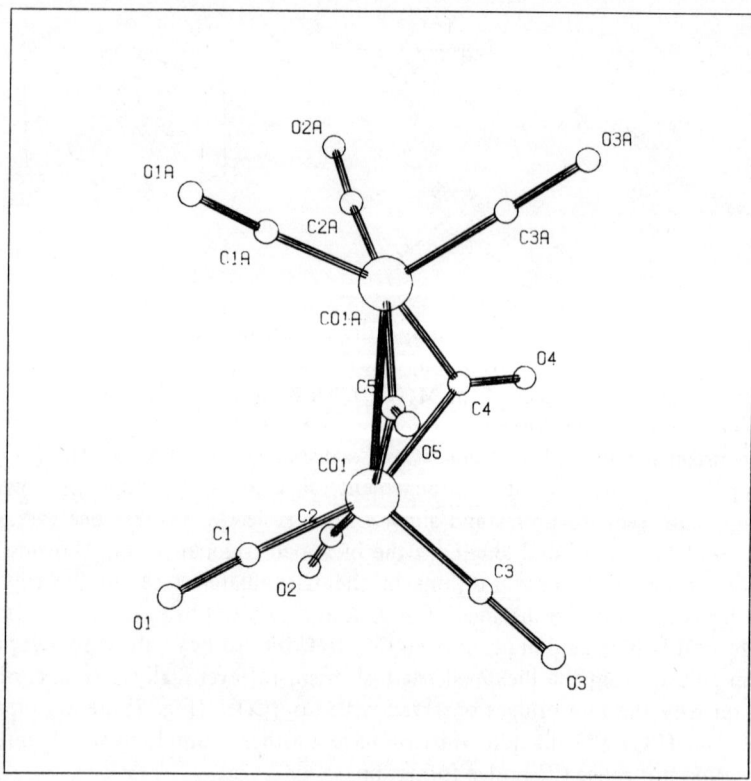

Fig. 7. Molecular structure of $Co_2(CO)_8$.

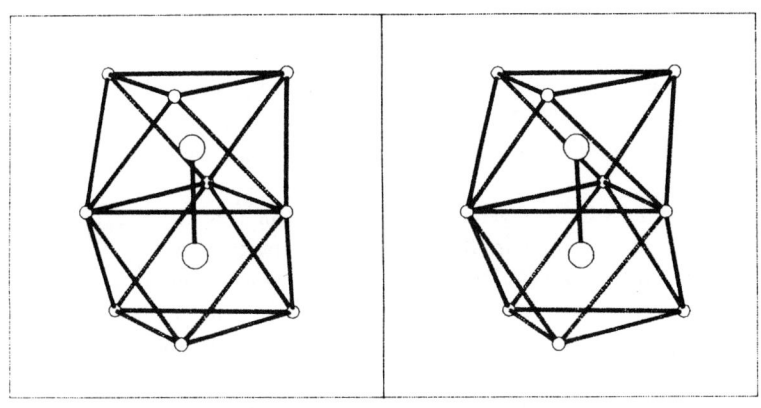

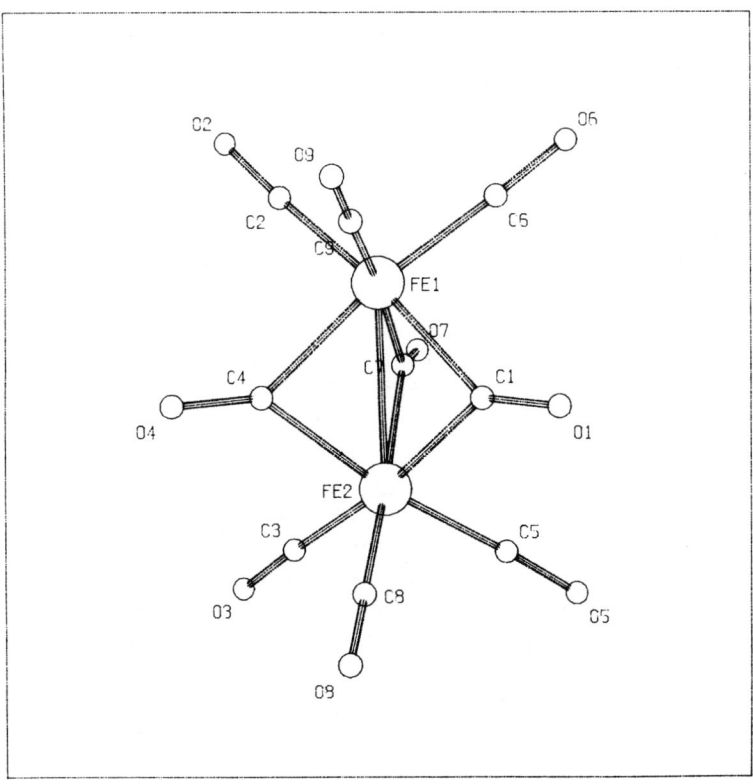

Fig. 8. Molecular structure of $Fe_2(CO)_9$.

found in $Fe_2(CO)_9$ (Fig. 8) (8). Placing the Fe_2 unit within this nine-vertex polyhedron such that it lies along the C_3 axis leads to the known structure and an Fe—Fe distance of 2.50 Å. It is no accident that the M—M distance is close to that found for Co—Co in $Co_2(CO)_8$; both compounds are based on the same fundamental CO polyhedron, namely, the trigonal prism.

$n = 10$. The CO arrangements in $Mn_2(CO)_{10}$ (Fig. 9), $Tc_2(CO)_{10}$, and $Re_2(CO)_{10}$ (9) correspond to the "best" 10-vertex polyhedron, the bicapped square antiprism (Table 14). The "long" M—M distances may be reconciled with the large polyhedral hole within the square antiprism; that is, it is the result of CO—CO interactions. The absence of CO bridges in these molecules is a direct consequence of the CO polyhedron.

$n = 12$ (Table 15 and Figs. 10–12). This problem has been discussed in detail for $M_m(CO)_{12}$ species. One additional molecule is worthy of consideration, namely, $V_2(CO)_{12}$. The existence of this dimer is in dispute. If a V_2 unit is placed within an icosahedron of CO groups, no CO bridges are expected and an exceptionally long V—V bond (~3.50 Å) is necessary.

$n = 13$. The seven most favorable polyhedra for $n = 13$ are listed in Table 16. Two, the edge-bridged icosahedron and the face-capped icosahedron, are found to be the most favorable forms, there being little energy difference between them. These are depicted in Figures 13 and 14. Closely related are the 1 : 5 : 6 : 1, the 3 : (3) : 3 : 4, and the 1 : 4 : (4) : 4 systems. A well-established binary carbonyl with 13 CO groups is the anion $Fe_4(CO)_{13}^{2-}$ (Fig. 15). A tetrahedron of iron atoms is enveloped by 13 carbonyls to give a structure with nine terminal CO ligands: three asymmetric edge CO bridges, and one three-center CO bridge; this is shown in Figure 15. The polyhedron defined by the oxygen atoms in this structure is a face-capped icosahedron, with the face cap corresponding to the face-center CO bridge lying beneath the base of the Fe_4 tetrahedron. Quantitatively, the ligand repulsion energy approximates well to the ideal value.

It should be noted that the CO polyhedron matches the symmetry of the Fe_4 unit; the alternative 1 : 2 : (2) : 2 : (2) : 2 :(2), edge-bridged icosahedron would match less well. A further consequence of packing an M_4 tetrahedron within the capped icosahedron is that the resulting structure will be closely similar to that derived by placing an M_4 tetrahedron within an icosahedron. Thus $Fe_4(CO)_{13}^{2-}$ and $Co_4(CO)_{12}$ have closely related structures, the only major difference being the additional three-center CO bridge in the iron system. The recently chacterized $[CoRu_3(CO)_{13}]^-$ possesses the same 1 : 3 : (3) : 3 : (3) ligand polyhedron (54). However, the metal atom tetrahedron is oriented differently, so that the unique "face-capping" carbonyl lies over its apex, the cobalt atom. Thus a different overall pattern of terminal and bridging carbonyls

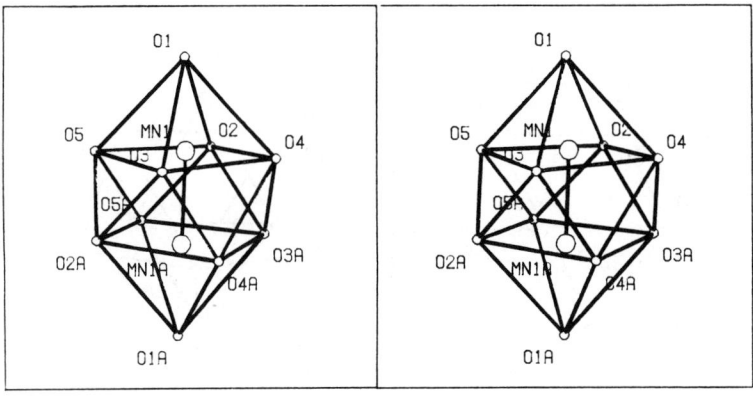
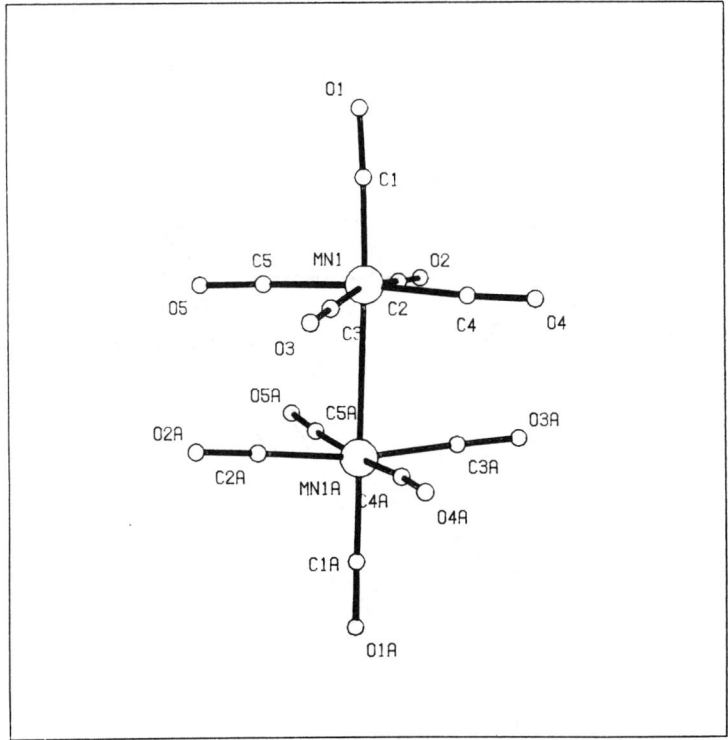

Fig. 9. Molecular structure of $Mn_2(CO)_{10}$.

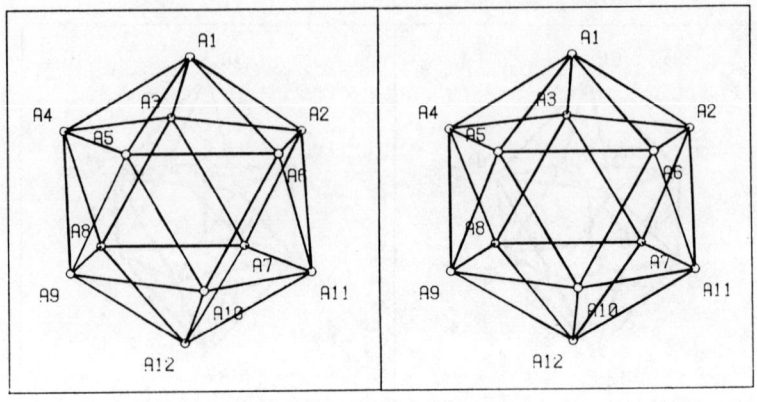
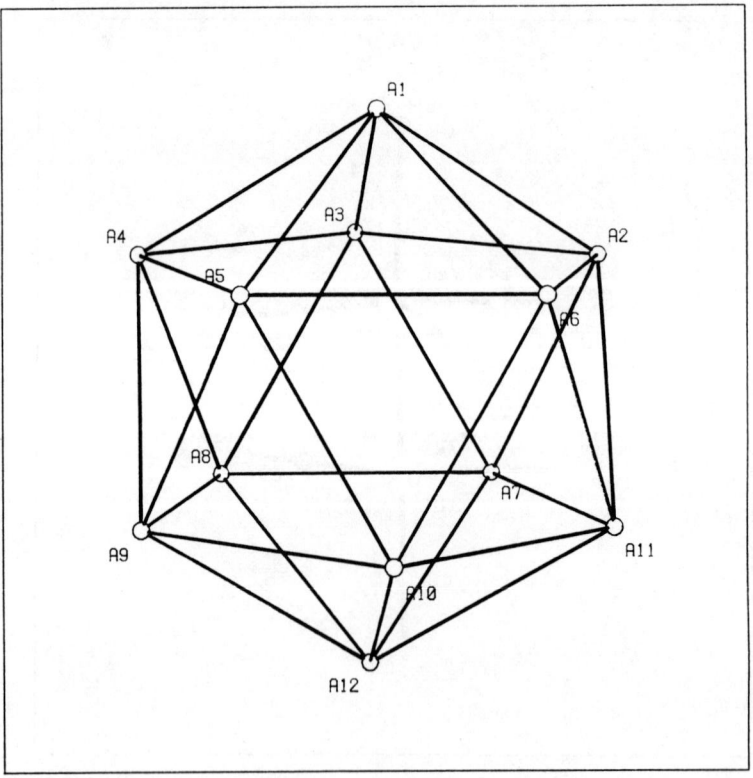

Fig. 10. Icosahedron.

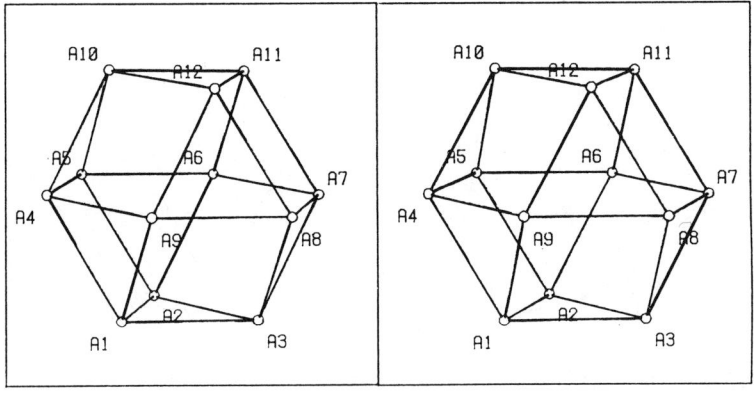

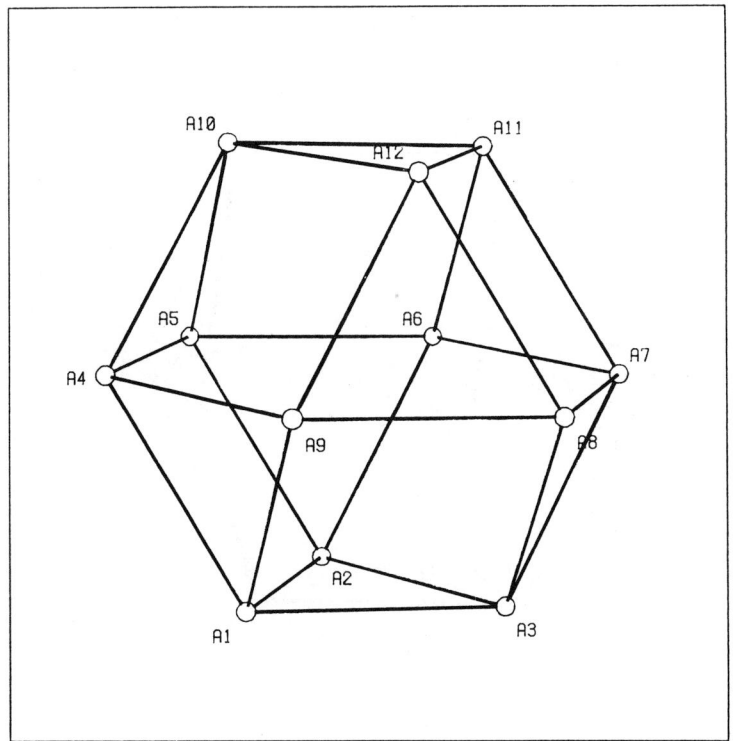

Fig. 11. Cuboctahedron.

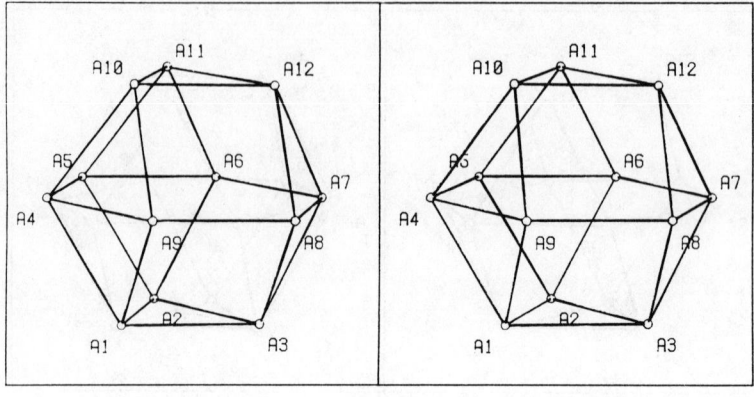

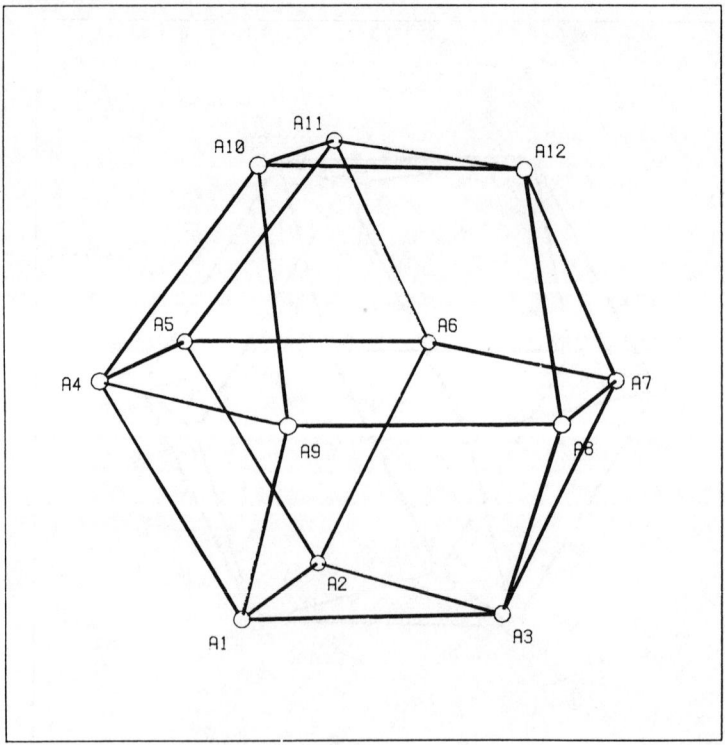

Fig. 12. Anticuboctahedron.

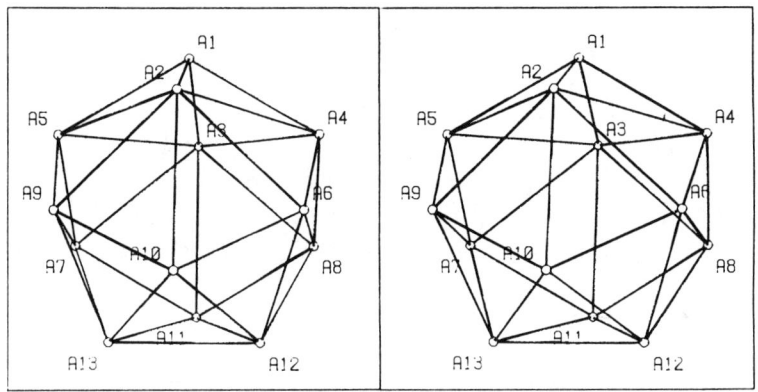
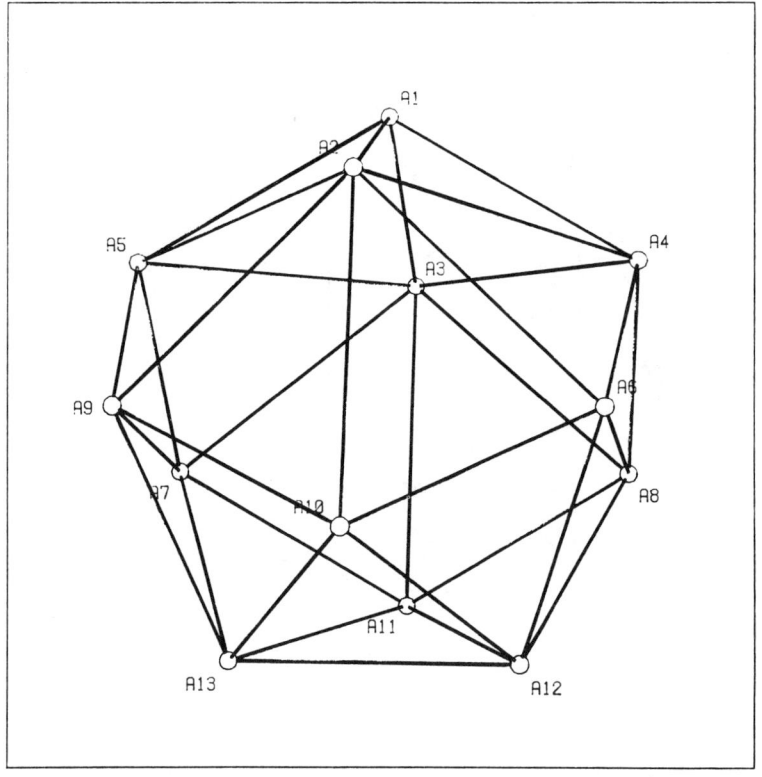

Fig. 13. Best 13-hedron 1 : 2 : (2) : 2 : (2) : 2 : (2).

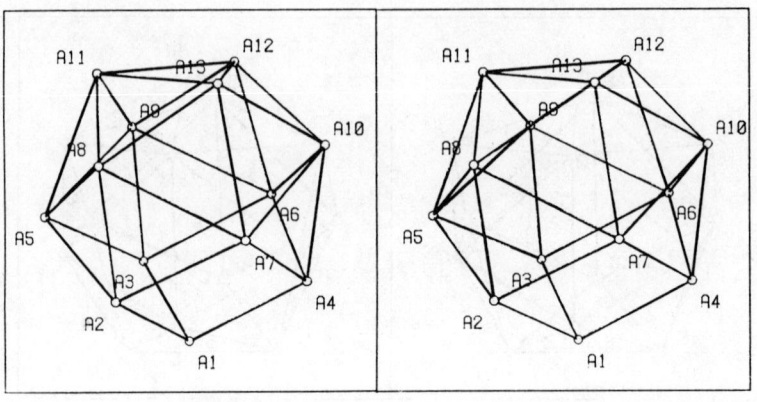
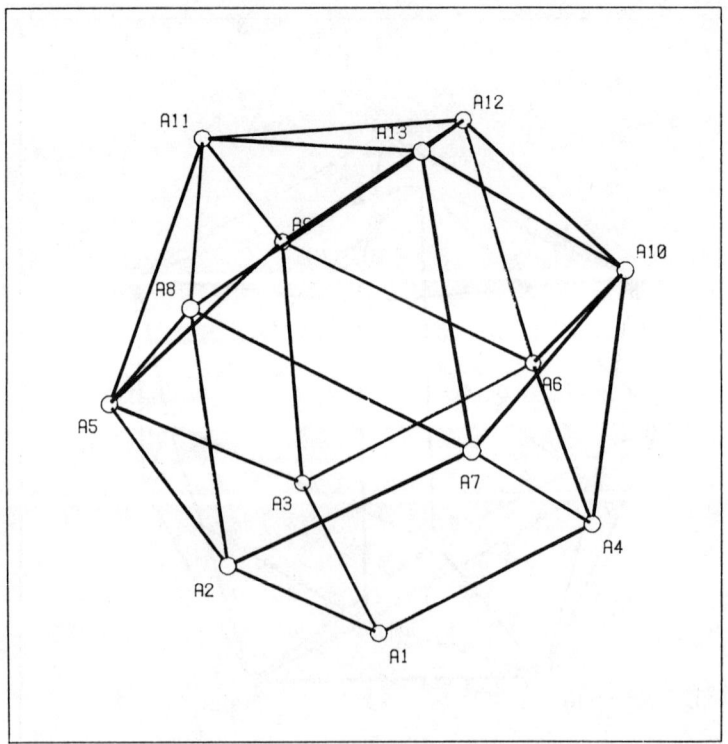

Fig. 14. Second-best 13-hedron 1 : 3 : (3) : 3 : (3).

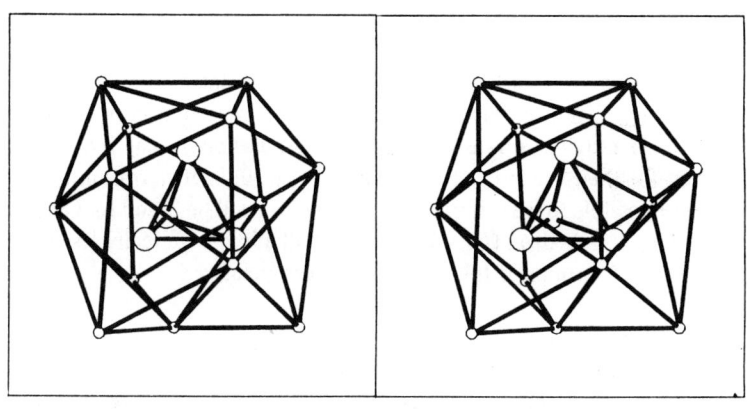

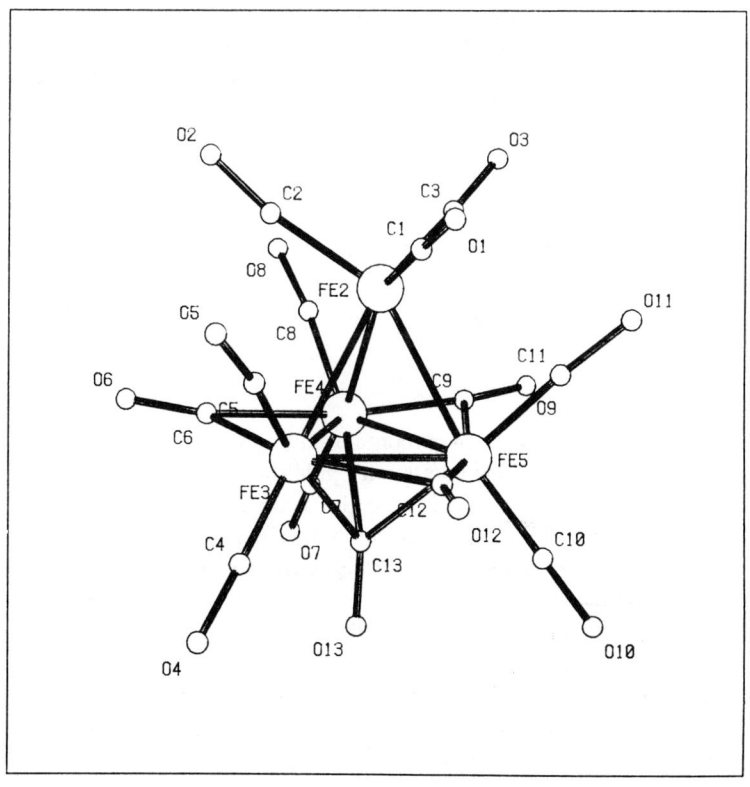

Fig. 15. Molecular structure of $[Fe_4(CO)_{13}]^{2-}$.

is generated. This may arise from the difference in size between the $CoRu_3^-$ and Fe_4^{2-} units, or from site preferences of individual metal atoms within the carbonyl polyhedron. In $[CoRu_3(CO)_{13}]^-$ the exact relative orientations of the metal atom and ligand polyhedra generate a chiral crystal structure. The possibility of optical activity in transition metal cluster carbonyls may have important implications for their use in catalysis involving optically active organic molecules.

It is quite possible that a different CO polyhedron will be found for $Os_4(CO)_{13}^{2-}$. The larger Os_4 tetrahedron (Os–Os ~ 2.9 Å) will allow a sterically less favorable distribution of CO groups, if this is favored electronically.

n = 14. The six best polyhedra for $n = 14$ are presented in Table 17.

The most favorable polyhedron corresponds to a bicapped hexagonal antiprismatic [1 : 6 : (6) : 1] arrangement with D_{6d} symmetry (Fig. 16). Slightly less favorable is the omnicapped cube with O_h symmetry (Fig. 17). Clearly for an O_h octahedral arrangement of metals the latter $(CO)_n$ polyhedron is preferred. Placement of an octahedral M_6 unit with the omnicapped cube produces a species $M_6(CO)_{14}$, which will have six terminal M—CO bonds and *eight* face-centered CO bridges (**10**).

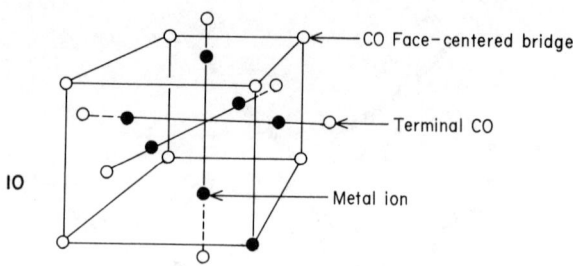

Two cluster carbonyls with 14 carbonyl ligands have been crystallographically characterized and their atomic coordinates published. These are $[Co_6(CO)_{14}]^{4-}$ (41) and $[Co_4Ni_2(CO)_{14}]^{2-}$. Both have the same structure in which an octahedron of metal atoms is bound to *six* terminal CO groups (one per metal) and eight CO face-centered bridges; the carbonyl envelope corresponds to an omnicapped cube (Fig. 18).

Inserting an M_6 octahedral unit within the bicapped trigonal antiprism, 1 : 6 : (6) : 1, would yield a structure with *six* terminal, six-edge bridges and two face-bridging carbonyls. The M_6 unit would also show strong trigonal distortion. No binary carbonyl with this structure has been observed. Real values of $\Sigma_{i \neq j} \Omega_{ij}^{-6}$ for the oxygen atom polyhedra found in $[Co_6(CO)_{14}]^{4-}$ and $[Co_4Ni_2(CO)_{14}]^{2-}$ show that they are almost exact approximations of the ideal form.

n = 15. The most favorable 15-vertex polyhedron is the 3 : (3) : 3 : (3) : 3 arrangement (Table 18, Fig. 19).

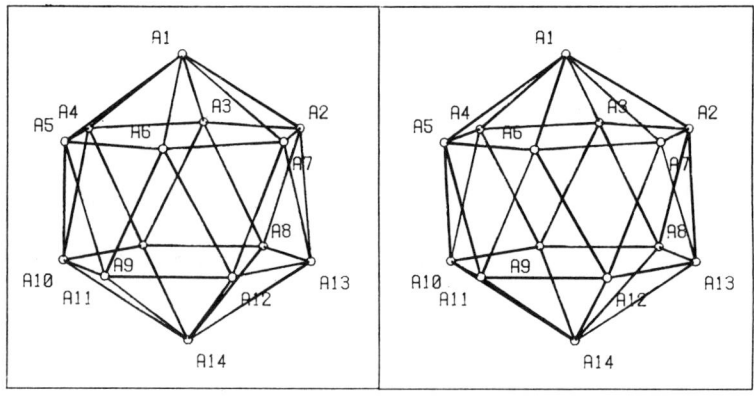

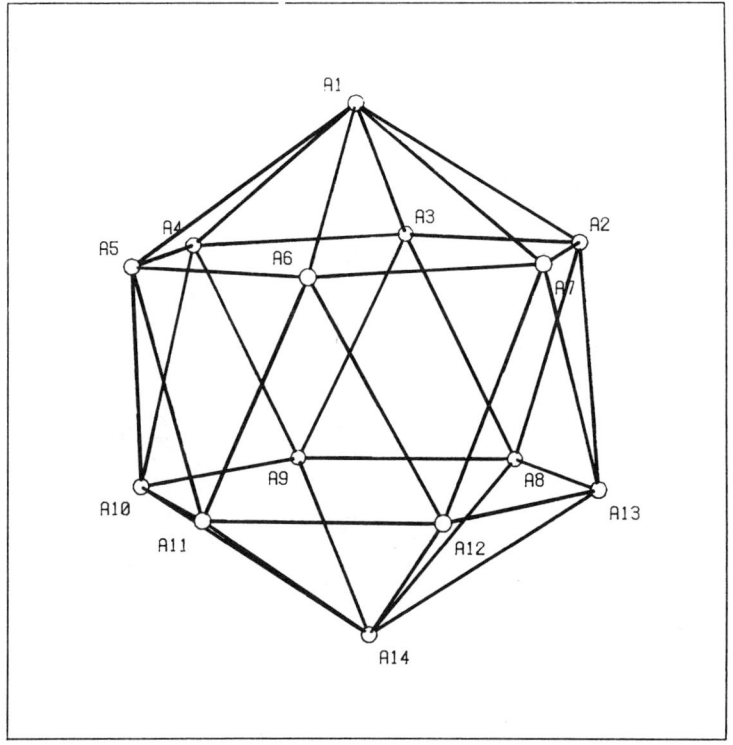

Fig. 16. Best 14-hedron 1 : 6 : (6) : 1.

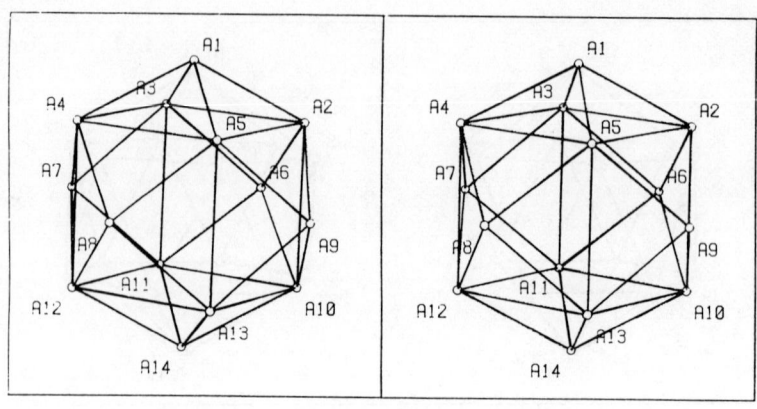
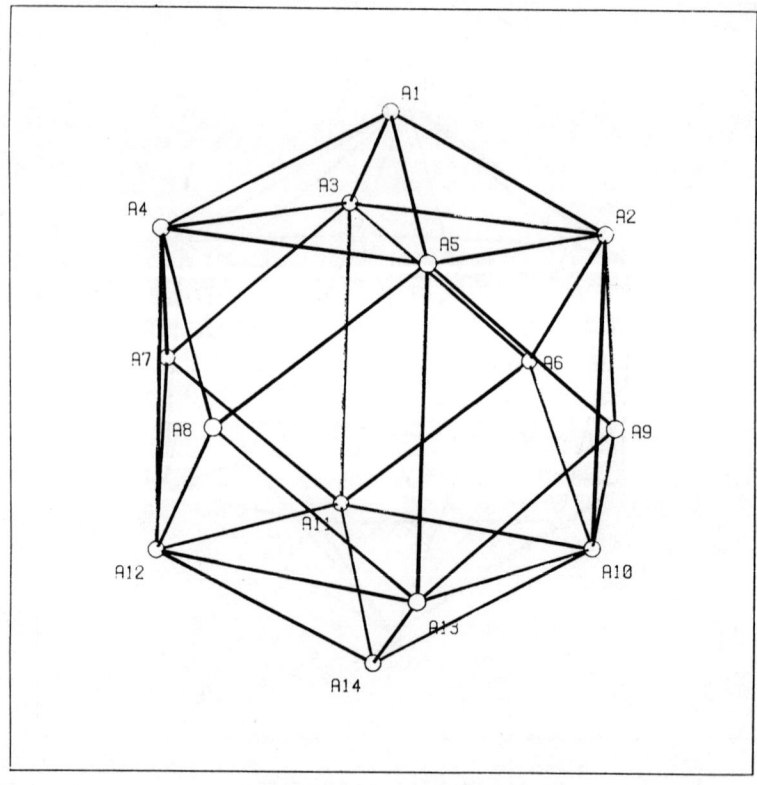

Fig. 17. Second-best 14-hedron 1 : 4 : (4) : 4 : 1.

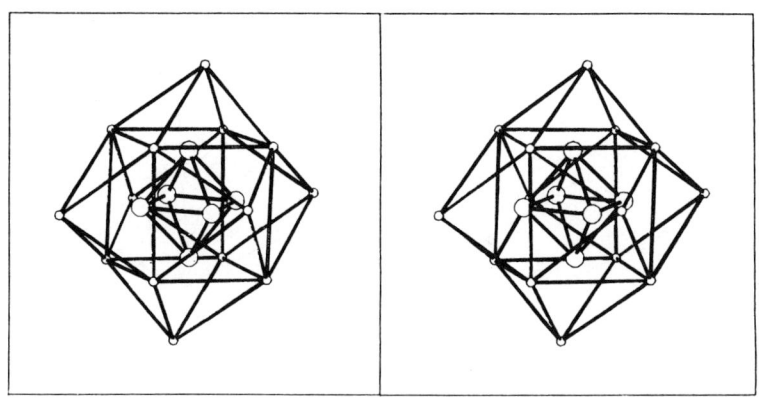

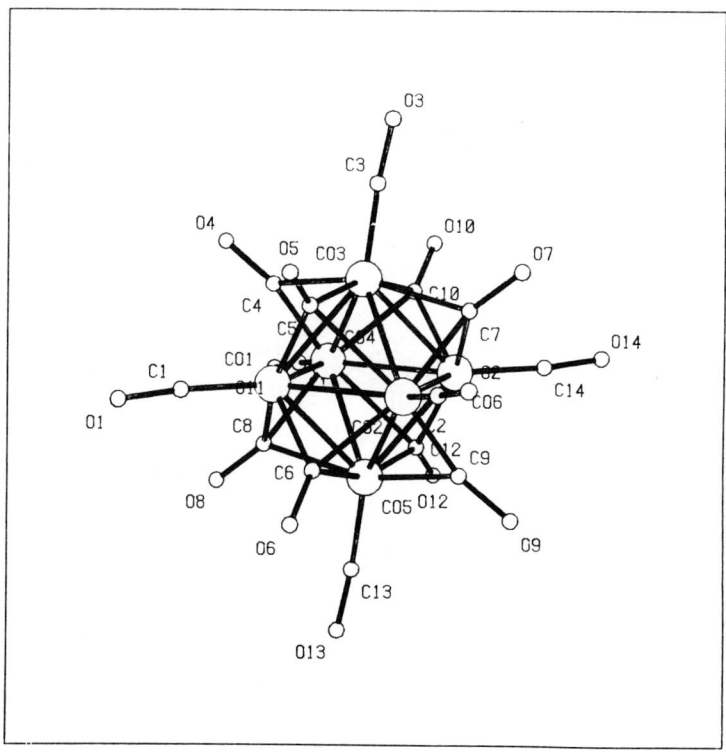

Fig. 18. Molecular structure of $[Co_6(CO)_{14}]^{4-}$.

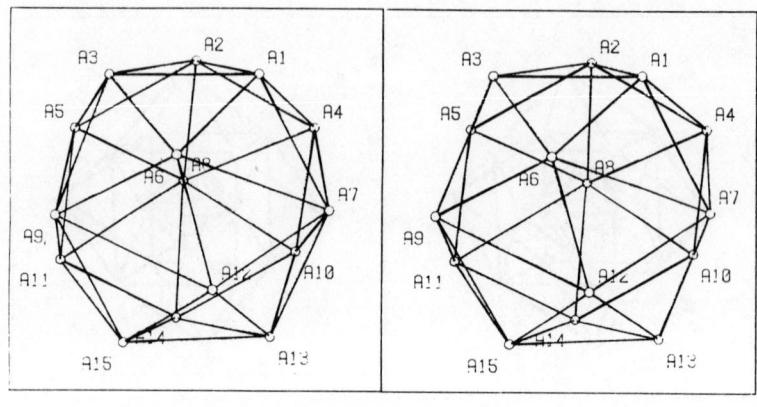
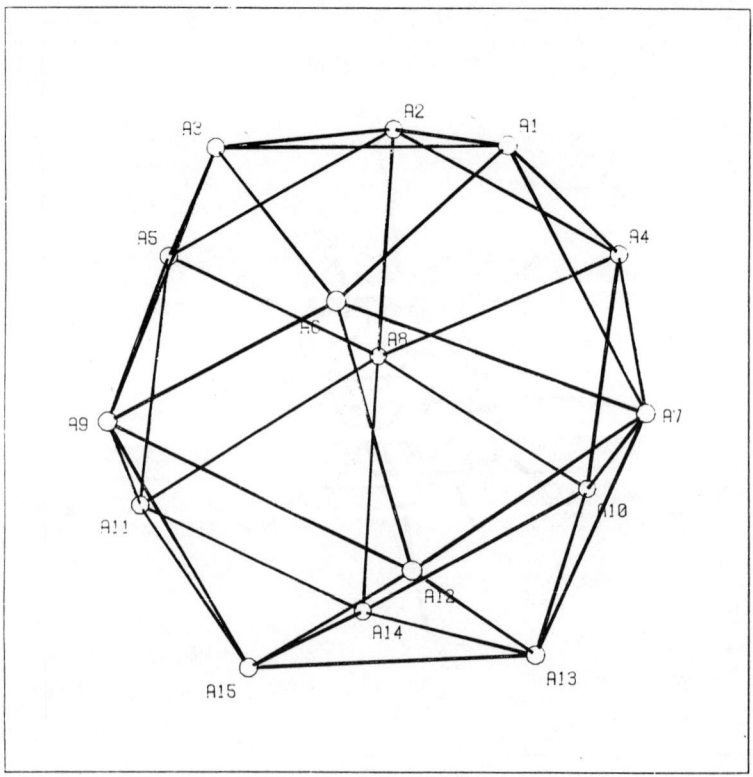

Fig. 19. Best 15-hedron 3 : (3) : 3 : (3) : 3.

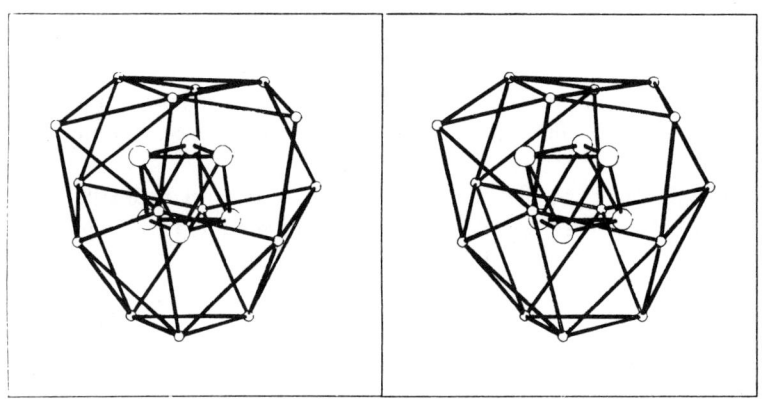
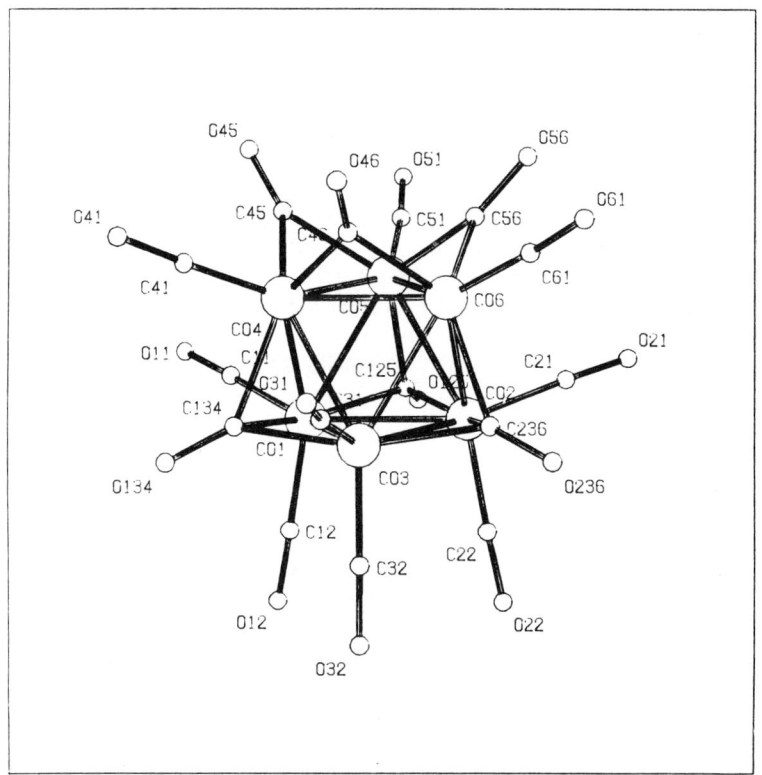

Fig. 20. Molecular structure of $[Co_6(CO)_{15}]^{2-}$.

Analysis of the oxygen atom coordinates for $[Co_6(CO)_{15}]^{2-}$ (Fig. 20) indicates that the 15 carbonyl groups adopt this arrangement with a high real $\Sigma_{i \neq j} \Omega_{ij}^{-6}$ value (42). Insertion of an M_6-O_h octahedron in the 3 : (3) : 3 : (3) : 3 polyhedron such that the C_3 axes of the polyhedrons coincide gives the established structure with three face bridges, three edge bridges, and nine terminal groups.

In $[Rh_6(CO)_{15}C]^{2-}$ the Rh_6 unit adopts a trigonal prismatic geometry (Fig. 21) (55). (See Sec. II-C-4). The insertion of the Rh_6 trigonal prism within a basically 3 : (3) : 3 : (3) : 3 ligand polyhedron gives a structure with nine terminal CO groups and nine edge bridges, although the ligand polyhedron in this highly aspherical cluster is quantitatively distorted.

It is interesting to explore the relationship between the Co_6 and Rh_6 species. If we fix the $(CO)_{15}$ polyhedron and allow one metal Δ face to rotate by 60° (i.e., converting the octahedron into a trigonal prism), it follows that the structure moves away from the three-edge-bridged, three-face-centered bridge system to the nine-edge-bridged configuration. The number and type of CO ligands in $[Co_6(CO)_{15}]^{2-}$ and $[Rh_6(CO)_{15}C]^{2-}$ reflect the consequence of placing either an octahedral or a trigonal prismatic M_6 unit within the same 15-vertex polyhedron.

The mixed cluster $[Rh_5Pt(CO)_{15}]^-$ adopts a ligand polyhedron resembling the next most favorable 1 : 6 : (6) : 2 arrangement (Fig. 22) (56). If the quasioctahedral Rh_5Pt unit is placed in this polyhedron such that the Pt atom is in association with *one* terminal CO group (because of its electronic requirements, see Sect. II), a structure with 11 terminal and 4 three-centered carbonyls is found. The Co_6^{2-} and Rh_5Pt^- metal clusters are both nearly spherical, but the latter is rather larger.

As noted above ($n = 6, 8$, or 12) as the size of the M_m unit increases, a progression from one $(CO)_n$ polyhedron to another is observed.

$[HCo_6(CO)_{15}]^-$, the subject of a recent neutron diffraction study, possesses a carbonyl polyhedron not recognizable as one of the standard polyhedral forms, but quantitatively the inter-ligand repulsions are almost the optimum value for a 15-carbonyl species (57).

n = 16. The three most favorable arrangements within this class are 1 : 3 : (3) : 3 : (3) : (3), 4 : (4) : 4 : (4) (Fig. 23) and 1 : 6 : 3 : (3) : 3 (Fig. 24), which all have very similar energies (Table 19).

The 1 : 6 : 3 : (3) : 3 arrangement, the tetracapped, truncated tetrahedron, is well known in metallurgy and mineral chemistry. Truncation of the tetrahedron by removal of its vertices results in a 12-vertex polyhedron with four hexagonal faces: capping these four faces produces the 1 : 6 : 3 : (3) : 3 polyhedron. Allowing for the hexagonal layer to pucker slightly produces the 1 : 3 : (3) : 3 : (3) : 3 form, of marginally lower energy. These four capping vertices are

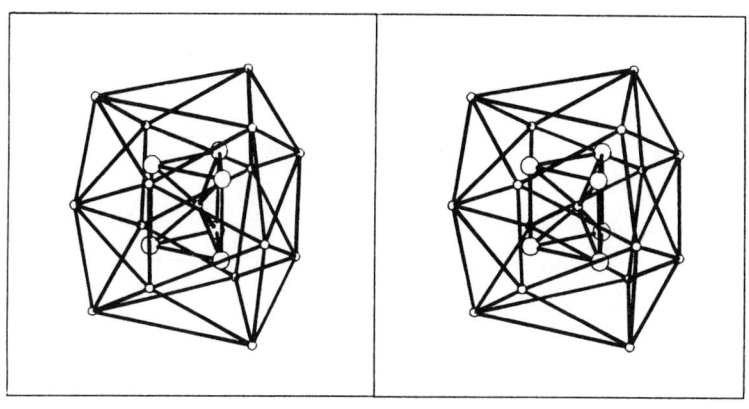

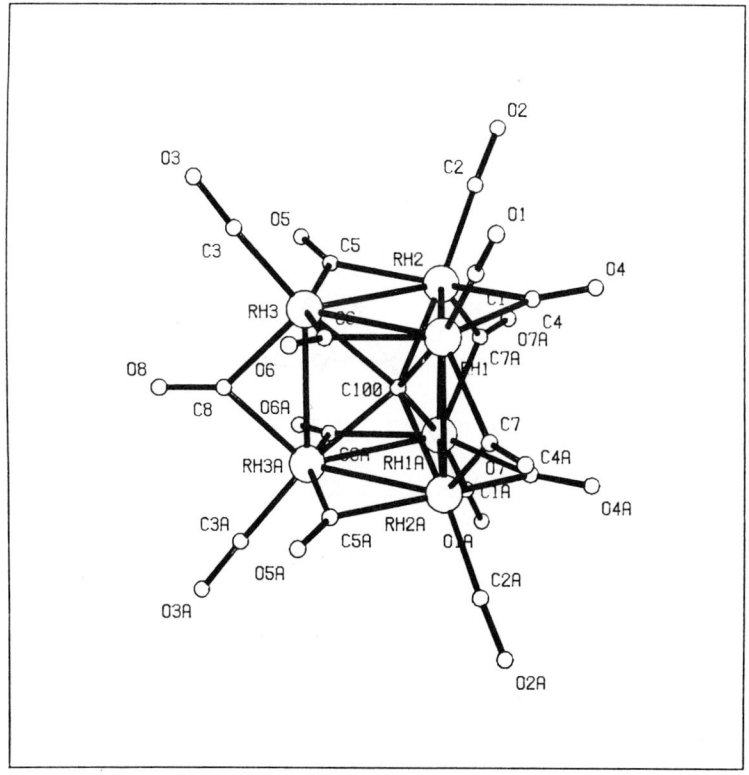

Fig. 21. Molecular structure of $[Rh_6C(CO)_{15}]^{2-}$.

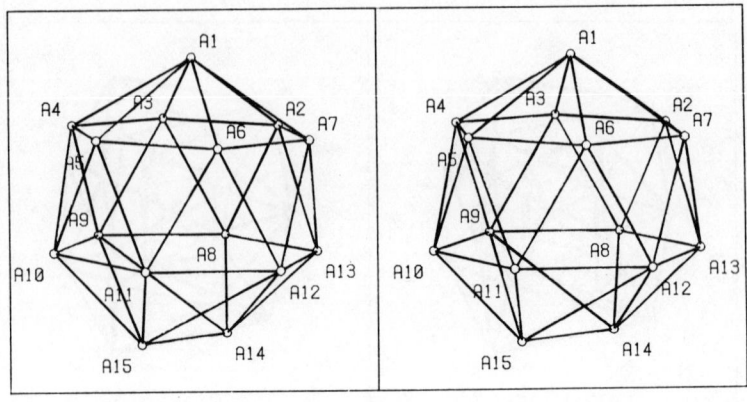

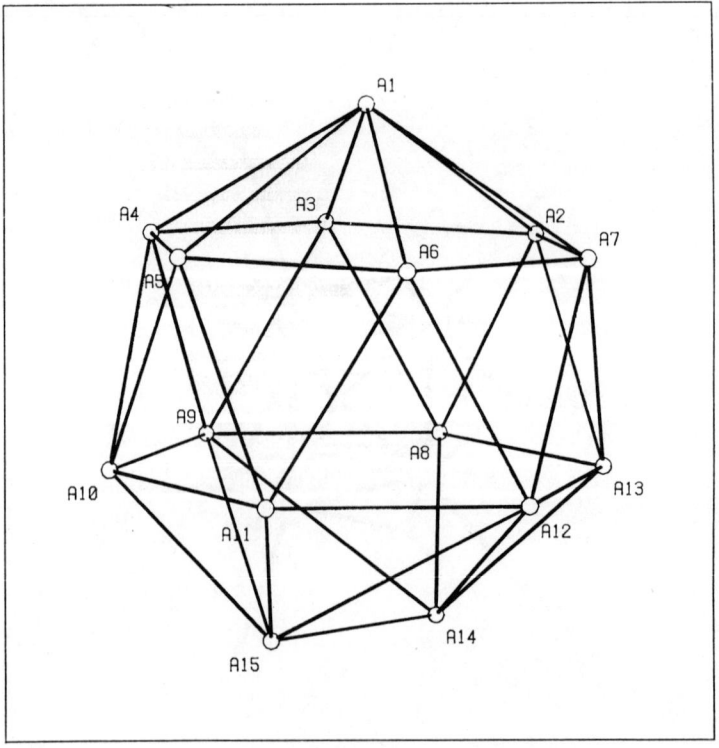

Fig. 22. Second-best 15-hedron 1 : 6 : (6) : 2.

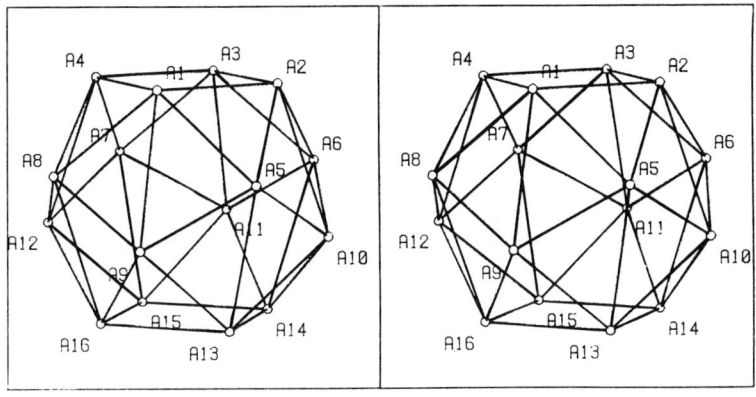

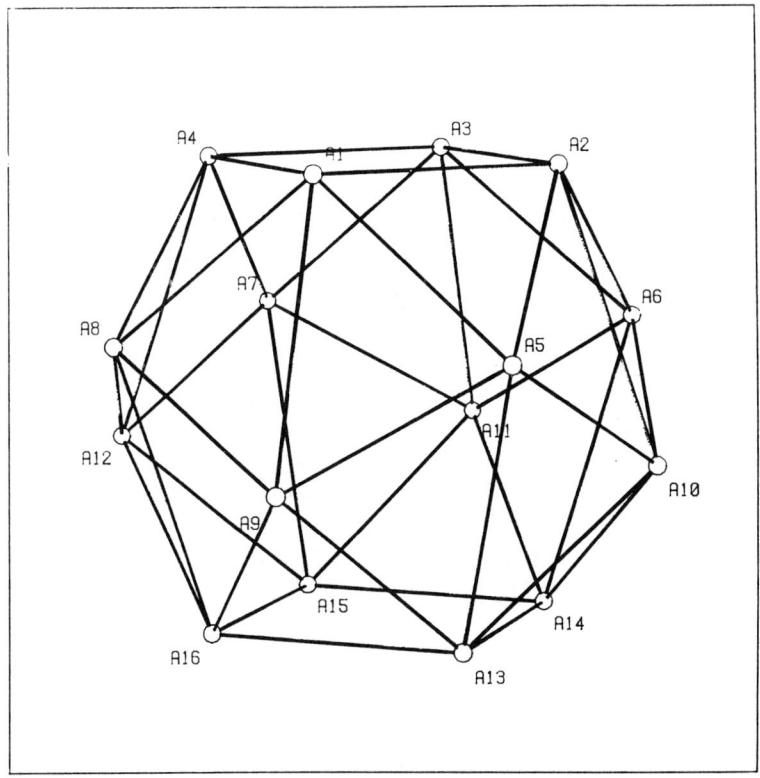

Fig. 23. Best 16-hedron 4 : (4) : 4 : (4).

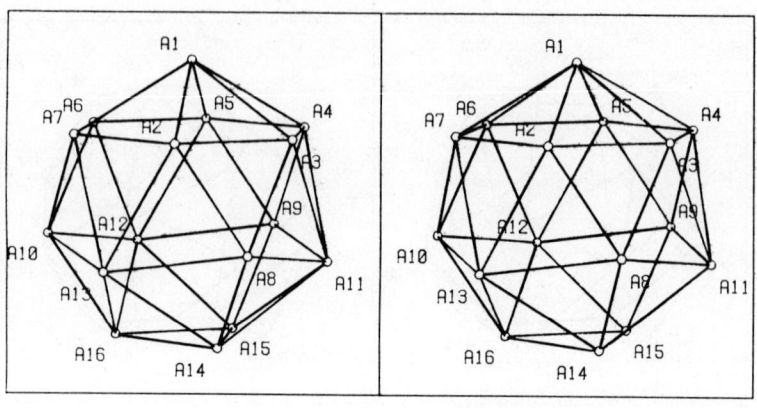
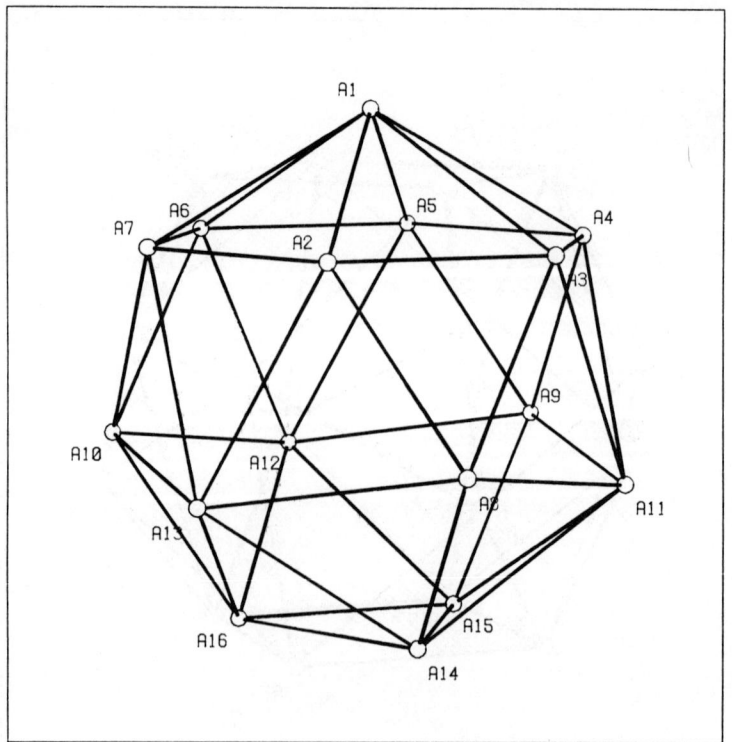

Fig. 24. Second-best 16-hedron 1 : 6 : 3 : (3) : 3.

326

Table 19 Favorable Polyhedra for $n = 16$

Order of Favorability	Description: Föppl Notation
1	1 : 3 : (3) : 3 : (3) : 3
2	4 : (4) : 4 : (4)
3	1 : 6 : 3 : (3) : 3
4	1 : 3 : (3) : 6 : 3
5	1 : 5 : 7 : 3
6	1 : 5 : (5) : 5

distinguished by a connectivity of 6 rather than 5, as shown by all other vertices. Several clusters with 16 carbonyl ligands have been structurally characterized. In $Rh_6(CO)_{16}$ (Fig. 25) (21), $[Fe_6C(CO)_{16}]^{2-}$ (Fig. 26) (58), and $Os_5(CO)_{16}$ (Fig. 27) (19) the CO polyhedron corresponds to this 1 : 6 : 3 : (3) : 3 arrangement.

On symmetry grounds the most reasonable orientation of an octahedral M_6 unit within the tetracapped truncated tetrahedron is one with an M atom lying beneath the center of each tetrahedron edge. This gives rise to six $M(CO)_2$ units; the four capping CO groups, which sit above the four hexagonal faces of the truncated tetrahedron, then correspond to face-centered bridges. This configuration corresponds to the structure of $Rh_6(CO)_{16}$ (Fig. 28).

In this structure each Rh atom occupies a "butterfly" interstitial site. For smaller M atoms such a structure is less desirable; long M—M distances are required. In $[Fe_6C(CO)_{16}]^{2-}$ the Fe atoms migrate away from these butterfly sites towards the smaller triangulated sites. This generates three CO edge bridges.

The pentanuclear osmium cluster $Os_5(CO)_{16}$ has a trigonal bipyramidal Os_5 unit (see Sec. II-C-3). Its structure, which is difficult to understand in terms of orthodox electron counting (18-electron rule), is geometrically a consequence of accommodating a trigonal bipyramid within the 1 : 6 : 3 : (3) : 3 polyhedron.

The real 1 : 6 : 3 : (3) : 3 polyhedra are quantitatively distorted (see below) but the $Os_5(CO)_{16}$ polyhedron is the most greatly distorted because of the asphericity of the metal cluster unit.

In addition, the crystal structures of the iodo derivatives $[Rh_6(CO)_{15}I]^-$ and $[Os_5(CO)_{15}I]^-$ are known. These structures are almost identical to those of the parent carbonyl molecules, one terminal CO being replaced by the iodide ligand. This has little effect on the rest of the ligand polyhedron.

The anion $[Re_4(CO)_{16}]^{2-}$ provides an example where the model of mutually repelling ligands on the surface of a sphere breaks down. This planar, rhombic metal atom cluster is extremely aspherical, and the need of the carbonyls to bind to the cluster unit with sensible M—C distances is more important than the optimization of non-bonding repulsions between ligands.

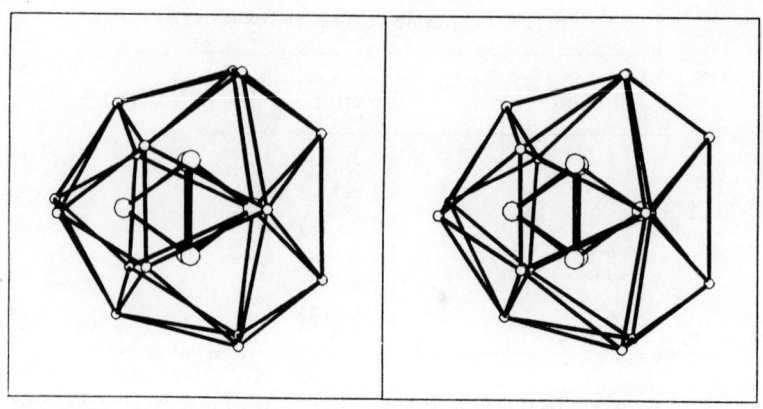

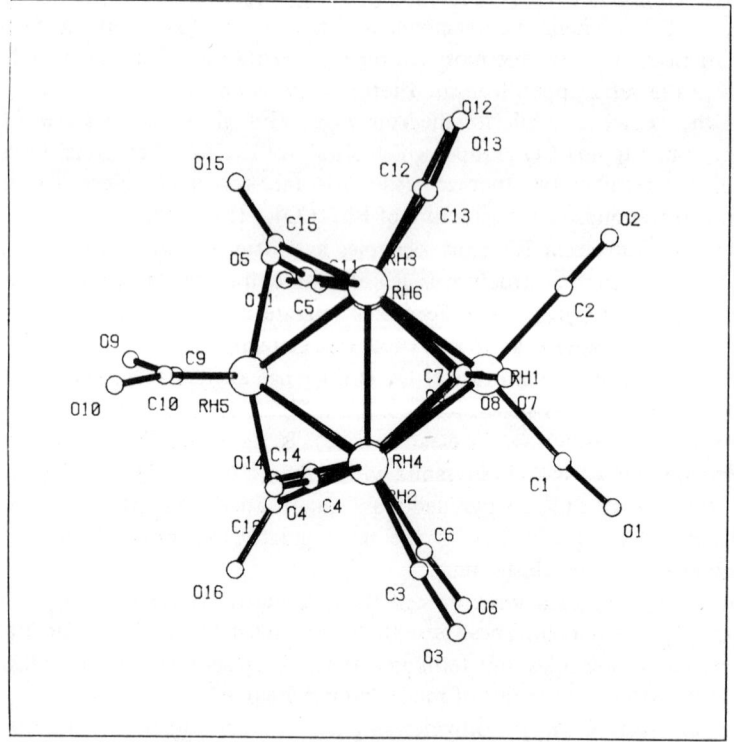

Fig. 25. Molecular structure of $Rh_6(CO)_{16}$.

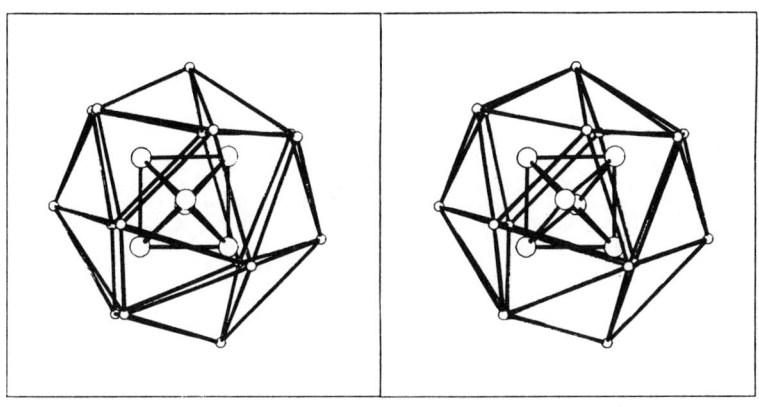

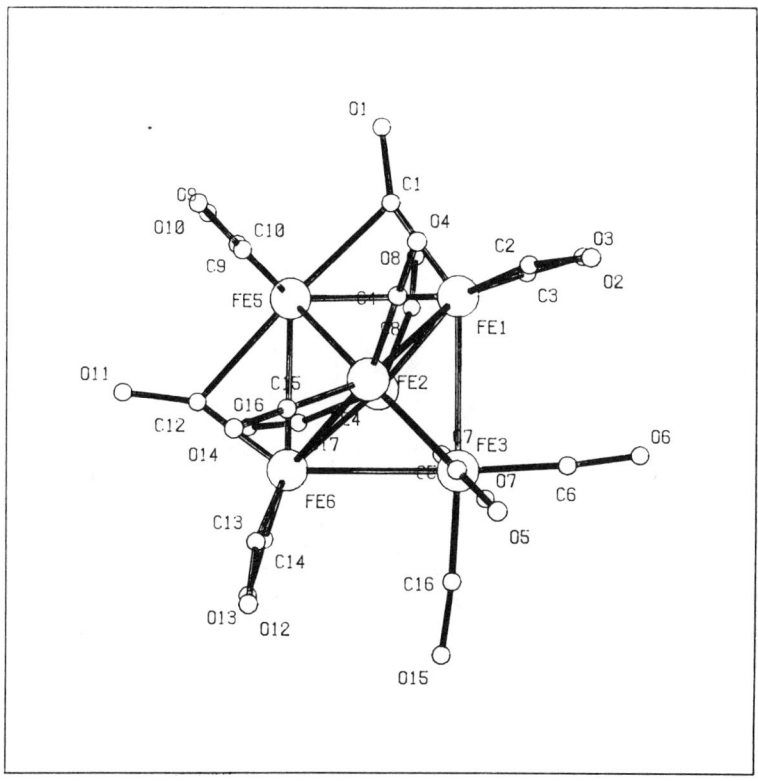

Fig. 26. Molecular structure of $[Fe_6C(CO)_{16}]^{2-}$.

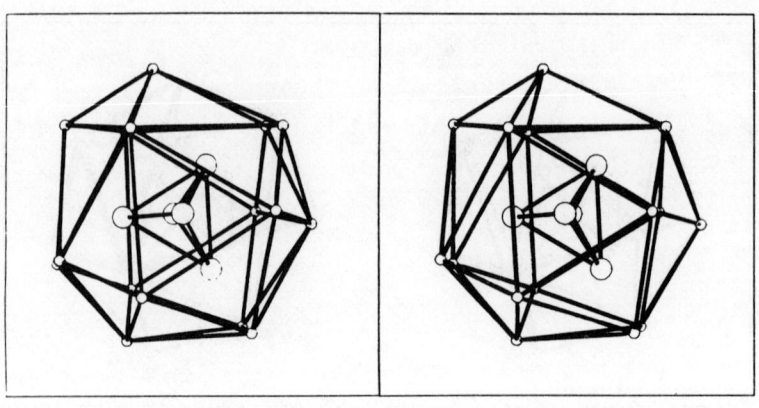

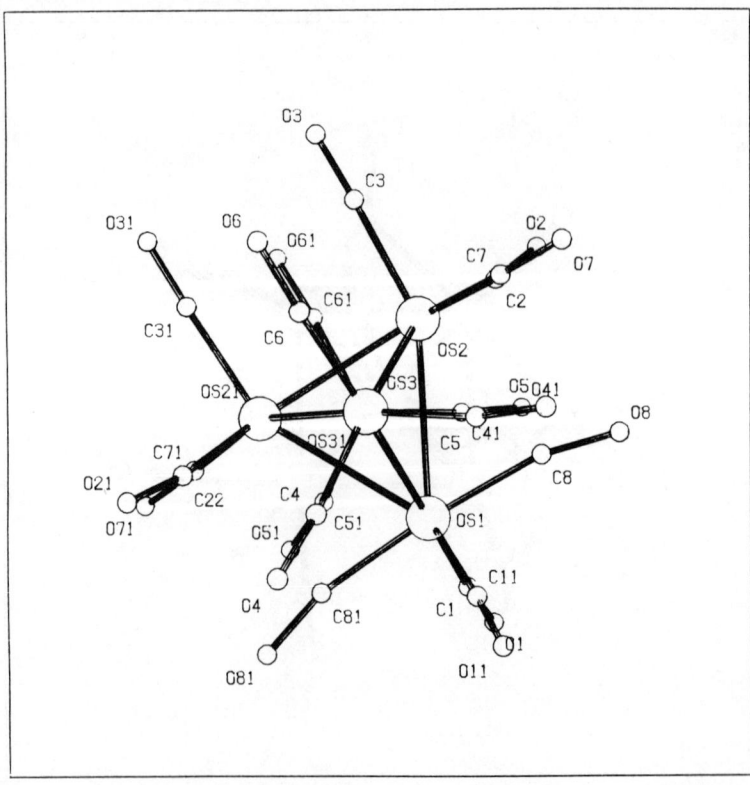

Fig. 27. Molecular structure of $Os_5(CO)_{16}$.

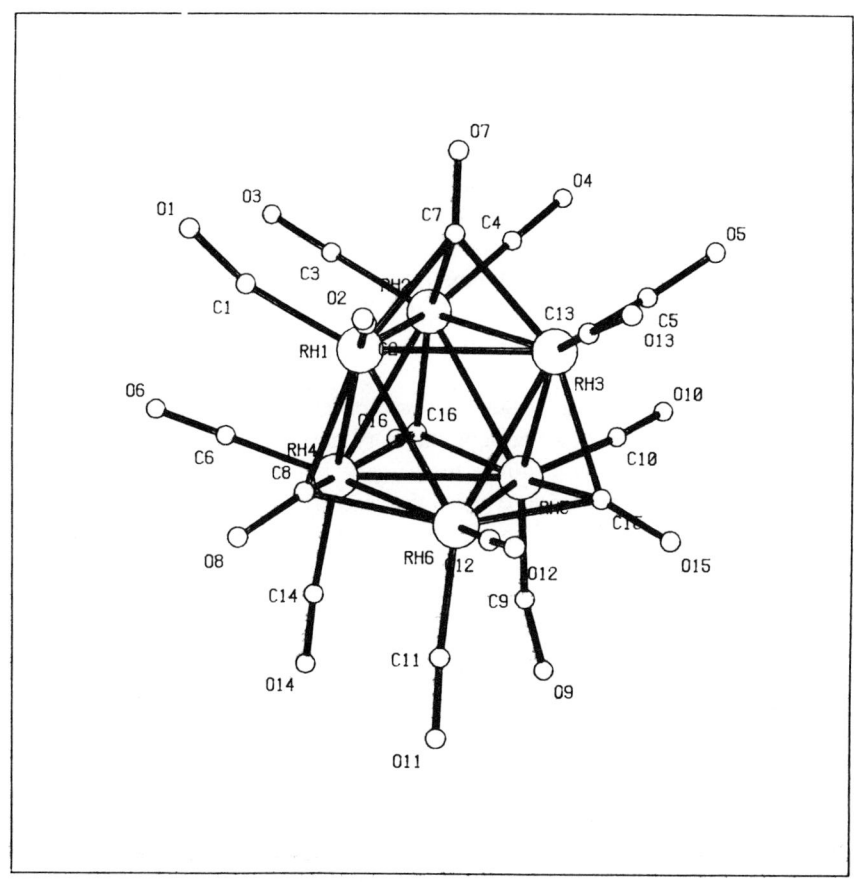

Fig. 28. Molecular structure of $Rh_6(CO)_{16}$.

C. Summary

We have shown that the numbers and distribution of μ_1, μ_2, and μ_3 carbonyl groups may be rationalized in terms of a simple model. This model postulates that the carbonyls are arranged in space in such a manner as to minimize their nonbonded repulsions. The ways in which this may be achieved are found by calculation of repulsion potentials between points of the surface of a sphere. This presupposes that the metal cluster enveloped by the ligand polyhedron is reasonably spherical, so that sensible M—CO bonding distances may be maintained. The success of the model for the range of metal geometries as diverse as the triangle, the tetrahedron, the trigonal bipyramid, and the octahedron is noteworthy. For less symmetric metal clusters the model also provides a clue to the mode of distortion of the ligand envelope, allowing sensible predictions of carbonyl structures to be made.

As mentioned above, the real values of $\Sigma_{i \neq j}\, \Omega^{-6}$ are, in general, higher than would be expected. An explanation must be sought if this result is to be reconciled with the proposition that the CO ligands adopt a configuration that minimizes nonbonded repulsions between them. First, the carbonyl polyhedra are distorted as a consequence of the cluster shape within them. This aspect has been discussed above, and only when there is a perfect symmetry match will the CO polyhedron achieve its most symmetric form. It must be remembered that in a sum of inverse sixth powers of angles, it will require only a few of the many interligand angles to deviate from their ideal values for a high value of $\Sigma_{i \neq j}\, \Omega^{-6}$ to be obtained.

More important, these molecules consist of a cluster of heavy transition metals enveloped by carbonyl groups; the very nature of the X-ray diffraction experiment means that the positions of the carbonyl ligands can be determined with only relatively low precision; hence the carbonyl atom coordinates in many clusters are not accurately known. The uncertainties in the oxygen coordinates will be greatly magnified (by a factor of $\sqrt{6}$) on calculation of $\Sigma_{i \neq j}\, \Omega^{-6}$. Bearing these factors in mind, the high values of $\Sigma_{i \neq j}\, \Omega_{ij}^{-6}$ in some real structures does not present too great a problem. It is significant that in molecules with a high degree of symmetry, such as $[Co_6(CO)_{14}]^{4-}$, agreement between real and ideal values is good.

Finally, the effect of crystal packing forces on the molecular structures should not be overlooked. These forces are difficult to quantify or to analyze qualitatively, but their magnitude can be quite large. In addition, the large, symmetric polyhedra defined by the CO groups have shapes inconsistent with the requirements of translational symmetry; to fill space efficiently within the crystal, therefore, they must become quite distorted.

CONCLUSION

Transition metal cluster carbonyls pose many interesting stereochemical problems both in their metal cluster geometries and in the distribution of the different modes of bonding of the CO ligands. Much progress has recently been made in understanding their structures, which have presented a challenge to generalised bonding theories in much the same way as did the boron hydrides. The steric importance of the carbonyl ligands in forming more or less close-packed arrays, and their mobility over the cluster surface in solution, have important implications for the reactivity of transition metal cluster carbonyls. It is their catalytic activity for the controlled reduction of carbon monoxide which has recently excited so much interest, particularly as hydrogen atoms may also be ligands in cluster carbonyls.

Index of Subjects

Metal cluster geometries	263
Metal-metal bonding in clusters	263
MO approaches	272
18-electron rule	264
Application of 18-electron rule	265
"Magic number" of electrons	266
Electron-rich clusters	266, 284
Electron-poor clusters	266
Skeletal Electron Pair Theory	268
Arachno- and nido-polyhedra	271
Capped polyhedra	271
Electron and Orbital Balance	277
Electron-precise clusters	273
$M(CO)_x$ fragments	270
Fragment Orbitals	274
d^{10} metal clusters	276
Mixed metal clusters	276
Electron-acceptor, electron-precise and electron-donor fragments	278
M_3 arrangements	278
M_4 arrangements	279
M_5 arrangements	282
M_6 arrangements	286
The CO Ligand envelope	290
CO-bonding modes	290
Ligand polyhedra $(CO)_n$	292, 302
Favorable polyhedra $n = 6 - 16$	301

REFERENCES

1. L. Mond, C. Langer, and F. Quincke, *J. Chem. Soc.*, **1890**, 749.
2. N. V. Sidgwick and R. W. Bailey, *Proc. R. Soc.*, *(London), Ser. A*, **144**, 521 (1934).
3. See, e.g., L. F. Dahl and R. E. Rundle, *J. Chem. Phys.*, **26**, 1751 (1957).
4. J. W. Lauher, *J. Am. Chem. Soc.*, **100**, 5305 (1978).
5. M. Elian and R. Hoffmann, *Inorg. Chem.*, **14**, 1058 (1975).
6. See, e.g., J. N. Murrell, S. F. A. Kettle, and J. M. Tedder, *Valence Theory*, Wiley, London, 1969.
7. G. G. Sumner, H. P. Klug, and L. E. Alexander, *Acta Crystallogr.*, **17**, 732 (1964).
8. F. A. Cotton and J. M. Troup, *J. Chem. Soc., Dalton Trans.*, **1974**, 800, and references therein.
9. L. F. Dahl and R. E. Rundle, *Acta Crystallog.*, **16**, 419. (1963).
10. C. H. Wei and L. F. Dahl, *J. Am. Chem. Soc.*, **91**, 1351 (1969).
11. M. R. Churchill and B. G. DeBoer, *Inorg. Chem.*, **16**, 878 (1977).

12. J. C. Calabrese, L. F. Dahl, P. Chini, G. Longoni, and S. Martinengo, *J. Am. Chem. Soc.*, **96**, 2614 (1974).
13. F. H. Carre, F. A. Cotton, and B. A. Frenz, *Inorg. Chem.*, **15**, 380 (1976).
14. V. G. Albano, G. Ciani, M. Freni, and P. Romiti, *J. Organomet. Chem.*, **96**, 259 (1975).
15. G. R. Wilkes, Ph.D. thesis, University of Wisconsin, Madison, 1965.
16. R. J. Doedens and L. F. Dahl, *J. Am. Chem. Soc.*, **88**, 4847 (1966).
17. M. R. Churchill and R. Bau, *Inorg. Chem.*, 7, 2606 (1968).
18. G. Longoni, P. Chini, L. D. Lower, and L. F. Dahl, *J. Am. Chem. Soc.*, 97, 5034 (1975).
19. B. E. Reichert and G. M. Sheldrick, *Acta Crystallogr.*, **B33**, 173 (1977).
20. P. F. Jackson, B. F. G. Johnson, J. Lewis, M. McPartlin, and W. J. Nelson, to be published.
21. E. R. Corey, L. F. Dahl, and W. Beck, *J. Am. Chem. Soc.*, **85**, 1202 (1963).
22. R. Mason, K. M. Thomas, and D. M. P. Mingos, *J. Am. Chem. Soc.*, **95**, 3802 (1973).
23. C. R. Eady, B. F. G. Johnson, J. Lewis, and M. McPartlin, *J. Chem. Soc., Chem. Commun.*, **1976**, 883.
24. D. M. P. Mingos, *Nature (Phys. Sci.)*, **236**, 99 (1972).
25. K. Wade, *Adv. Inorg. Radiochem.*, **18**, 1 (1976).
26. B. F. G. Johnson and R. E. Benfield, to be published.
27. R. Mason and J. A. Zubieta, *J. Organomet. Chem.*, **66**, 289 (1974).
28. G. Raper and W. S. McDonald, *J. Chem. Soc. A*, **1971**, 3430.
29. M. Manassero, M. Sansoni, and G. Longoni, *J. Chem. Soc., Chem. Commun.*, **1976**, 919.
30. C. J. Commons and B. F. Hoskins, *Aust. J. Chem.*, **28**, 1663 (1975).
31. F. A. Cotton and J. M. Troup, *J. Am. Chem. Soc.*, **96**, 1233 (1974).
32. G. F. Stuntz and J. R. Shapley, *J. Am. Chem. Soc.*, **99**, 607 (1977); V. Albano, P. Bellon, and V. Scatturin, *J. Chem. Soc., Chem. Commun.*, **1967**, 730.
33. D. A. Brown, W. J. Chambers, N. J. Fitzpatrick, and R. M. Rawlinson, *J. Chem. Soc. A.*, **1971**, 720.
34. L. F. Dahl and J. F. Blount, *Inorg. Chem.*, **4**, 1373 (1966).
35. L. J. Guggenberger and E. L. Muetterties, *J. Am. Chem. Soc.*, **96**, 1748 (1974).
36. B. F. G. Johnson, *J. Chem. Soc., Chem. Commun.*, **1976**, 211.
37. See, e.g., E. Band and E. L. Muetterties, *Chem. Rev.*, **78**, 639 (1978); R. E. Benfield and B. F. G. Johnson, in "Transition Metal Clusters", ed. B. F. G. Johnson, Wiley (London) 1980, p. 472.
38. B. F. G. Johnson and R. E. Benfield, *J. Chem. Soc., Dalton Trans.*, **1980**, 1743.
39. T. A. Claxton and G. C. Benson, *Can. J. Chem.*, **44**, 157 (1966).
40. J. C. Calabrese, L. F. Dahl, A. Cavalieri, P. Chini, G. Longoni, and S. Martinengo, *J. Am. Chem. Soc.*, **96**, 2616 (1974).
41. V. G. Albano, P. L. Bellon, P. Chini, and V. Scatturin, *J. Organomet. Chem.*, **16**, 461 (1969).
42. V. Albano, P. Chini, and V. Scatturin, *J. Organomet. Chem.*, **15**, 423 (1968).
43. R. B. King and D. H. Rouvray, *J. Am. Chem. Soc.*, **99**, 7834 (1977).
44. D. M. P. Mingos, *J. Chem. Soc., Dalton Trans.*, **1974**, 133.
45. E. W. Plummer, W. R. Salaneck, and J. S. Miller, *Phys. Rev.*, **18B**, 1673 (1978).
46. J. C. Green, D. M. P. Mingos, and E. A. Seddon, *J. Organomet. Chem.*, **185**, C20 (1980).
47. R. Hoffmann, B. E. R. Schilling, R. Bau, H. D. Kaesz, and D. M. P. Mingos, *J. Am. Chem. Soc.*, **100**, 6088 (1978).
48. J. K. Ruff, R. P. White, and D. F. Dahl, *J. Am. Chem. Soc.*, **93**, 2159 (1971).

49. E. H. Braye, L. F. Dahl, W. Hubel, and D. L. Wampler, *J. Am. Chem. Soc.*, **84**, 4633 (1962).
50. M. I. Bruce, G. Shaw, and F. G. A. Stone, *J. Chem. Soc., Dalton Trans.*, **1972**, 2094.
51. E. L. Muetterties and L. J. Guggenberger, *J. Am. Chem. Soc.*, **98**, 7221 (1976).
52. R. E. Benfield and B. F. G. Johnson, *J. Chem. Soc., Dalton Trans.*, **1978**, 1554.
53. R. Bau, H. B. Chin, M. B. Smith, and R. D. Wilson, *J. Am. Chem. Soc.*, **96**, 5285 (1974).
54. P. C. Steinhardt, W. L. Gladfelter, A. D. Harley, J. R. Fox, and G. L. Geoffroy, *Inorg. Chem.*, **19**, 332 (1980).
55. V. G. Albano, M. Sansoni, P. Chini, and S. Martinengo, *J. Chem. Soc., Dalton Trans.*, **1973**, 651.
56. A. Fumagalli, S. Martinengo, P. Chini, A. Albanati, S. Bruckner, and B. T. Heaton, *J. Chem. Soc., Chem. Commun.*, **1978**, 195.
57. D. W. Hart, R. G. Teller, C.-Y. Wei, R. Bau, G. Longoni, S. Campanella, P. Chini, and T. F. Koetzle, *Angew. Chem. Intl.*, **18**, 80 (1979).
58. M. R. Churchill and J. Wormald, *J. Chem. Soc., Dalton Trans.*, **1974**, 2410.
59. V. G. Albano and G. Ciani, *J. Organomet. Chem.*, **66**, 311 (1974).
60. R. G. Vranka, L. F. Dahl, P. Chini, and J. Chatt, *J. Am. Chem. Soc.*, **91**, 1574 (1969).
61. L. J. Farragia, J. A. K. Howard, P. Mitrprachachan, J. L. Spencer, F. G. A. Stone, and P. Woodward, *J. Chem. Soc., Chem. Commun.*, **1978**, 260.

Index

A (stereochemical descriptor), 17
Absolute configuration, see Configuration, absolute
Acetaldehyde, as reactant in asymmetric synthesis, 150, 151
Acetoacetates, as substrates for asymmetric hydrosilylation, 148
(3-Aceto-1,2-dideuteropentan-4-ono) cyclopentadienyl(triphenylphosphine) palladium(II), 70, 78
3-Acetoxy-5-carbomethoxycyclohexene, 136
Acetoxypalladation, 67
(Acetylacetonato)[N-alkylephedrinato] dioxomolybdenum(VI), 138
Acetylenes, see Alkynes
1-Acetyl-2-methylferrocene, 32
α-Acylaminoacrylic acid, as ligand, 123
(Acyl)dichlorobis(triphenylphosphine) rhodium(III), 77, 87
η^2-Acyls, 90
1-Adamantyl bromide, 43
Alkenes, addition to: of acetate, 67
 of amines, 68
 of chloride, 68
 of methanol, 66
 of unstabilized carbon nucleophiles, 70
 of water, 63
(Alkenyl)chlorobis(cyclopentadienyl) zirconium(IV), 104
Alkenylpentafluorosilicates, 62
Alkenyl sulfide coupling, 55
Alkyl hydroperoxides, 138
Alkylmetal hydride, stable, 76
Alkynes: additions of carbon nucleophiles to, 73
 oxidative addition of, 81
Allothreonine, 150, 151
Allylic alcohols, asymmetric epoxidation of, 138, 140
Allylic alkylation, 135, 136
Allyl ligand, 203
Allylnitrosylbis (triphenylphosphine) iridium tetrafluoroborate, 205

Allylnitrosylbis (triphenylphosphine) ruthenium, 205
Allylpalladium complexes, 135, 138
Aluminum alkenyls, 54, 61
Aluminum alkyls, 61
Amine (3-chloro-4-methoxyphenolato) (3-chloro-5-methoxyphenolato) (N,N-dimethylcyclobutylamine) (N,N-dimethylcyclopropylamine) (pyridine) cobalt, 13
A-Aminoacidatobis(ethylenediamine) cobalt(III), 151
Anisochronus magnetic sites, 239
Anisylmethyl, see 3-(P-Anisylmethyl)
Anticuboctahedron, 312
Apical attack, in trigonal bipyramid, 229
Arachno-trigonal bipyramid, 271
Asymmetric codimerization, 132
 of 1,3-cyclooctadiene and ethylene, 133
 nickel (II) hydride as catalyst for, 133
 nickelocyclopentane complex in, 134
 of norbornene and ethylene, 133
Asymmetric cyclopropanation, 127, 128
 of cis[^{2}H$_2$]-styrene, 130
 of conjugated olefins, 128
 deuterium scrambling in, 132
 of 1,1-diphenylethylene, 131
 of olefins, 130
 reactions, table of, 129
Asymmetric epoxidations, 139
Asymmetric hydroesterification, 145, 146
Asymmetric hydroformylation, 140, 141, 143, 144
Asymmetric hydrogenation:
 enantiomeric excesses in, 123
 homogeneous, 121
 of ketones, 125
 ^{31}P NMR studies of, 125
Asymmetric hydrosilylation, 147
 asymmetric induction in, 149
 catalysts for, 147
 of carbonyl compounds, 148

Asymmetric induction, 127, 144, 150
  in hydroformylation, 142
  in hydrosilylation, 149
  in platinum system, 144
Asymmetric synthesis: amino acids as substrates in, 150
  of chiral diphosphines, 121
  coordination complexes, using, 150
  coordination compounds of cobalt(III) in, 149
  definition of, 120
  Knoevenagel mechanism in, 152
  oxazolidine ring in, 151
Azidonitrosylbis(triphenylphosphine)nickel, 181

Baeyer-Villiger oxidation, 41
(η^6 - Benzene)tricarbonylchromium, 31
1-Benzoyl-2-phenylcyclopropane, 130
Benzyl-α-d, ligand, 49, 108
(Benzyl-α-d), chlorobis(triphenylphosphine)platinum(II), 47, 108
A-N-Benzylglycinatobis(ethylenediamine)cobalt (III), 150
N-Benzylglycine, 150, 151
Benzylic halides, oxidative addition of, 48
Benzylmethylphenylphosphine (BMPP), 148
Berry mechanism, 232
5H-Bibenzophospholyl, 144
Binary carbonyls, molecular structures of, 254
Binding constants, in hydrogenation, 123
Bipyrimidal structures (TB-5,PB-7, HB-8,HB-9), 22
Bis(acetonitrile)dichloropalladium(II), 91
Bis[N-(2-aminoethyl)-1,2, ethanediamine-N,N', N'''] bromochloropraseodymium(1 +), 30
Bis[camphorquinone-α-dioximato]cobalt(II) hydrate, 128
Bis(cyclohexylglyoximato)(pyridine)cobaltate (I) anion, 58
Bis(1,5-cyclooctadiene)nickel(0), 53
Bis(cyclopentadienyl)deuterio[dimethyl α-(β-deuterio)succinato]molybdenum(IV), 96
Bis(cyclopentadienyl)(dimethyl 2-succinato-3-d) deuteriomolybdenum(IV), 77
Bis(cyclopentadienyliodonitrosylmolybdenum)-1,1-dimethylhydrazide, 185
Bis(cyclopentadienyl)(methyl-β-acrylato)rhenium(III), 81
Bis(cyclopentadienyl) nitrosylmethylmolybdenum, 185

Bis(dimethyldithiocarbamato)nitrosylcobalt, 178
Bis(dimethyldithiocarbamato)nitrosyliron, 175
Bis(dimethylglyoximato)(neohexyl-1,2-d_2) (pyridine)cobalt(III), 105
Bis(dimethylglyoximato)(pyridine)cobaltate(I) anion, 42, 43, 81, 102
1,1'-Bis(diphenylphosphino)ferrocenyl-2-ethyldimethylamine, 148
4,5-Bis(diphenylphosphinomethyl)-2,2-dimethyl-1,3-dioxolane(diop), 136, 137, 141, 145, 147, 148
Bis(propenyl)mercury alkenylates, 72
[1,2-Bis(trifuoromethyl)vinyl] pentacarbonylmanganese(I), 104
Bistrimethylsilylamide ligand, 180
Bistriorganogermylmercury, 226
Boron alkenyls, 59
Boron aryls, 71
β-Bromoacrylates, oxidative addition to, 53
2-Bromobutane, coupling reaction of, 43
1-Bromo-4-t-butylcyclohexane, 43
Bromochloro(2-ethyl-13-methyl-3,6,9,12,18-pentaazabicylco[12.3.1]octadeca-1 (18),2,12,14,16-pentaene-N^3,N^6,N^9, N^{18})iron(1 +), 29
Bromodicarbonylbis(triphenylphosphine) iridium, 190
Bromodinitrosylbis(triphenylphosphine)iridium, 190
6-Bromo-1-hexene, 48
Bromo(p-methoxybenzyl)bis(triethylphosphine) palladium(II), 50
Bromo(methyl β-acrylato)bis (triphenylphosphine)palladium(II), 92
Bromo(methyl β-acrylato)bis (triphenylphosphine)platinum(II), 92
2-Bromooctane, 102
1-Bromo-1-octene, 51, 104
β-Bromostyrene, 51, 53, 54, 104
Bromo(β-styryl)bis(triphenylphosphine) nickel(II), 100
Bromo(β-styryl)bis(triphenylphosphine) platinum(II), 54
Bromo(tri-n-butylphosphine)copper(I), 60
Butadiene, 130
1-Butene, 68, 142
2-Butene, 66, 100, 142
Butenolide preparation, 55
2-Butenylsilver, 100
t-Butylethylene, 77
sec-Butyl ligand, 40

C, 17
Cahn-Ingold-Prelog (CIP) sequence rule, 7, 9, 11, 17, 31, 32
Camphorquinone α-dioximate, 132
CANON algorithm, 9
Carbene-transition metal complexes, 126
Carbido (heptadecacarbonyl)hexaruthenium, 268, 286
Carbido (hexadecacarbonyl)hexairon dianion, 327, 329
Carbido (pentadecacarbonyl)hexarhodium dianion, 286, 288, 322, 323
Carbon-carbon coupling, catalytic, 53, 55, 61
Carbon monoxide, insertion of, 38, 40, 41, 49, 57, 61, 64, 65, 66, 77, 83, 88
Carbonylation oxidatively induced, 41
Carbonylchlorobis(triethylphosphine)rhodium(I), 47
Carbonylchlorobis [tri(p-methoxyphenyl)phosphine]rhodium(I), 85
Carbonylchlorobis(triphenylphosphine)iridium(I), 53, 81
Carbonylchlorobis(triphenylphosphine)rhodium(I), 47
Carbonylchloro(methyl)bisphosphine iridium(III)cation, 47
Carbonyl compounds, asymmetric hydrosilylation of, 148
Carbonyl(cyclopentadienyl)(2,3-dideuterio-4,4-dimethyl-pentan-1-ono)(triphenylphosphine)iron(II), 83
Carbonyl(cyclopentadienyl)(2,3-dideuterio-3-phenylpropan-1-ono)(triphenylphosphine)iron(II), 83
Carbonyldichloro(β-phenethyl-1,2-d_2)bis(triphenylphosphine)iridium(III), 84
Carbonyl(dimethyl α-fumarato)bis(triphenylphosphine)rhodium(I), 81
Carbonyl(dimethyl α-maleato)bis(triphenylphosphine)rhodium(I), 100
Carbonyl(dimethyl-α-maleato)iodo(methyl)bis(triphenylphosphine)rhodium(III), 95
μ-Carbonyl-μ-nitrosylbis(cyclopentadienylcobalt), 198
Carbonyltrinitrosylmanganese, 190
Carbonyltris(triphenylphosphine)palladium(0), 50
Catalyst, for asymmetric codimerization, 132
3-Center bond, in transition metal carbonyl clusters, 292

4-Center bond, in transition metal carbonyl clusters, 292
Central atom chirality symbol, 17
Chelating ligands, multidentate, notation for, 4
Chirality labels, 17
Chirality symbols, assignment of, 25
Chiral rhodium-phosphine complexes, 147
μ-Chloro-abefh-pentachloro-μ-nitrosylnitrosyldiplatinate, 198
β-Chloroacrylate, oxidative addition of, 51
Chlorobis(cyclopentadienyl)(2,3-dideuterio-4,4-dimethylpentan-1-ono)zirconium(IV), 84
Chlorobis(cyclopentadienyl)(3,3-dimethylbut-1-eno)zirconium(IV), 100
Chlorobis(cyclopentadienyl)(neohexyl-1,2-d_2-sulfinato)zirconium(IV), 93
Chlorobis(cyclopentadienyl)(neohexyl-1,2-d_2)zirconium(IV), 84, 93, 104, 108, 109
Chloro(dinitrogen)bis(triphenylphosphine)iridium(I), 53
Chlorodinitrosylbis(triphenylphosphine)ruthenium, 187, 189, 190
Chlorodinitrosyltriphenylphosphineiron, 190
3-Chloro-5-methoxyphenol, 13
Chloromethyl(trimethylsilyl)stannanamine, 10
Chloro(neohexyl-1,2-d_2)mercury(II), 108
m-Chloroperbenzoic acid, 107, 108
Chlorotris(triphenylphosphine)rhodium(I), 76, 77, 85, 87
3-Cholestanyl ligand, 44
Chrysanthemic acid, asymmetric synthesis of, 128
CIP, see Cahn-Ingold-Prelog sequence rule
Cluster compounds, 34, 254, 263
Cobalt(II), alkylation by Co(III), 58
Cobalt(III) alkenyls, 104, 105
Cobalt alkyls, as alkene dimerization catalysts, 134
Cobalt(III) alkyls, autoxidation of, 109
Cobalt(III) alkyls, halogenolysis of, 102
Cobalt(III) alkyls, HgX_2 cleavage of, 105
Cobalt(I) alkyls, isomerization of, 78
Cobalt(I) anion: acetylenes, addition to, 51, 81
  alkylation by Co(III), 58
  oxidative addition to, 43, 51
Cobalt carbene complexes, 132
Cobalt(I) complex, oxidative addition to, 42
Cobalt(III), coordination compounds of, in asymmetric synthesis, 149

Cobalt Schiff base complexes, as catalysts for asymmetric hydroformylation, 140
Cobalt trichloride tetraammoniate, 2
Codimerization, asymmetric, 132
Configuration, absolute: glyceraldehyde, 7
  tartaric acid, 7
  tris(ethylenediamine)cobalt(3 +), 8
Configurational analysis: of metal alkyls, 40
  of vinylplatinum complexes, 54
Configuration numbers, 17, 20, 23, 24
Coordination compounds, stereochemical notation for, 17
Copper alkenyls, reductive elimination, 95
Copper alkyls, protonolysis, 99
Copper(I) alkenyls, halogenolysis, 104
Copper(I) alkyls, acetylene insertion, 75
Copper complexes, 60, 128
Copper hydride, 60, 98
Cubane, oxidative addition of, 57
Cuboctahedron, 311
Cyclobutylamine, 13
Cyclohexene, 61
Cyclohexene-d_4, addition to, 67, 73
Cyclohexene oxide, 42
Cyclohexylisonitrile, 89
1,3-Cyclooctadiene and ethylene, codimerization of, 133
Cyclopentadienyldinitrosylchromium chloride, 187
Cyclopentadienyl(η^2-ethylene-1,2-d_2)(triphenylphosphine)palladium(II) cation, 70
Cyclopentadienyliodonitrosyl(phenylhydrazine)molybdenum cation, 185
Cyclopropanation, 126, 127, 130
Cyclopropylamine, 13
Cyclopropylcarbinyl ligand, 44
Cyclopropyllithium, 99

Decacarbonylbis(triphenylphosphine)tetrairidium, 291
Decacarbonyl(dihydrido)(trialkylphosphine)triosmiumplatinum, 281
Decacarbonyl(dihydrido)triosmium, 266
Decacarbonyl dimanganese, 265, 291, 292, 309
Decacarbonyldirhenium, 292, 308
Decacarbonylditechnetium, 292, 308
Decacarbonylpentaplatinum dianion, 276
Δ(right-handed helix), 8, 17
Diastereotopic groups, 238

Diastereotopic interactions, 120, 121, 123, 125
Diazoacetophenone, 130
Diazoalkane, 128
Dibenzosemibullvalene, oxidative addition of, 57
Di-butyl mercury, 226
(1,2-Dicarbomethoxy-3,4-dideuterio-5,5-dimethylhex-1-eno)dicarbonyl(cyclopentadienyl)iron(II), 92
Dicarbonylcyclopentadienylferrate(0) anion, 40, 44
Dicarbonyl(cyclopentadienyl)methyl(tri-n-butylphosphine)molybdenum(II), 106
Dicarbonyl(cyclopentadienyl)methyl(triphenylphosphine)molybdenum(II), 106
Dicarbonyl(cyclopentadienyl)(neohexyl-1,2-d_2)iron(II), 83, 92, 101, 105
Dicarbonyl(cyclopentadienyl)(neohexyl-1,2-d_2-sulfinato)iron(II), 92
Dicarbonyl(cyclopentadienyl)(β-phenethyl-1,1-d_2)iron(II), 101
Dicarbonyl(cyclopentadienyl)(β-phenethyl-1,2-d_2)iron(II), 42, 93, 101, 105
Dicarbonyl(cyclopentadienyl)(β-phenethyl-1,2-d_2-sulfinato)iron(II), 92
Dicarbonyl(cyclopentadienyl)β-phenethyl-1,2-d_2)(triethylphosphine)tungsten(II), 92, 105
Dicarbonyl(cyclopentadienyl)(triethylphosphine)tungstate(0) anion, 42
Dicarbonylnickel fragment, 282
Dichlorobis(benzoylacetonato)tin, 236, 243
Dichloro [1,3-bis(diphenylphosphino)propane]nickel(II), 55
Dichloro(1,5-cyclooctadiene)palladium(II), 107
Dichloroethylene, 54
Dichloropentafluoroniobate, 6
Dideuterioethylene, 65
2,3-Dideuteriopropiolactone, 65
3-(1,2-Dideuteriovinyl)acetylacetone, 78
1,2-Dihalobenzocyclobutene, 44
Dihydridobis(benzylmethylphenylphosphine)bis(solvent)rhodium(I) perchlorate, 148
(Dimethyl n^2-acetylenedicarboxylate)bis(triphenylphosphine)platinum(0), 81
[(Dimethylamine)(methylamine)(methylamino)(N-methylethylamine)(pyridine)(pyrimidine)metal] complex, 11

INDEX

Dimethyl 2-butynedioate, 81, 92
(μ^2-Dimethylcarbamido)
  tetraisocyanatonitrosylmolybdenum, 184, 185
1,4-Dimethylcyclohexane, 41
[Dimethyl(*N,N*-dimethylethanolamino)
  3,5-dimethylpryazolyl)gallato(*N*(2), *N*(3),0)] dinitrosyliron, 190
[Dimethyl(*N,N*-dimethylethanolamino)
  (3,5-dimethylpyrazolyl)gallato
  (*N*(2),*N*(3),0]nitrosylnickel(I), 181
Dimethyl α-methylmaleate, 95
Dimethylphenylsilane, 148
α-[2-Dimethylphosphinoferrocenyl]
  ethyldimethylamine(MPFA), 148
(μ-3,5-Dimethylpyrazolyl)bis(nitrosylnickel), 184
Dimethyl succinate, 96
m-Dinitrobenzene, 51, 53
Di-μ-nitrosylbis(cyclopentadienylcobalt), 198
Di-μ-nitrosylbis(cyclopentadienyl)iron, 198
Dinitrosylbis(dimethylphenylphosphonite)
  manganese chloride, 187, 189, 190
Dinitrosyl complexes, 185
  five-coordination in, 187
  four-coordination in, 190
  six-coordination in, 185
  $[M(NO)_2]^6$, structural data for, 186
  $[M(NO)_2]^8$, structural data for, 188
  $[M(NO)_2]^9$, structural data for, 191
  $[M(NO)_2]^{10}$, structural data for, 192
Dinitrosyltris(dimethylphenylphosphonite)
  manganese, 187, 189
Diop (4,5-Bis(diphenylphosphinomethyl)-2,2-dimethyl-1,3-dioxolane), 136, 137, 141, 143, 147, 148
2,3-Diphenylbutanoyl chloride, 77
2,2-Diphenylcyclopropylcarboxaldehyde, decarbonylation of, 84
1,1-Diphenylethylene, 131
Diphenylphosphine, 144
1,2-Diphenylpropene, 72
1,2-Diphenylpropyl ligand, 78
Diphosphine-rhodium complexes, cationic, 122
Diphosphines, 121
Disjoint subgraphs, 231
Disodium tetracarbonylferrate(-II), 41
Dodecacarbonyldicobaltdiridium, 293
Dodecacarbonyl(dihydrido)triosmium, 266
Dodecacarbonyldivanadium, 308
Dodecacarbonyl(hexahydrido)rhenium dianion, 279, 281
Dodecacarbonylhexanickel dianion, 293, 297, 299
Dodecacarbonylhexaplatinum dianion, 276, 286, 288, 293, 297, 300
Dodecacarbonylpentanickel dianion, 282, 293, 297, 298
Dodecacarbonyltetracobalt, 265, 290, 291, 292, 293, 297, 308
Dodecacarbonyl(tetrahydrido)tetraosmium, 281
Dodecacarbonyl(tetrahydrido)tetrarhenium, 279, 281
Dodecacarbonyltetrairidium, 281, 291, 292, 293, 297
Dodecacarbonyltetrarhodium, 291, 292, 293
Dodecacarbonyltriiron, 265, 276, 292, 293, 295, 301
Dodecacarbonyltriosmium, 291, 292, 293, 296
Dodecacarbonyltriruthenium, 291, 292, 293
(Dodecachloro)hexaniobium dication, 286

Effective Atomic Number (EAN) rule, 175
Eighteen electron rule, 264, 266, 267, 273
Electron pair theory, skeletal, 268
Enantiomeric excess, 120, 123, 124
Enantiomers, locants for, 5
Epimerization: in oxidative additions, 45
  via transmetalation, 58
Expoxidation, asymmetric, 139, 140
Equatorial attack in trigonal bipyramid, 229
Ethyl α-bromophenylacetate, 47
α-Ethylcinnamaldehyde, 85
Ethyl diazoacetate, 224
Ethyldimethyl(triphenylphosphine)gold(III), 96
Ethylene, 133, 135
Ethylene, oxidation of, 63
Ethylene-1,2-d_2, addition to, 65, 66, 70
[[*N,N'*-Ethylenebis(glycinato)]
  (2-)][oxalato(2-)]cobalate(1-), stereodescriptor for, 28
N,N'-Ethylenebis(salicylideneiminato)
  nitrosylcobalt, 178
N,N'-Ethylenebis(salicylideneiminato)
  nitrosyliron, 177
3-Ethylhexanal, 79
Ethyl isobutyrate, lithium enolate of, 54
Ethylisopropylphenylgermanium-*d*
  α-bromocamphorsulfonate, 223
Ethyl-1-naphthylphenylgermanium hydride, 223

Ethyl-1-naphthylphenylgermyllithium, 225
Ethyl-1-naphthylphenylsilicon chloride, 219

Facial, 3
Feist's esters, ring opening of, 91
Ferrocene, 31
Fifteenhedron, 320, 324
Five-coordinate structures, 22
Fluorosilanes, 219
Fourteenhedron, 317, 318

Galvanoxyl, 46
Germyl radicals, 224
Glyceraldehyde, absolute configuration of, 7
Glycine, 150
Gold(III) alkyls, reductive elimination, 95

α-Haloesters, oxidative additions of, 45, 47
Hapto complexes, 31
Helicity, 236, 237, 238, 239
Hexacarbonyltrinickel dianion, 302
Hexacarbonyltriplatinum, 288, 289
Hexacarbonyltriplatinum dianion, 277, 289, 302
Hexacyanonitrosylvanadate, 184
Hexadecacarbonyldimolybdenumtrinickel dianion, 282
Hexadecacarbonylhexacobalt, 291
Hexadecacarbonylhexarhodium, 268, 270, 286, 290, 293, 327, 328, 331
Hexadecacarbonylpentaosmium, 265, 266, 271, 282, 327, 330
Hexadecacarbonyltetrarhenium dianion, 301, 327
Hexaorganoditin compounds, optically active, 224
Hexylcyclopropane-2,3-d_2, 57
Homotopic groups, 238
α-[$^2H_1$]-styrene, see Styrene
β-Hydride elimination, 67, 73
Hydridocarbonyltris(triphenylphosphine)rhodium(I), 141, 142
Hydridonitrosyltris(triphenylphosphine)iridium perchlorate, 178
Hydridonitrosyltris(triphenylphosphine)ruthenium, 178
Hydroalumination, 61
Hydroesterification, asymmetric, 145, 146
Hydroformylation reaction, 77, 78
  asymmetric, 140, 141, 142, 143, 144

Hydrogenation, asymmetric: homogeneous, 121, 123
  of ketones, 125
  of olefins, enantiomeric excess in, 123, 124
  stereospecific, 122
Hydrosilylation, asymmetric, 147, 148
Hydroxamic acids, as ligands in asymmetric epoxidation, 139

Icosahedron, 310
Iminoacyl complexes, 90
Inversion, minimum constraints for, 231
3-Iodocholestane, 44
1-Iodohexene, 53
(Iodomethyl)methylneophylphenyltin, 221
2-Iodooctane, coupling reaction, 51
Iodo(tri-n-butylphosphine)copper(I), 60
Iodotrimethylbis(dimethylphenylphosphine)platinum(IV), 96
Iridium(III) alkyl, CO insertion stereochemistry, 88
Iridium(I) complexes, oxidative addition to, 45, 53
Iridium hydride, acetylene insertion in, 81
Iridium(IV) oxidation of M-C, 102
Iron(II) alkyls: acetylene insertion in, 92
  CO insertion in, 83
  halogenolysis of, 101, 104, 105, 106
  HgX_2 cleavage of, 104
  β-hydribe abstraction from, 80
  optically active, 88, 93, 106
  SO_2 insertion in, 92
Iron (0) anion, oxidative addition to, 40, 44
Iron catalysis, 56
Iron complexes, chiral: CO insertion, 88
  electrophilic cleavage, 106
  SO_2 insertion, 93
Isochrony, accidental, 248
Isocyanatocyclopentadienyldinitrosylchromium, 187
Isomer notation, permutational, 9
Isonitrile induced CO insertion, 90
Isonitrile insertion, 92
Isoprene, 130
Isopropyldimenthylphosphine, 133
Isopropyl-1-naphthylphenylgermanium hydride, 223
Isopropyl-1-naphthylphenylgermyllithium, 226
(Isothiocyanato)nitrosylbis(triphenylphosphine) nickel, 181, 182

INDEX 343

α-Ketoesters, as subtrates for asymmetric hydrosilylation, 148
Knoevenagel mechanism, in asymmetric syntheses, 152

Λ (left-handed helix), 8, 17
Levulinates, as substrates for asymmetric hydrosilylation, 148, 149
Ligand: degeneracy in atom ranking, 19
  envelope, 290
  index numbering, 8
  stereochemical label of, 17
Lithium alkenyls, transmetalation of, 60
Lithium alkyls, 59, 73
Lithium dialkylcuprates, 43, 55, 59
Lithium enolates, 54
Locant designators, 5

M (configurational symbol), 8
McDonnell Notation, class symbol, 5
Magic number, 266, 268
Manganese(I) alkenyls, 104
Manganese(I) alkyls: acetylene insertion in, 74
  CO insertion in, 85
  halogenolysis of, 102
  HgX_2 cleavage of, 105
  SO_2 insertion into, 92
Manganese(-1)anion, oxidative addition to, 41, 42
Manganese hydride, acetylene insertion into, 81
Maximum difference subrule, 19
Menshutkin reaction, 46
Mercury alkenyls, 60, 72
Mercury alkyl: $CuCl_2$ cleavage of, 108
  oxidative addition to Pt(0), 59
  transmetallation of, 60, 108
Mercury aryls, 71
Meridional, 3
Metal alkyls, configurational analysis of, 40
Metal-metal bonds, in clusters, 263, 264
Metal-nitrogen bond distances, 169
Metal-nitrogen-oxygen bond angles, 169
Metal nitrosyls, 156
Metal stereochemistry: of CO insertion, 85
  in electrophilic cleavage, 106
  of oxidative additions, 46, 55
  in reductive C-C formation, 95
  in SO_2 insertion, 94
Methoxypalladation, 66
1-Methyl-1-acetoxymethylcyclopentene, 136

N-α-Methylbenzylsalicylaldimine, 140
Methyl β-bromoacrylate, 54, 92
2-Methylbutanal, 144
2-Methylbutanoic acid, 40
1-Methyl-4-t-butylcyclohexane, 40, 43
4-Methylcyclohexyl ligand, 40, 83
Methyldimenthylphosphine, 133
1-Methyl-2,2-diphenylcyclopropane, 44
1-Methyl-2,2-diphenylcyclopropyl bromide, 43
1-Methyl-2,2-diphenylcyclopropyl ligand, 44
3-Methyl-1-hexene, 79
Methyl iodide, oxidative addition of, 46
Methyl-1-naphthylphenylgermanium chloride, 218
Methyl-1-naphthylphenylgermanium hydride, 223
Methyl-1-naphthylphenylgermyllithium, 226
Methyl-1-naphthylphenylsilane, 224
Methyl-1-naphthylphenyltin hydride, 220
Methylneophylphenyl(phenylethynyl)tin, 221
Methylneophylphenyltin deuteride, 228
Methylneophylphenyltin halide, 218, 219
Methylneophylphenyltin hydride, 220, 224
Methylneophyltrityltin bromide, 219
Methylpentacarbonylmanganese(I), 74
Methylphenyl(2-phenylpropyl)stannylcyclopentadienyldicarbonyliron, 221
Methylphenyl(2-phenylpropyl)stannylpentacarbonylmanganese, 221
Methylphenyl(2-phenylpropyl)stannyltricarbonyl triphenylphosphine)cobalt, 221
Methylphenyl-2-phenylpropyltin hydride, 220
Methylphenyl-t-butyltin hydride, 220
Methyl propynoate, 81
α-Methylstyrene, 126, 127, 146
Molybdenium(II) alkyl, CO insertion into, 87
Molybdenium hydride, alkene insertion into, 77
Molybdenum(II) alkyls, metal cleavage, stereochemistry of, 106
Molybdenum (0) anion, oxidative addition to, 42
Molybdenum complex, as catalyst for asymmetric epoxidation, 138
Molybdenum(IV) hydridoalkyl, reductive elimination of, 96
Molybdenum-oxo-complexes, 138
Mononitrosyl complexes, see Nitrosyl complexes

Monophosphines, as ligands in asymmetric hydrosilylation, 148
Morgan algorithm, 9
Multidentate ligands in ligand atom ranking, 19

α-Naphthylphenylsilane, 148
Naphthylphenylvinylsilicon fluoride, 228
Neohexyl-1,2-d_2 alcohol, 109
Neohexyl bromide, 40
Neohexyl-1,2-d_2 ligand, 39, 42, 61, 77, 83, 91, 101, 104, 105, 108
Neohexyl-1,2-d_2 trifluoromethanesulfonate, 42
Nickel(II) alkenyls: CO insertion into, 85
  protonolysis of, 100
Nickel(II) alkyls, acetylene insertion into, 75
Nickel π-allyl complex, 51, 53, 72
Nickel(II) aryls, acetylene insertion into, 74
Nickel(0) complexes, oxidative addition to, 50, 53
Nickel hydride, 134
Nickelocyclopentane complexes, 134, 135
Nitratonitro(pyridine-1-oxide)(1,1,1-trifluoro-2,4-pentanedionato)metal complex, 13
Nitrosylbis[1,2-bis(diphenylphosphino)ethane] ruthenium tetraphenylborate, 179
Nitrosylbis[1,3-bis(diphenylphosphino)propane] ruthenium tetraphenylborate, 179
Nitrosylbis(maleonitriledithiolato)iron, 175
Nitrosylbis(methyldiphenylphosphine) dichlorocobalt, 179
μ-Nitrosyl [bis(N,N'-dimethyl-N,N'-bis(2-mercaptoethyl)ethylenediamine)] dicobalt tetrafluoroborate, 198
Nitrosylbis(o-phenylenebis(dimethylarsine) cobalt diperchlorate, 179
Nitrosylbis(triphenylphosphine)rhodium, 206
Nitrosyl bridges, 195
Nitrosylcarbonylbis(triphenylphosphine) iridium, 181
Nitrosyl cluster compounds, 195
Nitrosyl complexes, 156, 158
  allyl groups, involving, 203
  ambidentate ligands in, 203
  bond lengths in, 172
  bridging and polynuclear, 195
  containing SO_2 and C_3H_5 ligands, 200, 202
  doubly bridging, 196
  electron deficient, 175
  equatorial NO groups in, 179
  five-coordinate, 174, 179, 180
  four-coordinate, 180, 181
  of higher coordination numbers, 184
  infrared spectra of, 175
  M-N-O bending vibration in, 183
  [MNO]^n geometry of, 159
  polynuclear metal, 200, 201
  seven-coordinate, 184
  six-coordinate, 169, 170, 174
  SO_2 ligands, involving, 203
  static structures of metal in, 206
  stereochemical dynamics of metal in, 206
  structural data for, 173
    [MNO]^3, 159
    [MNO]^4, 160
    [MNO]^5, 161
    [MNO]^6, 162
    [MNO]^7, 164
    [MNO]^8, 165
    [MNO]^9, 167
    [MNO]^{10}, 167
  tetradentate ligands in, 178
  tetragonal pyramidal, 174, 175, 177
  tetrahedral, 181
  thermal parameters for, 175
  three-coordinate, 184
  trigonal bipyramidal, 178
  triply bridging, 198, 199
  vibronic coupling in, 183
  see also Dinitrosyl complexes
Nitrosyldicarbonylbis(triphenylphosphine) manganese, 179
Nitrosyl groups: equivalent, 187
  geniculation of, 90
  slightly bent, 187
  thermal motion of, 171
Nitrosyl(N,N'-dimethyl-N,N'-bis(2-mercaptoethyl)ethylenediamino)iron, 175
Nitrosylpenta amminecobalt, 172
Nitrosylsulfonylbis(triphenylphosphine)cobalt, 203, 204
Nitrosylsulfonylbis(triphenylphosphine) rhodium, 203, 204
Nitrosyltetracarbonylmanganese, 179
Nitrosyl(1,4,8,11-tetramethyl-1,4,8,11-tetraazacyclotetradecane)iron, 177
Nitrosyl-α,β,γ,δ-tetraphenylporphinatocobalt, 178
Nitrosyl-α,β,γ,δ-tetratolylporphinatoiron, 176
Nitrosyl-α,β,γ,δ-tetratolylporphinatomanganese, 175

Nitrosyltris(bis(trimethylsilyl)amido)chromium, 180
Nitrosyltris(2-diphenylphosphinoethyl)phosphineiron tetraphenylborate, 178
Nitrosyltris(N,N'-di-n-butyldithiocarbamato)molybdenum, 184
Nitrosyl(1,1,1-trisdiphenylphosphinomethylethane)nickel, 181
Nitrosyltris(triphenylphosphine)rhodium, 182, 183, 206
Nomenclature, stereochemical, 2, 3, 6, 8, 11, 17, 18, 32
Nonacarbonyldiiron, 290, 292, 308
Nonacarbonyltris(dimethylphenylphosphine)triiron, 290
Norbornadiene, 61, 134
Norbornane-2-d_1, 98
Norbornene, 133, 134, 135
1-Norbornyl bromide, 43
Norbornyl ligand, 60
2-Norbornylmagnesium bromide, 60
Norbornylmercuric bromide, 60
2-Norbornyl(tri-n-butylphosphine)copper(I), 98
Nucleophilic transmetalation, 45

Octacarbonyldicobalt, 290, 292, 305, 306, 308
Octacarbonyldiiron dianion, 305
Octacarbonylmonocobaltmonoiron anion, 305
Octacarbonyltetraplatinum, 276
Octacarbonyltriplatinum dianion, 276, 288
Octachlorohexamolybdenum tetracation, 273
Octadecacarbonylhexaosmium, 268, 271, 286, 288, 292
  dianion, 268, 286, 288, 292
Octahedral complexes (OC-6), 20
  nomenclature of, 4
2-Octyl, optically active ligand, 41
Optically active compounds, nomenclature for, see Nomenclature
Optical purity, definition of, 120
Organocuprates, reductive elimination in, 95
Organogermanium compounds, optically active, 223
Organometallic complexes, transition metal, 264
Ortho-metalation, 81
Oxazolidine ring, in asymmetric synthesis, 151
Oxidative addition, 39

Oxidative cleavage: by Cu(II), 66, 68, 107
  by Ir(IV), 102
  by m-chloroperbenzoic acid, 108
  by Pb(OAc)$_4$, 107
Oxidative elimination, 39
μ-Oxido-bis[chlorotriphenylphosphinenitrosyliridium(I)], 180
Oxiranes, chiral, 139
"Oxo" reaction, 140

P (configurational symbol), 8
Palladium(II) alkene complexes, 68, 71, 72
Palladium π-alkyl complexes, 72
Palladium(II) alkyls:
  β-hydride elimination in, 78
  CO insertion of, 83
  CuCl$_2$ cleavage of, 108
  halogenolysis of, 103
  rearrangement of, 67
  reduction of, 97
Palladium(0) complexes, oxidative addition to, 47, 54
Palladium(II) complexes, SO$_2$ extrusion from, 93
Palladium(II) diene complexes: addition of stabilized carbanions to, 70
  hydration of, 64
  methanol addition to, 66
  oxidative cleavage of, 107
Palladium(II), methylenecyclopropane ring opening by, 91
Palladium(II) oxidation, of ethylene, 63
Palladium(0) phosphine complex, as catalyst in allylic alkylation, 135
Palladium(II) phosphine complexes for asymmetric hydroesterification, 145
3-(p-Anisylmethyl-1-naphthylstannyl)-1,1-dimethyl-1-propanol, 221
Pentacarbonyl(dimethyl α-fumarato)manganese(I), 104
Pentacarbonyl(ethyl α-propionato)manganese(I), 45, 58, 102
Pentacarbonylmanganate(-I) anion, 41, 42, 45, 58
Pentacarbonylmolybdenum fragment, 284
Pentacarbonyltetra(diphenylmethylphosphine)tetraplatinum, 277
Pentacarbonyltris(triphenylphosphite)irondiplatinum, 276

Pentacyanonitrosylchromate, 172
Pentadecacarbonylhexacobalt dianion, 268, 286, 291, 321, 322
Pentadecacarbonyl(hydrido)hexacobalt anion, 322
Pentadecacarbonyl(iodo)hexarhodium anion, 327
Pentadecacarbonyl(iodo)pentaosmium anion, 327
Pentadecacarbonylpentametal, 284
Pentadecacarbonylpentaosium dianion, 282
Pentadecacarbonylpentarhodiumplatinum anion, 322
Perchloratonitrosylbis(ethylenediamine)cobalt, 171
Permutation group, 236
Phantom atoms, 11
β-Phenethyl-1,2-d_2 ligand, 40, 42, 83, 92, 101, 102, 105
α-Phenethyl optically active ligand, 41
Phenonium ion, formation in halogenolysis, 101
Phenylacetylene, 51
Phenylchorobis(benzoylacetonato)tin, 236, 245
1-Phenyl-1-chloro-1-silaacenaphthene, 228
2-Phenylcyclopropanecarboxylic esters, 130
2-Phenyl-1,1-dicyanocyclopropane, 130
3-Phenyl-3-hexene, 74
2-Phenyl-2-methylbutanal, decarbonylation of, 84
Phenylpentacarbonylmanganese(I), 74
5-Phenyl-4-penten-2-one, 53
1-Phenylpropene, 72
α-Phenylpropionic acid, 41
Phenyl t-butyl ketone, 148
Phosphines: chiral unidentate, 121
  as ligands in asymmetric hydroesterification, 146
Photochemical decarbonylations, 88
Planar chirality, 32
Platinum(II) alkene complex, amine addition to, 68
Platinum(IV) alkyls, reductive elimination, 96
Platinum(0) complexes, oxidation additions to, 47, 54
Platinum(II) complexes, oxidative addition to, 57
Platinum(II) diene complexes, 66, 70
Platinum hydride, 80, 91
Platinum phosphine complexes, as catalysts for asymmetric hydroformylation, 141

Polyhedra, numbering and nomenclature for, 5
Polynitrosyl complexes, 190
Polynuclear compounds, 34
Polypropylene, 91
Potassium bis(N-salicylideneglycinato)cobaltate, 151
Prochiral ketones, 148
Λ-[Prolinatobis(ethylenediamine)cobalt(III), 151
Proline, as ligand in coordination complexes, 151
Propargyl bromide, 226
Propylene-d_1 polymerization, 91
n-Propyl pyruvate, 148
Propynal, 81
Pseudo-atoms, 32
p-Toluoisonitrile, see Toluoisonitrile

Quadricyclane, oxidative addition of, 57

R (symbol), in octahedral compounds, 8
Racemization: via nucleophilic transmetalation, 45, 47
  in oxidative additions, 44, 48
Radical intermediates, in oxidative additions, 43, 46, 48, 50
Rearrangements: combinations of modes, 240-248
  graphical topological representations of, 242
  threshold mode of, 243
Reduction:
  of M-C by HCuL, 98
  of M-C by HMn(CO)$_5$, 97
  of M-C by LiAlH$_4$, 61, 68, 97
  of M-C by NaBH$_4$, 97
  see also under compound
Reductive elimination, to form a C-C bond, 52
Retention, minimum constraints for, 232
Rhenium hydride, acetylene insertion into, 81
Rhodium(I), in decarbonylation, 84
Rhodium(I) alkenyls, 100
Rhodium(I) alkyls, β-hydride elimination in, 78
Rhodium(I) catalyzed hydrogenation, 76
Rhodium(I) complexes: acid chloride decarbonylation, 98
  oxidative addition to, 47, 52, 57
Rhodium hydride, acetylene insertion into, 81
Rhodium-hydride bond, 142
Rhodium phosphine complexes: in asymmetric hydrogenation, 121

as catalysts for asymmetric hydroformylation, 141
Rotationally equivalent permutations, 237
Roussin's black salt, 200

S (symbol), for octahedral compounds, 8
Septacarbonyl(diphenylphosphino)(ethane) dimanganese, 290
SEP(skeletal electron pair) theory, 268, 271, 274
Sequence rule, CIP (Cahn, Ingold, and Prelog), 7, 9, 11, 17, 31, 32
Silicon, substitutions at, 227
Silver alkenyls, reductive elimination in, 95
Silver(I) alkenyls, protonolysis of, 100
Silver(I) catalysis, 56, 57
Sixteenhedron, 325, 326
Sodium methylphenylsulfonylacetate, 136
Solvent effect, on CO insertion stereochemistry, 87
Square based pyramid, *nido*-octahedron, 271
Square planar complexes (SP-4), 20
Square pyrimidal structure (SP-5), 22
Stannyl radical, optically stable, 224
Stereochemical descriptor, 17
Stereochemical nomenclature, *see* Nomenclature
Stereochemistry of metals, *see* Metal stereochemisty; *and individual metals and compounds*
Styrene, 100, 130
α-[^{2}H$_1$]-Styrene, asymmetric hydroformylation of, 143
Sulfur dioxide insertion, 92
Symmetry equivalent permutations, 237
Symmetry site term, 17

Tartaric acid, absolute configuration of, 7
Tetraaqua(2,4,5,6,(1H,3H)-pyrimidinetetrone 5-oximato-N^5,O^4)(2,4,5,6,(1H,3H)-pyrimidinestetrone-5-oximato-O^4,O^5) strontium, 30, 31
Tetracarbonylcobaltate(-I) anion, 42
Tetracarbonylhydridocobalt(I), 42
Tetracarbonylhydrido(triphenylphosphine) manganese(I), 81
Tetracarbonylosmium, 282
Tetracarbonyl(β-phenethyl-1,2-d_2) (triethylphosphine)manganese(I), 92, 102, 105

Tetracarbonyl(triethylphosphine)manganate(-I) anion, 42, 102
(η^2-Tetrachloroethylene)bis(triphenylphosphine) platinum(0), 55
Tetracyanonitrosylferrate, 175, 176
Tetradecacarbonylhexacobalt tetraanion, 268, 286, 291, 316, 319, 332
Tetradecacarbonyltetracobaltdinickel dianion, 316
Tetrahedral complexes (T-4), 20
Tetrahedron, parity of, 229
Tetrakis(dimethylphenylphosphine) palladium(0), 54
Tetrakis(methyldiphenylphosphine) palladium(0), 54
Tetrakis(triphenylphosphine)nickel(0), 50, 53
Tetrakis(triphenylphosphine)palladium(0), 48, 50, 54, 136
Tetranitrosylchromium, 190
Tetraorganogermanium compounds, 226
Tetraorganotin compounds, optically active, 221, 224
Thirteenhedron, 313, 314
Three-center bridges, 270
Threonine, 7, 150, 151
Threose, 7
Tin, substitution at, 228
Tin aryls, 71
Titanium(IV) alkyls, optically active, 93
p-Toluoisonitrile, 92
Trans effects in CO insertion, 88
Transition metal-carbene complexes, 126
Transition metal carbonyl clusters, 254, 273, 301
  multi-center bonds in, 292
  M_3 unit, 278
  M_4 unit, 279
  M_5 unit, 282
  M_6 unit, 286
Transition-metal organometallic complexes, 264
Transmetalation, 58-60
Transmetalations, alkenyl, 62
Tricarbonyl(η^5-cyclopentadienyl)manganese, 31
Tricarbonyl(cyclopentadienyl) methylmolybdenum(II), 74
Tricarbonyl(cyclopentadienyl) methyltungsten(II), 74
Tricarbonylcyclopentadienylmolybdate(0) anion, 42
Tricarbonylosmium fragment, 282, 288

Tridecacarbonyl(hydrido)tetrairon anion, 290
Tridecarbonyltetrairon dianion, 290, 308, 315
Tridecarbonyltetraosmium dianion, 316
1,1,1-Trifluoro-2,4-pentanedione, 13
β,β,β-Trifluoro-α-phenethyl chloride, 48
Trigonal bipyramid, 22, 229
Trigonal prism (*TP*-6), 22
Trigonal prism square face tricapped structure (*TPS*-9), 22
Triicosacarbonyloctaosmium, 271
Trimethylsilane, 148
Triorganogermanes, 218, 225, 226
Triorganogermyllithium compounds, 225, 226
Triorganosilyllithium compounds, 225
Triorganostannylgermane, 224
Triorganostannyl-transition metal complexes, 221
Triorganotin halides, 218
Triorganotin hydrides, 220, 224
Triorganotin methoxides, reduction of, 224
Triphenylphosphine, 50, 291
Triphenylphosphine ligand, 221
Tris(cyclopentadienyl)nitrosylmolybdenum, 185
Tris(dimethylphenylphosphine)platinum(0), 54
Tris(di-*o*-substituted aryl)germanes, 248
Tris(ethylenediamine)cobalt (3+), absolute configuration of, 8
Tris(1,2-propanediamine-*N,N'*)cobalt(3+), 18, 27
Tris(*tert*-butylisonitrile)platinum(0), 92
Tris(triethylphosphine)palladium(0), 50
Tris(triethylphosphine)platinum(0), 48
Tris(triphenylphosphine)nitrosyliridium, 183
Tungsten(II), alkyls: HgX_2 cleavage, 105
  SO_2 insertion, 92
Tungsten(0) anion, oxidative addition to, 42

Undecacarbonyl(hydrido)triiron anion, 292
Unicosacarbonylheptaosmium, 271

Vanadium-oxo-complexes, 138, 139
Vinyl halide π complexes, 54
Vinyl-1-cyclooctene, 133
Vinyl metal complexes, configurational analysis of, 51
Vinyl radical intermediates, 52

Wacker process, 63, 66
Walden cycle, at a chiral germanium atom, 223
Wilkinson's catalyst, 76

Ziegler polymerization, 91
Zirconium(IV) alkenyls, 61, 100, 104
Zirconium(IV) alkyls: acetylene insertion into, 75
  autoxidation of, 108
  CO insertion into, 84
  halogenolysis of, 104, 108
  SO_2 insertion into, 93
  transmetaltion of, 61
Zirconium hydride: acetylene insertion into, 80
  alkene insertion into, 77

Cumulative Index, Volumes 1-12

	VOL.	PAGE
Absolute Configuration of Planar and Axially Dissymmetric Molecules *(Krow)*	5	31
Absolute Stereochemistry of Chelate Complexes *(Saito)*	10	95
Acetylenes, Stereochemistry of Electrophilic Additions *(Fahey)*	3	237
Analogy Model, Stereochemical *(Ugi* and *Ruch)*	4	99
Asymmetric Synthesis, New Approaches in *(Kagan* and *Fiaud)*	10	175
Asymmetric Synthesis Mediated by Transition Metal Complexes *(Bosnich* and *Fryzuk)*	12	119
Atomic Inversion, Pyramidal *(Lambert)*	6	19
Axially and Planar Dissymmetric Molecules, Absolute Configuration of *(Krow)*	5	31
Barriers, Conformational, and Interconversion Pathways in Some Small Ring Molecules *(Malloy, Bauman,* and *Carreira)*	11	97
Barton, D.H.R., and Hassel, O.--Fundamental Contributions to Conformational Analysis *(Barton, Hassel)*	6	1
Carbene Additions to Olefins, Stereochemistry of *(Closs)*	3	193
Carbenes, Structure of *(Closs)*	3	193
sp^2-sp^3 Carbon-Carbon Single Bonds, Rotational Isomerism about *(Karabatsos* and *Fenoglio)*	5	167
Carbonium Ions, Simple, the Electronic Structure and Stereochemistry of *(Buss, Shleyer* and *Allen)*	7	253
Chelate Complexes, Absolute Stereochemistry of *(Saito)*	10	95
Chirality Due to the Presence of Hydrogen Isotopes at Noncyclic Positions *(Arigoni* and *Eliel)*	4	127
Chiral Lanthanide Shift Reagents *(Sullivan)*	10	287
Classical Stereochemistry, The Foundations of *(Mason)*	9	1
Conformational Analysis, Applications of the Lanthanide-induced Shift Technique in *(Hofer)*	9	111
Conformational Analysis–The Fundamental Contributions of D. H. R. Barton and O. Hassel *(Barton, Hassel)*	6	1
Conformational Analysis of Intramolecular Hydrogen-Bonded Compounds in Dilute Solution by Infrared Spectroscopy *(Aaron)*	11	1
Conformational Analysis of Six-membered Rings *(Kellie* and *Riddell)*	8	225
Conformational Analysis and Steric Effects in Metal Chelates *(Buckingham* and *Sargeson)*	6	219
Conformational Analysis and Torsion Angles *(Bucourt)*	8	159
Conformational Barriers and Interconversion Pathways in Some Small Ring Molecules *(Malloy, Bauman* and *Carreira)*	11	97

	VOL.	PAGE
Conformational Changes, Determination of Associated Energy by Ultrasonic Absorption and Vibrational Spectroscopy *(Wyn-Jones* and *Pethrick)*	5	205
Conformational Changes by Rotation about sp^2-sp^3 Carbon-Carbon Single Bonds *(Karabatsos* and *Fenoglio)*	5	167
Conformational Energies, Table of *(Hirsch)*	1	199
Conformational Interconversion Mechanisms, Multi-step *(Dale)*	9	199
Conformations of 5-Membered Rings *(Fuchs)*	10	1
Conjugated Cyclohexenones, Kinetic 1,2 Addition of Anions to, Steric Course of *(Toromanoff)*	2	157
Crystal Structures of Steroids *(Duax, Weeks* and *Rohrer)*	9	271
Cyclobutane and Heterocyclic Analogs, Stereochemistry of *(Moriarty)*	8	271
Cyclohexyl Radicals, and Vinylic, The Stereochemistry of *(Simamura)*	4	1
Double Bonds, Fast Isomerization about *(Kalinowski* and *Kessler)*	7	295
Electronic Structure and Stereochemistry of Simple Carbonium Ions *(Buss, Schleyer* and *Allen)*	7	253
Electrophilic Additions to Olefins and Acetylenes, Stereochemistry of *(Fahey)*	3	237
Enzymatic Reactions, Stereochemistry of, by Use of Hydrogen Isotopes *(Arigoni* and *Eliel)*	4	127
1,2-Epoxides, Stereochemical Aspects of the Synthesis of *(Berti)*	7	93
EPR, in Stereochemistry of Nitroxides *(Janzen)*	6	177
Five-Membered Rings, Conformations of *(Fuchs)*	10	1
Foundations of Classical Stereochemistry *(Mason)*	9	1
Geometry and Conformational Properties of Some Five- and Six-Membered Heterocyclic Compounds Containing Oxygen or Sulfur *(Romers, Altona, Buys* and *Havinga)*	4	39
Hassel, O. and Barton, D. H. R. – Fundamental Contributions to Conformational Analysis *(Hassel, Barton)*	6	1
Helix Models, of Optical Activity *(Brewster)*	2	1
Heterocyclic Compounds, Five- and Six-Membered, Containing Oxygen or Sulfur, Geometry and Conformational Properties of *(Romers, Altona, Buys* and *Havinga)*	4	39
Heterocyclic Four-Membered Rings, Stereochemistry of *(Moriarty)*	8	271
Heterotopism *(Mislow* and *Raban)*	1	1
Hydrogen-Bonded Compounds, Intramolecular, in Dilute Solution, Conformational Analysis of, by Infrared Spectroscopy *(Aaron)*	11	1
Hydrogen Isotopes at Noncyclic Positions, Chirality Due to the Presence of *(Arigoni* and *Eliel)*	4	127
Infrared Spectroscopy, Conformational Analysis of Intramolecular Hydrogen-Bonded Compounds in Dilute Solution by *(Aaron)*	11	1
Intramolecular Hydrogen-Bonded Compounds, in Dilute Solution, Conformational Analysis of, by Infrared Spectroscopy *(Aaron)*	11	1

CUMULATIVE INDEX, VOLUMES 1–12

	VOL.	PAGE
Intramolecular Rate Processes *(Binsch)*	3	97
Inversion, Atomic, Pyramidal *(Lambert)*	6	19
Isomerization, Fast, About Double Bonds *(Kalinowski* and *Kessler)*	7	295
Ketones, Cyclic and Bicyclic, Reduction of, by Complex Metal Hydrides *(Boone* and *Ashby)*	11	53
Lanthanide-induced Shift Technique – Applications in Conformational Analysis *(Hofer)*	9	111
Lanthanide Shift Reagents, Chiral *(Sullivan)*	10	287
Mass Spectrometry and the Stereochemistry of Organic Molecules *(Green)*	9	35
Metal Chelates, Conformational Analysis and Steric Effects in *(Buckingham* and *Sargeson)*	6	219
Metal Hydrides, Complex, Reduction of Cyclic and Bicyclic Ketones by *(Boone* and *Ashby)*	11	53
Metallocenes, Stereochemistry of *(Schlögl)*	1	39
Metal Nitrosyls, Structures of *(Feltham* and *Enemark)*	12	155
Multi-step Conformational Interconversion Mechanisms *(Dale)*	9	199
Nitroxides, Stereochemistry of *(Janzen)*	6	177
Non-Chair Conformations of Six-Membered Rings *(Kellie* and *Riddell)*	8	225
Nuclear Magnetic Resonance, ^{13}C, Stereochemical Aspects of *(Wilson* and *Stothers)*	8	1
Nuclear Magnetic Resonance, for Study of Intra-Molecular Rate Processes *(Binsch)*	3	97
Nuclear Overhauser Effect, Some Chemical Applications of *(Bell* and *Saunders)*	7	1
Olefins, Stereochemistry of Carbene Additions to *(Closs)*	3	193
Olefins, Stereochemistry of Electrophilic Additions to *(Fahey)*	3	237
Optical Activity, Helix Models of *(Brewster)*	2	1
Optical Circular Dichroism, Recent Applications in Organic Chemistry *(Crabbé)*	1	93
Optical Purity, Modern Methods for the Determination of *(Raban* and *Mislow)*	2	199
Optical Rotatory Dispersion, Recent Applications in Organic Chemistry *(Crabbé)*	1	93
Overhauser Effect, Nuclear, Some Chemical Applications of *(Bell* and *Saunders)*	7	1
Phosphorus Chemistry, Stereochemical Aspects of *(Gallagher* and *Jenkins)*	3	1
Phosphorus-containing Cyclohexanes, Stereochemical Aspects of *(Maryanoff, Hutchins* and *Maryanoff)*	11	186
Piperidines, Quaternization Stereochemistry of *(McKenna)*	5	275
Planar and Axially Dissymmetric Molecules, Absolute Configuration of *(Krow)*	5	31

	VOL.	PAGE
Polymer Stereochemistry, Concepts of *(Goodman)*	2	73
Polypeptide Stereochemistry *(Goodman, Verdini, Choi* and *Masuda)*	5	69
Pyramidal Atomic Inversion *(Lambert)*	6	19
Quaternization of Piperidines, Stereochemistry of *(McKenna)*	5	75
Radicals, Cyclohexyl and Vinylic, The Stereochemistry of *(Simamura)*	4	1
Reduction, of Cyclic and Bicyclic Ketones by Complex Metal Hydrides *(Boone* and *Ashby)*	11	53
Resolving-Agents and Resolutions in Organic Chemistry *(Wilen)*	6	107
Rotational Isomerism about sp^2-sp^3 Carbon-Carbon Single Bonds *(Karabatsos* and *Fenoglio)*	5	167
Small Ring Molecules, Conformational Barriers and Interconversion Pathways in Some *(Malloy, Bauman* and *Carreira)*	11	97
Stereochemical Aspects of ^{13}C Nmr Spectroscopy *(Wilson* and *Stothers)*	8	1
Stereochemical Aspects of Phosphorus-containing Cyclohexanes *(Maryanoff, Hutchins* and *Maryanoff)*	11	186
Stereochemical Nomenclature and Notation in Inorganic Chemistry *(Sloan)*	12	1
Stereochemistry, Classical, The Foundations of *(Mason)*	9	1
Stereochemistry, Dynamic, A Mathematical Theory of *(Ugi* and *Ruch)*	4	99
Stereochemistry of Chelate Complexes *(Saito)*	10	95
Stereochemistry of Cyclobutane and Heterocyclic Analogs *(Moriarty)*	8	271
Stereochemistry of Germanium and Tin Compounds *(Gielen)*	12	217
Stereochemistry of Nitroxides *(Janzen)*	6	177
Stereochemistry of Organic Molecules, and Mass Spectrometry *(Green)*	9	35
Stereochemistry of Reactions of Transition Metal-Carbon Sigma Bonds *(Flood)*	12	37
Stereochemistry of Transition Metal Carbonyl Clusters *(Johnson* and *Benfield)*	12	253
Stereoisomeric Relationships, of Groups in Molecules *(Mislow* and *Raban)*	1	1
Steroids, Crystal Structures of *(Duax, Weeks* and *Rohrer)*	9	271
Structures, Crystal, of Steroids *(Duax, Weeks* and *Rohrer)*	9	271
Torsion Angle Concept in Conformational Analysis *(Bucourt)*	8	159
Ultrasonic Absorption and Vibrational Spectroscopy, Use of, to Determine the Energies Associated with Conformational Changes *(Wyn-Jones* and *Pethrick)*	5	205
Vibrational Spectroscopy and Ultrasonic Absorption, Use of, to Determine the Energies Associated with Conformational Changes *(Wyn-Jones* and *Pethrick)*	5	205
Vinylic Radicals, and Cyclohexyl, The Stereochemistry of *(Simamura)*	4	1
Wittig Reaction, Stereochemistry of *(Schlosser)*	5	1

JUN 1 8 1981